Storey's Guide to

RAISING HORSES

Storey's Guide to
RAISING
HORSES

Heather Smith Thomas

STOREY BOOKS

Schoolhouse Road
Pownal, Vermont 05261

The mission of Storey Communications is to serve our customers
by publishing practical information that encourages personal independence
in harmony with the environment.

Edited by Deborah Burns and Marie Salter
Copyedited by Alice Vigliani
Cover design by Meredith Maker
Front cover photograph © Larry Lefever, Grant Heilman Photography, Inc.
Back cover photograph © Gemma Giannini, Grant Heilman Photography, Inc.
Text production by Erin Lincourt
Production assistance by Jennifer Jepson Smith
Photographs by Heather Smith Thomas, except those on pages 10, 12 (bottom), 23 by
June T. Campbell; 12 (top) by Margaret Thomas; and 15 and 74 by Lisa Helfert
Line drawings by Elayne Sears, except that on page 48 by James E. Dyekman
Indexed by Susan Olason, Indexes & Knowledge Maps

Printed in the United States by R. R. Donnelley
10 9 8 7 6 5 4 3 2

Library of Congress Cataloging-in-Publication Data

Thomas, Heather Smith, 1944–
     Storey's guide to raising horses / Heather Smith Thomas.
          p. cm.
     ISBN 1-58017-127-3
          1. Horses. I. Title.
SF285.3 .T48 2000
636.1—dc21

                                                                    99-057461

DEDICATION

To Velma Ravndal — my 4-H leader and mentor — who shared her love and knowledge of horses, and her high standards of horsemanship, with many eager young riders. We learned from her that good horsemanship always has a purpose: to be in the best interests of the horse and to promote unity of horse and rider. My commitment to horses and to sharing that commitment with others in my writing is a direct result of her inspiration and guidance during my youth.

CONTENTS

Preface ix

Acknowledgments xi

Part I: Basic Horsekeeping

Chapter 1 Facilities 2

Chapter 2 Feeding and Nutrition 32

Chapter 3 Seasonal Care 60

Chapter 4 Foot Care 80

Chapter 5 Horse Handling 97

Part II: Health Care

Chapter 6 Reading the Signs of Health or Sickness 126

Chapter 7 Disease Prevention 138

Chapter 8 Infectious Diseases of Horses 154

Chapter 9 Non-Infectious Diseases of Horses 179

Chapter 10 Parasites 197

Chapter 11 Toxic Plants and Poisons 223

Chapter 12 Skin Problems 240

Chapter 13 Dental Care and Mouth Problems 250

Chapter 14 Eyes and Ears 260

Chapter 15 Digestive Tract Problems and Colic 270

Chapter 16 First Aid and Medical Treatments 279

Part III: Breeding Horses

Chapter 17 Selecting Breeding Stock 302

Chapter 18 Genetics 318

Chapter 19 Keeping a Stallion 331

Chapter 20 Breeding the Mare 343

Chapter 21 Care of the Broodmare 360

Chapter 22 Foaling 372

Chapter 23 Care of the Newborn Foal 400

Chapter 24 Care of the Young Horse 430

Epilogue 451

Appendixes

A. Horse Anatomy 454

B. Resources 455

C. Glossary 460

Index 474

PREFACE

For thousands of years, the horse has served humankind; our history is inextricably linked with his. Until recently, he has been our chief means of transportation and our source of power for farming. After the invention of tractors and cars, horses' numbers dwindled, but since the 1950s there's been renewed interest in horses — for sport and pleasure and for working small farms.

The horse of the twenty-first century has a distinct advantage over his forebears. In the early days nearly everyone owned a horse, the principal means of conveyance. Today, he's no longer needed for transportation, and people who own horses enjoy and appreciate them. Now, perhaps as in no other time in history, the horse in the United States is owned by folks who respect him, seek to understand him, and want to do their best to take good care of him.

Our biggest stumbling block as horse owners comes from our Computer Age thinking, wanting to accomplish everything quickly, whether training a horse or learning to ride. Contemporary conveniences and timesaving devices have spoiled us; we want shortcuts. There is no shortcut to becoming a good horseman, a knowledgeable horse owner, and a good horse breeder. There's no substitute for time invested and the desire to learn all we can. It can take a lifetime.

Dedication makes a horseman. Only one who seeks to understand a horse can truly enjoy him. The casual horse enthusiast may admire a horse from a distance or ride one on weekends, but that person misses a great deal. One who lives with horses, cares for them daily, and works with them, sincerely trying to improve horsemanship and understanding, continually thinking about the horses, their problems, their personalities — that person will get the most satisfaction from horses.

Enjoying horses involves more than admiring them in pictures or watching one gallop across a pasture. The key to real enjoyment of horses is *doing*, whether feeding, grooming, breeding, showing, moving cattle, racing, jumping, or riding for pleasure. We all need a sense of identity, of belonging, a realm in which we feel at home and with which we can identify. Working with horses and accomplishing something with them can give us this purpose.

Understanding and experience make a good horseman. An intuitive "feel" for horses — a sense of what a horse needs at a certain time and how best to care for him — makes a responsible horseman. Part of enjoying a horse is caring for him as best we can, being able to tell when he's fit and healthy and when he isn't, and if he isn't, being able to tell what he needs and whether he needs veterinary attention. Your horse needs proper care — the right amount and right kind of feed, enough of the right kind of exercise to keep him fit and in condition for the work he is to do. The good horseman cares for his horses as regularly and responsibly as possible, as he would care for a child or anything else he loves. The dedicated horseman is a perfectionist as far as horses are concerned, for he will do anything and everything within his power to take proper care of the horses and to learn what proper care entails.

Horses are unique creatures. The horseman must come to understand each horse in his care, learning what's best for each individual in a given situation. There is no substitute for the "eye of the master," whether feeding, doctoring, or training a horse. The person who understands a horse will give him the proper amount of work and rest, reward and discipline. Horse owners have an enormous responsibility, for these animals are totally dependent on us for their existence and well-being.

The horse is a challenge, and there is much to learn. The serious student of horses is fortunate, for today there are many sources of help and advice. This book is one of many and does not attempt to cover all aspects of horse care and breeding but will be useful to the reader who wants to learn more about good horsekeeping, the raising of horses, and their health and soundness.

My aim in this book is to provide useful information to the horse owner with special emphasis on horse care and "intelligent horsekeeping." This is not a book on training (although every time you handle a horse you are training it, for better or worse) but on *horse husbandry* — the art of keeping a horse fit and healthy in body and spirit. Learn from your horses, and they will learn from you.

Author's Note: Throughout the text, except when referring to mares or fillies, I use the masculine pronoun to refer to horses. This is done for simplicity and reader ease.

ACKNOWLEDGMENTS

I owe special thanks to many people who helped make this book possible. I want to thank the many horsemen who have given me good advice and encouragement over the years. This is a wonderful aspect of the world of horses — other horsemen are always willing and happy to help you; we share a common bond. Our love and respect for horses drives and unites us in a special way.

I want to thank my husband, Lynn, for his support, patience, help, and encouragement in all my endeavors with horses. His partnership and contributions have been indispensable. He has helped me with everything — from building foaling pens and hay sheds, to holding a sick horse for treatment, to accompanying me on a "baby-sitter horse" while I started a youngster under saddle on first rides into the mountains surrounding our cattle ranch.

I must also thank several generations of horses that have grown up here on Withington Creek over the past 40 years. My first-born foal, Khamette (1959), was raised as a 4-H project when I was in high school, and we still have grandsons of hers on the ranch. My father's old Thoroughbred mare Nell blessed us with many half-Arab offspring and their descendants; some have been the best cowhorses I've been privileged to know.

Over the years we've had numerous horses of various breeds — Arabian, Thoroughbred, Quarter Horse, Appaloosa, Morgan — and appreciated the talents of all. Lynn and I raise beef cattle on a ranch that depends on 4 months of summer grazing in the mountains, and we need our horses to help us take care of the cattle. We appreciate athletic and sure-footed horses with the stamina to travel day after day in rough country, working cattle.

Our horses are an integral part of our ranch; we could not raise cattle without them. Yet they are a joy and pleasure as well as a necessity for the job. We delight in good horses that enjoy their work, and we respect the amazing partnership that we have with them. We travel many miles with our horses, and in some years we've done endurance riding, enjoying comradeship with other horsemen and taking advantage of the fact that our horses were already fit for this demanding sport just from doing the range riding.

Working with a variety of horses over the past 45 years has been one of my greatest satisfactions and accomplishments. They have been my teachers as well as my pupils, and I owe a large debt of gratitude to them. I cannot imagine my life without horses. They are the reason for my writing and the inspiration for this book.

PART I

Basic Horsekeeping

ONE

FACILITIES

Good horsekeeping involves two important, interrelated components — dedicated and conscientious personal care of each animal (i.e., regular daily feed and care if horses are confined, and daily inspection if they're at pasture) and good facilities. If you don't have good facilities your horses may become ill or injured, even with the best of care. For example, a poorly ventilated barn may cause respiratory problems; a stall that's too small may get a foaling mare into trouble; a poorly designed or run-down fence may invite injury; and a weedy pasture containing toxic plants may cause colic or fatal illness. If you strive to make sure your facilities are adequate, safe, and in good repair, you can greatly reduce the risk of injury to your horses. And you make your own job of caring for each animal much easier.

A barn is extremely useful, providing out-of-the-weather storage for all your horse-related items and shelter for your horses.

Barns

What you need will depend on the climate where you live and your situation. Many horses are at pasture year-round, being fed hay if snow gets deep. They never see a barn and are none the worse for it, but they appreciate a windbreak, natural or man-made. In such cases, a grove of trees or brushy draw can provide a lot of shelter from wind or driving rain and snow.

If you work with your horses during the winter or have mares that might foal during bad weather, it's nice to have a good barn. In a cold or rainy climate, horses in pen or pasture should have a run-in shed as a place to get out of the weather. Shelter for a sick or injured horse is also helpful.

Consider Your Needs

If there is already a barn on your property, it may just need some repair and maintenance to be hazard free and safe for horses, or it may need a little remodeling to suit your purposes. If you are building a new barn, consider visiting other horsekeeping facilities to get ideas for design and layout that will work well for you and your horses. There are many good designs, and some not so good. A barn is expensive and should last your lifetime, so make sure it is well designed before you begin construction.

Building a Barn

In a cold climate, your horses need a barn that is not drafty and retains heat. In a climate where summers are hot and humid, they benefit from a barn that stays cool. Choose the site carefully, considering climate as well as present and possible future needs.

Location

Choose a well-drained site. The barn should be on higher ground than surrounding pasture areas, not in a swale or floodplain that gets wet with every major rain or snowmelt. You don't want a sea of mud encircling the barn, moisture seeping through stall floors, or spring runoff coming in the doors. If the ideal barn location is not in a good spot for adequate drainage, arrange for a contractor to haul in enough fill to build up the area so it *does* have good drainage.

Convenience is key. The barn should be within reach of pens and pastures and your house. It should be far enough away so flies and odors aren't a problem, but close enough that you can get to it quickly for chores and emergencies.

The barn should be located where there's access to electricity and water. After all, you'll want lights, outlets for electrical equipment or a heater, and a source of water for horses confined in stalls. Consider also whether there's room for additional barn area if in the future you decide to add more stalls or an indoor work area.

Lights in the barn (in the aisle, stalls, and feed areas) should be at least 8 feet (2.4 m) high or higher and covered with a wire cage to protect them from breakage.

A safe light fixture has a wire framework to protect it from breakage, and a glass jar inside the wire cage to contain bulb fragments if the light shatters.

Permits

You may need a permit to build a barn if you live on the edge of town or in a subdivision. Check the local zoning regulations before you start building. Your barn also must be constructed in accordance with local building codes.

Conversion Caveats

If you are converting an existing building (chicken house, garage, or some other type of shed) into a horse barn, make sure it is high enough — at least 8 or 10 feet (2.4 or 3.1 m) for large horses — inside so that your horse will never hit the ceiling with his head, even if he rears. Also make sure the building does not have a concrete floor, which has poor drainage and offers poor footing. A horse may be injured while slipping and falling. Wood has some give but can be slippery; dirt makes the best floor.

Building Site and Related Issues

Prepare a good building site before you start construction. The ground underneath should be solid with some rock so nothing will shift, sink, or sag. If the existing soil will not support a solid building with good drainage, the area should be excavated to a depth of at least 4 feet (1.2 m), and large coarse rocks put into the hole, with crushed rock on top of that layer. Fine gravel can be spread on top of the crushed rock. This will provide good drainage. A foot (0.3 m) of soil or tamped clay (or a mixture of clay and sand — usually about three parts clay to one part sand) can be put on top of the gravel for the floor base, unless you choose to have a wooden or synthetic floor.

The barn should have a foundation of concrete, cinder blocks, or pressure-treated wood. An inexpensive pole structure can be built using tall, well-treated posts set deeply into firm ground as the main supports for roof and walls.

Dirt floors are often used for stalls because they are easy to keep dry (they provide some drainage) and not too firm or unforgiving if a horse lies down a lot or crashes down hard on the floor. You may want rubber mats on stall floors in some situations — as for a foaling mare, young horse, or any horse that might spend time lying down.

Windows should be covered with bars so they cannot be broken. They should also let in light and sunshine but keep out wind and moisture, as should doors. Openings should face east and south for solar warmth from sun in winter. The barn should be designed to stay warm in winter but cool in summer; take summer sun into consideration when placing windows.

rubber mat
fine gravel
crushed rock
coarse rocks

Base material for the barn floor consists of a layer of coarse rocks, topped with crushed rock, then fine gravel, all of which is covered with a rubber mat. (Redrawn from Cherry Hill, Horsekeeping on a Small Acreage. Pownal, VT: Storey Books, 1990.)

Stall windows should be covered by bars so they aren't broken accidentally.

Ventilation

Ventilation is very important. You do not want an airtight barn that holds in humidity or stall dust. Dust from hay and bedding can cause respiratory problems in horses, as can an environment that stays too warm and humid. The barn shouldn't be drafty, but it must have adequate ventilation. In designing your barn to suit your site in terms of ventilation requirements, consider the following guidelines. In warm climates, a row of covered stalls that open to outside paddocks will do. In cold climates, two rows of enclosed stalls separated by a center aisle work well. The aisle should be at least 8 to 10 feet (2.4–3.1 m) wide; 12 feet (3.7 m) is even better if you have the space.

The main reason a horse is better off outside than in a barn — even in cold or stormy weather — is that it's unnaturally damp and dusty indoors. If your horse must be stalled in a barn, good ventilation can minimize the problems of dampness and dust.

Whenever an animal exhales, moisture accompanies air expelled from the lungs. This is why when you breathe in an enclosed space (such as a car on a cold day) condensation is produced — droplets of water "steam up" the

*Three methods to improve barn ventilation. **A**. Air enters the upper wall louvers and is pulled out the top of the roof by a turbine vent. **B**. Air enters through soffit louvers under the eaves and exits through a continuous ridge vent. **C**. Air enters and exits through gable louvers. (Redrawn from Cherry Hill,* Horsekeeping on a Small Acreage. *Pownal, VT: Storey Books, 1990.)*

windows. With a large animal an even greater volume of moisture is exhaled with every breath. The average 1,000-pound (453.6 kg) horse puts out about 2 gallons (7.6 L) of moisture daily with air expelled from its lungs. A barn with six horses can have 12 gallons (45.4 L) of water vapor put into the air daily, as well as moisture created by the evaporation of urine.

Damp air (along with irritation of air passages from ammonia fumes produced by urine and manure) can contribute to respiratory problems and encourage growth of fungus, molds, and certain bacteria. Also, when weather is cold, moisture condenses on the underside of a metal or uninsulated roof, dripping down like gentle rain or freezing and making icicles that drip when they melt.

Air movement and ventilation are important to take out damp air and replace it with dry fresh air. Condensation does not occur as much if the air is slowly moving, taking the moisture with it. Ventilation can be obtained by:

- Opening windows, barn doors, and tops of stall doors, although in winter this may cause a cold draft
- Installing vents, such as louvers under the eaves or roof ridge
- Installing a spinning ventilator on the roof to pull air out of the barn, to supplement the ridge vents
- Using a fan to force air through the barn, making sure this does not create a draft

Health Tip

Ammonia fumes near the stall floor are a prime cause of lung irritations in foals (they spend a lot of time lying down), so make sure stagnant air from the lower part of the stall is moved out. In addition to the ventilation methods listed above, you may need louvers in the sides of solid stall walls to help clear stagnant air.

Heated Barns

You probably don't need a heated barn; it requires good insulation and a source of heat, and it is very expensive. Moreover a heated barn is an unnatural environment for horses. An unheated barn is usually much healthier for horses; they are less prone to develop colds and respiratory problems. You won't need a source of heat except in unusual situations — such as for a seriously sick horse or new foal during very cold weather. At these times you can usually supply enough warmth with a space heater. Consider cost, durability, maintenance needed, and fire resistance.

Building Materials

When choosing building materials, consider cost, durability, and insulating quality. Different materials have various advantages and disadvantages.

Advantages and Disadvantages of Common Building Materials

MATERIAL	ADVANTAGES	DISADVANTAGES
Construction materials		
Metal	• Quick to build • Less expensive than wood • Fireproof • Minimal maintenance	• Noisy during rain, hail, wind • Cold in winter, hot in summer (low insulating ability) • Can be damaged by kicking or rubbing
Wood	• More forgiving (less apt to cause injury) than metal or stone	• Materials and labor are expensive • Wood must be treated to prevent chewing • Not fireproof
Brick, cinder block, concrete, stone masonry	• Cool in warm climate • Fireproof	• Damp and cold in winter • Require good ventilation • Expensive
Roofing materials		
Wooden shakes or shingles	• Warm in winter, cool in summer (good insulating ability) • Do not cause condensation in barn	• Materials and labor are expensive • Fire hazard • Cannot be used on flat roofs (water won't run off well) • Prone to weather and rot
Asphalt shingles, rolled roofing	• Durable (20 years or more if properly installed) • Less expensive than wooden shingles • Easy to install • Less noisy than metal • Better insulating quality than metal	• Prone to damage by heat, wind, or ice build-up • Become brittle in temperatures below freezing
Metal	• Quick and easy to install • Relatively inexpensive and maintenance free • Translucent fiberglass sections can be installed to let in light	• Hot in summer • Noisy during storms • Condensation forms inside and drips down • Moisture leaks around screws • Snow may slide off and build up in front of doorways and walls

---

Wet and Wintry Climates

Install rain gutters and downspouts to handle water flow off the roof. If you get snow, build overhangs to shelter the doorways and keep entryways less muddy from snowmelt. If exercise paddocks are attached to the barn, locate them so rain and snow coming off the roof won't make mudholes.

Barn Stalls

Most horses can be comfortable in a stall 12 by 12 feet (3.7 x 3.7 m), but for adequate room foaling stalls should be at least 14 by 14 or 14 by 16 feet (4.3 x 4.3 m *or* 4.3 x 4.9 m). A narrow tie stall may be fine for working with a horse but should never be used for keeping a horse more than a few hours. Many horses are reluctant to lie down when tied and cannot get any exercise. A box stall is much more comfortable.

The stall should be large enough so the horse has a clean space in which to lie down and to eat. Many horses accustomed to living in a stall urinate and defecate in the same area, leaving the rest of the stall somewhat clean. Because no one size is perfect for every horse, it's wise to have several stall sizes in your barn. A 10 by 10 foot (3.1 x 3.1 m) stall may be adequate for a pony or a horse that must be strictly confined after injury. A large stall is needed for a large horse, a foaling mare, a horse that rolls frequently and gets cast against the wall in a small stall, or a recuperating horse that cannot be turned out but needs room to move around. It's handy to have a double stall divided by a movable partition, so you can use it as two separate stalls or as one large stall for a foaling mare or a mare with a young foal.

Stalls should be sturdy and durable to withstand rubbing, kicking, chewing, and the corrosive effects of urine and manure. Their walls should be smooth and free of hazardous projections that might injure a horse (e.g., wood splinters, protruding nails, bolt ends, or sharp edges). Stall doors should be at least 4 feet (1.2 m) wide and 8 feet (2.4 m) high so a horse cannot run into the edge or hit his head on the top of the opening.

The top portion of the front wall can be pipe or metal bars so the horse can see out and have proper ventilation. Spaces between bars should be small, no more than 3 to 4 inches (7.6–10.2 cm) so the horse can never get a foot or nose through, nor teeth or lower jaw caught when trying to nibble and chew. Dividers between stalls should always be solid so horses in adjacent stalls cannot smell each other and start an argument.

vent

water

vent

12' x 14'
or 14' x 24'
STALL

swinging
partition

water

vent   12' x 14'
STALL

12' x 14'
STALL

vent

water

vent

10' AISLE
road base

feed
barrels

hay drop
from loft

shelf

vent
water

vent

10' x 12'
STALL

vent

10' x 8'

10' x 16' TACK
ROOM
insulated floor,
ceiling, walls

tool
hooks

fence
charger

4'
scale

This sample barn floor plan shows feed and tack rooms and stalls of various sizes, including a double stall with a movable partition. A movable partition facilitates creation of one large stall if needed. (Modified from Cherry Hill, Horsekeeping on a Small Acreage. Pownal, VT: Storey Books, 1990.)

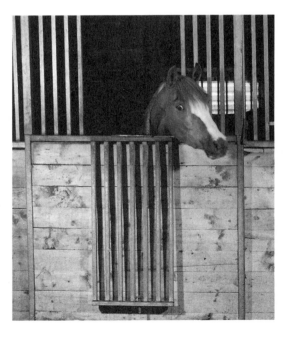

The top of the stall wall or stall door should be made of pipe or metal bars so the horse can see out. This door top has a drop panel that can be closed if needed.

Flooring

Flooring should be a material easily kept clean, since the average horse produces up to 50 pounds (22.7 kg) of manure and more than 10 gallons (38 L) of urine daily (the actual amount depends on whether he is eating hay or pellets, and whether the weather is warm or cold — there's less water consumption and urine production in cold weather). A clay floor has good traction, some cushion, and a moderate degree of warmth, but moisture does not drain through it as readily as through a mix of clay and sand or crushed rock. However, sand or crushed rock may get mixed with the feed or bedding if the horse likes to paw.

Wooden floors are fine for tie stalls but not for box stalls; they are too slippery for good traction. Stall mats can be used over wood or dirt floors. Such mats are usually a combination of rubber, clay, and synthetic materials. If they have a textured surface they afford good traction and aren't slippery. They keep horses from pawing holes in the floor or eating dirt, and they have enough cushioning to be comfortable. Stall mats also can be easily cleaned.

Bedding

What you use for bedding in a barn stall or shed will depend on the materials available in your area, their cost, and your budget. Straw and wood shavings or chips are the two best options.

In a farming region, **straw** may be the cheapest and most plentiful material. Wheat straw is often the bedding of choice because it is less palatable than oat straw (horses won't eat the bedding as much) and less abrasive than barley straw. The sharp awns that protrude from the grain heads of barley — if there is any grain left in the straw — can become embedded in a horse's mouth.

If you live where wood products are processed, **shavings** or **wood chips** may be economical. Soft woods like pine and fir make good bedding, but hardwoods have less absorbancy and may even be toxic to horses (black walnut is especially toxic and should never be used). Sawdust is usually too fine and dusty for bedding, getting into eyes and nose. Wood products should not be used as bedding for foaling mares or newborns because shavings or chips can harbor *Klebsiella* bacteria (see chapter 22). Make sure bedding is not dusty or moldy. If it is, your horses may develop respiratory problems.

Manure Pile

Plan stalls and barn with ease of cleaning in mind, and have a place nearby to put soiled bedding and manure. The manure pile should be far enough away that flies aren't a problem, but close enough for easy access. If horses spend much time in the barn, stall cleaning must be a daily chore. You don't want your horses lying in wet bedding, at more risk for health problems.

Clean stalls at least once a day so horses have fresh, dry bedding.

Hay Feeders

Horses in stalls can be fed in mangers or hay racks. If stalls are kept clean, hay can be fed on the floor — the most natural position for a horse to eat. But if the floor is sandy or gravelly a horse may pick up sand when cleaning up the last wisps and leaves of hay, which can lead to sand colic. There is also some risk of parasitism if hay gets mixed with manure.

An alternative is a walled-off feed bunk where horses eat at floor level (kept clean!) or a good hay rack or wall feeder in a corner, with no protrusions that might cause injury. The rack should be high enough so a horse cannot get a foot caught in it, but low enough so he does not have to reach up for feed, because hay leaves and dust may fall into his eyes as he eats. As the horse tugs at the hay, falling particles and dust may be breathed in, causing irritation to respiratory passages and inviting lung problems.

Combination hay and grain feeder; this type opens into the barn aisle so you can feed the horse without having to enter his stall.

Tack Room

You might want a room for storing horse gear, medications, and first aid equipment. Bridles, halters, and other equipment can be hung neatly on walls, and saddles stored on saddle racks, to keep them in good condition. The door should be wide enough to allow access by a person carrying a large saddle. It is wise to have a strong door that can be locked to prevent theft. If there's no separate room for tack, a wide area in the aisle or entryway can be used for tack storage, with tack hangers and saddle racks mounted on the wall.

Feed Room

Many barns include a place where feed can be stored. Most of your hay should be stored somewhere else (in a hay shed or on well-drained ground with a good tarp over it), but it's handy to have a few bales in the barn along with any grain you are feeding. Have a horse-proof door and latch for the feed room. Be sure that all grain storage containers have lids. Do not store sacks of feed in a barn; not only will mice get into them, but there is always risk of a horse foundering (i.e., developing laminitis from overeating on grain) if it happens to get into the feed room.

Feed Room Tips

Make the feed room as rodent-proof as possible, and keep it dry and well ventilated. Moisture can cause grain to mold or ferment. Large plastic garbage cans work well for grain storage; make sure their lids are securely replaced after taking out grain.

Feed Storage

Large quantities of hay should *not* be stored in a horse barn; it takes up too much space and is a fire hazard. It's best if you can make a hay shed with a tall roof to protect hay from rain and melting snow, preventing weather damage and spoilage. An inexpensive hay shed can be made by using tall posts to support a metal roof. The shed should be on high ground that never floods, and the base should be of crushed gravel so bottom bales do not draw moisture from the ground. Build the hay shed as large as needed — for storing several tons of hay or a year's worth of hay for many horses.

Hay stored outside should be kept on pallets (so bottom bales do not draw moisture, which can cause mold or rot) and covered with a good tarp. It keeps the top bales from spoiling, but you may get occasional leakage and some ruined hay anyway.

One risk in storing hay in or near a barn is spontaneous combustion from the natural heating of hay that has been baled too "green" (not yet dried enough) or wet. In some instances fermentation and chemical changes can cause the hay to heat so much that it ignites. If you must store hay in the barn, make sure it is well cured. Open a few bales and check that there is no dampness, heat, or mold.

Fire Prevention and Safety

Fire is the greatest danger for a stabled horse, especially if the fire starts at night and you do not discover it in time to save the barn and horses. Always build barns with this in mind, using the most fireproof materials you can afford. Have electric wiring installed (or repaired) by a professional, make sure there are no fire hazards in the barn, and insist that everyone who comes in or uses electrical equipment is careful to follow basic safety rules.

◆ No smoking in the barn.
◆ Store a minimal amount of carefully checked hay in the barn.
◆ Be sure all light fixtures, switches, and wires are out of reach of horses and all lights are covered with protective cages to prevent breakage.
◆ Run all wires through conduit, and outfit switches with safety covers.
◆ Unplug all electrical appliances when not in use. Check all cords frequently, and keep them in good repair.
◆ Do not leave a heater running when you are not there to watch it.
◆ Have fire extinguishers in the barn in handy locations.
◆ Keep the barn clean and free of dust, cobwebs, and feed spillage.
◆ Hang a halter and rope by every stall door (but out of reach of the horse), and always replace it after use.
◆ Post the phone number of the fire department beside all telephones, along with easy directions to tell them how to get to your place, and the official name of your road or street.

An open-sided pole shed for hay storage protects the hay from moisture while allowing good ventilation and easy access.

Run-In Shed

An outdoor shelter for horses at pasture or in paddocks can be used instead of or in addition to a barn. A run-in (three-sided) shed gives adequate protection from cold wind, driving rain, or blizzards. It also gives some relief from flies (horse flies do not like to go into a shady area) and hot sun.

The open side should face south (away from cold winds coming from the north) or away from prevailing winds. Horses often use the shelter more as a windbreak than to get out of wet weather. The shed should be built on a high spot so horses have dry footing instead of mud, and with the slope of the roof away from the opening so moisture can't accumulate in front of it. There should be good drainage away from the shed. It should be large enough (about 140 square feet [42.7 m²] per horse) to accommodate all horses in that pen or pasture, to avoid injuries when horses jostle for space.

This run-in shed provides adequate protection from the elements and accommodates several horses.

If there is a natural windbreak of trees or brush, you won't need a shed. This saves the expense of building one and eliminates the risk of horses being injured when trying to get into the shed all at once or chasing one another through it.

Fences and Paddocks

Fences must be secure enough to hold horses and safe enough to prevent injury. Choice of fencing material will be influenced by what you can afford, type of horse you are enclosing (a paddock for a stallion must be more substantial than a pasture fence for a group of geldings, for instance), and time and money you can afford for fence maintenance.

A fence that won't hold a horse is no fence at all. You don't want your horse to get out on a highway to be hit by a vehicle or cause a fatal accident, escape into a neighboring pasture, or be injured during fence-crashing. You also don't want your horse in the neighbor's garden or flower beds. Fencing for horses should be 5 feet (1.5 m) high, even higher if you have a horse that likes to jump.

Avoid sharp corners in a pen, pasture, or paddock where a horse might get injured while running along the fence-line or being chased by another horse.

All pens, paddocks, and corrals should have secure fences because any confined horse will test them diligently. At pasture he has more to keep his mind occupied, room for grazing and exercise. In a small area he can become bored and unhappy, spending more time checking out the fence or trying to find a way out.

Be aware that a poorly maintained fence can become dangerous and unsightly or make you vulnerable to a lawsuit if your horse gets out and damages property or causes an accident. Check fences often. Tighten loose wires, replace splintered boards or poles, pound in loose nails or staples, restretch sagging net wire, and test electric fences periodically to be sure they are working. Keeping fences well maintained not only lengthens their life but also helps ensure horses won't be injured in encounters with their boundaries.

Types of Fencing

Many types of fencing can be used to confine a horse, but not all are safe. The materials you choose will depend on the situation — whether a roomy paddock for mares and foals, a pasture for a group of horses, a stallion pen, a yard for playful yearlings, or a boundary between other horses next door.

- Paddocks for mares and foals should be very safe and durable because foals are inquisitive and can get into trouble. You don't want them injured trying to put a head or foot through the fence. *Best bet:* Wood or net wire, possibly reinforced with electric wire, depending on size of paddock
- Any enclosure used for weaning should be completely hazard-proof. *Best bet:* Diamond mesh wire
- A stallion pen should be at least 5 feet (1.5 m) high or higher, and the materials stout and durable. *Best bet:* Wood or pipe fence, possibly reinforced with electric wire
- Fencing around small paddocks or pens should be chew-proof, paw-proof, kick-proof, and lean-over-proof because horses confined in small areas think up all sorts of fence-terrorizing activities to entertain themselves. If turned out in the paddock only for short periods, they may gallop and buck around the small enclosure, kicking up their heels — so the fence must be strong. *Best bet:* Diamond meshwire
- You may get by with less complicated fencing around large pastures where horses have room to roam and lots of grass to graze, but if part of the pasture borders another you'll need safety measures to prevent neighboring equines from rowdying each other at the fence. *Best bet:* Wood or wire, synthetic materials — possibly reinforced with electric wire

Barbed Wire

When it is tight, barbed wire makes a good barrier for cattle (which have thick, tough hides) or even mature ranch horses accustomed to this kind of fence. However, it can be a disaster for young horses learning about fences or for any horse not used to barbed wire. Even horses that respect barbed wire can be ruined by it — if pushed into it by another horse, pawing at it, or attempting to jump or lean over it.

Barbed wire can be safely used for horses, with modifications — if your place already has that type of fence and you cannot afford to replace it yet. Make sure the fence is in good repair, with wires stretched tight. Sagging wires are more apt to get a horse in trouble because he may try to reach through to grab a bite of grass on the other side or get a foot through it. To make sure horses won't be injured by a barbed-wire fence, put a strand of electric fence along the inside to teach the horses to respect the fence and stay well away from it. This eliminates the leaning, reaching, rubbing, or pawing that gets a horse in trouble.

Wood Fencing

Wood poles or boards make good fences for horses under some conditions. Strong and durable, wood is more forgiving and resilient than metal if a horse runs into it. Wood works best where horses have a lot of room. Confined, bored horses like to chew wood and can demolish a pole or board fence in a short time. As a general rule, horses that grow up in large areas and big pastures don't become wood chewers (unless they have mineral deficiencies or dietary imbalances). The chewing habit is learned early, however, in confined youngsters. Once a wood chewer, always a wood chewer — unless there is a deficiency that is later corrected.

In many instances, the only way you can get by with wood fences is to add an electrified wire or two, or electrified tape, to keep horses away from the fence or prevent their chewing it. An electric wire can also keep a horse from rubbing on the fence or trying to lean over it — habits that take their toll on fences.

Nail boards to the inside of the posts to create a smoother fence; horses running along it won't run into the posts.

Nail poles to the inside of the posts so they can't be knocked off by a horse kicking, rubbing, or leaning on the fence.

Wood fencing with an electric wire along the top and one out to the side to keep horses back from it will last a long time. In a dry climate, untreated poles or boards may last 20 to 30 years, even longer if treated. Wood preservatives are necessary in a wet climate and should be reapplied periodically.

Wood fences require a great deal of maintenance to be safe for horses. Board fences should be checked for splinters. All types of wood fence should be thoroughly checked at least once a year to see if more paint or preservative is needed. Broken, cracked, splintered, chewed, or rotten poles or boards must be replaced, and nails redriven if they start to work out. Nails that creep out are a serious hazard to any horse that runs into them.

Make sure poles or boards are nailed to the inside of posts, so a horse rubbing, leaning on, or kicking the fence cannot knock them off. This makes a smoother fence line; horses running along it will not run into the posts. Wood posts should always be thoroughly treated before being set in the ground, or they will rot.

PVC Tubing

Polyvinyl chloride (PVC) tubing doesn't rot or splinter and horses won't chew it. This material, which was originally developed for plumbing pipes and drain tiles, was first used for fencing in the early 1980s. The PVC tubing used for fencing today has been treated to resist color change caused by sunlight and has been modified so it won't crack in extremely cold weather.

Most companies that sell PVC fence give a 15- to 20-year warranty. PVC tubing is initially quite expensive (about twice as much as treated wood) but requires little maintenance and usually lasts longer. It never needs painting or preservatives, won't rot, and can't be damaged by horses or termites. The material is somewhat flexible and won't splinter if a horse hits it. PVC tubing will occasionally break if impact is great enough, but it is safer than metal or wood.

Polyethylene-Coated Wood

This type of fencing (planks and square posts covered with a white polymer resin) combines some of the advantages of wood and synthetic materials. The coating makes a nice white fence and protects the wood from moisture; the ends of boards and the tops of posts have protective caps to prevent moisture damage. This type of fencing requires less maintenance than wood and is highly visible, durable, and safe. Also, horses don't chew it as readily as wood.

Rubber and Nylon Strips

This fencing, made from reprocessed tires and conveyer belts, is stretched tightly between wood posts. A fence of this sort is durable, although strips may occasionally have to be tightened between posts. It has a lot of give if a horse runs into it. Newer versions don't fray or deteriorate as badly as earlier materials. But if a horse chews on this type of fence, he can get indigestible fibers in his stomach that could lead to impaction and colic. If you have a horse that chews, add an electric wire or tape to keep him from nibbling at the fence.

Net or Mesh Wire

This fencing material is much safer than barbed wire and more foolproof than smooth wire, although it's possible for a horse to get a foot caught if spaces are hoof size or larger. "No climb" netting (2 x 4 in. [5.1 x 10.2 cm] rectangular spaces) is fairly safe for adult horses, but a foal can get a foot through. Larger mesh, such as that used for cattle, pigs, and sheep, is not safe for horses. The best type of net wire for horses is diamond mesh, since even a small hoof rarely gets through it and a shoe is less likely to be caught.

Wire mesh will stay tight a long time if properly stretched and braced, and if it is topped by a pole, board, or pipe to keep horses from leaning over it. A pole at the bottom and a diagonal brace between posts help reinforce netting.

Diamond mesh is the safest type of net wire fence for horses.

Smooth Wire

Five to seven strands of smooth, 12- to 13-gauge barbless wire (two wires twisted together without the barbs) on wooden posts may work as boundary fencing for large pastures, but in smaller areas horses may lean over or reach through it unless the fence is topped with a pole, pipe, or board or used in conjunction with an electric fence.

Electric Fencing

By itself, electric fencing is not adequate for horses — it's more a psychological barrier rather than a physical one. Electric fencing is not very visible, and a horse may run into it before seeing it, unless you use brightly colored electric fence tapes. Electric fencing may break if a horse hits it at full speed or tries to jump it. Even the newer plastic tapes cannot hold a determined horse. Flagging an electric fence with bright pieces of cloth or using highly visible tapes will not stop a horse that is being chased by a pasturemate and seeking an easy way out.

Horses unaccustomed to electric fencing must learn about it; otherwise they may crash through. The horse can best learn about the fence in a situation where he is not upset and trying to go through it; if he crashes through he will get shocked but will not necessarily learn to respect it. When you first turn him out in a pasture with electric fence, watch to make sure he checks it (most horses try to smell the fence and get a shock on the nose). If he does not check it, tempt him to touch it by offering a bucket of grain or a bite of green grass from the other side of the fence. This may seem like a mean trick, but it will alert him to the hot wire and make him stay back from it after that; it may even prevent serious injury later because he will respect the fence.

Some horses know when the electricity is off or the fence is disconnected or shorting out in the rain. If the power goes off, the battery wears out, or the

In this solar-powered fence charger, sunlight provides the power to charge the wire. When a horse touches the wire, his body shorts it out, with the electricity taking a short-cut through him, back into the ground (completing the circuit), thereby giving the horse a shock.

fence is broken and shorting out, your horse may walk through it, or over it or crawl under it. If you live in a sunny climate you can use a solar-powered fence charger, but you still need to check the fence periodically to make sure it is not broken or shorting out. You also need to keep the charger area free of weeds or tree branches that might block the sun.

Electric fencing is best used in conjunction with a solid barrier. The hot wire teaches the horse to respect the fence; even a bold one won't keep trying to lean over, jump, or push on it after having been zapped by the electric wire a time or two. Used in conjunction with other fencing, an electric wire can do wonders in preventing over-the-fence battles and in keeping maintenance costs of the primary fence to a minimum.

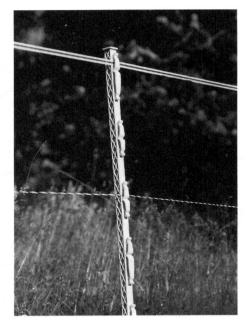

Lightweight push-in post with a cap on top protects horses from injury; clips hold electric tapes or wires. (Courtesy of Premier Fencing Company.)

An electric wire at the top of a wood fence can keep horses from chewing and leaning over it.

Maintaining Electric Fencing

Follow a program of regular checking and maintenance to make sure electric fencing is always working properly. Wet grass, weeds, or leaves may short it out, or deer jumping over it may push the hot wire or tape into the solid fence next to it. If the hot wire gets caught on a wire fence, nail, or staple, it will short out. Periodically trim grass and branches away from wires during summer when plants and bushes are growing rapidly; otherwise, wires may become choked with green foliage and short out. Many fence chargers have a blinking light to show the fence is working. If it's not plugged in or is shorting out, the light will be off. You can use an inexpensive fence tester to see if there's an electrical charge in the fence. The tester has a small rod that you stick in the ground and a wire that you touch to the fence. If the fence is working when you do this, the light on the tester goes on.

Fence tester with lights showing the strength of the electrical charge

Steel Pipe

This fencing was originally created from surplus or discarded drill rig pipe from the oil industry, but now some companies make steel pipe especially for horse fencing. Pipe rails are generally secured to metal (pipe) posts set in the ground in concrete. Often 2-inch (5.1-cm) drill-stem pipe is used for top and bottom rails, with three or four strands of cable or sucker rod in between. Because pipe has more structural strength than wood, pipe rails may be 10 to 20 feet (3.1–6.1 m) long, requiring fewer posts to be installed.

Pipe fence is durable and lasts forever, especially if coated with antirust paint. Because of its strength and durability, pipe makes good fence for stallion pens, road boundaries, or other places where you want a fence that never fails. However, it can cause injury if a horse runs into it at full speed or if the spacing of pipes enables a horse to get a leg or head caught. There have been a few deaths when horses put their heads through the pipes and tried to chew on the next pipe down — getting mouth and jaw caught on the lower pipe while the upper pipe pressed into the windpipe.

Fence Posts

Make sure posts are set properly; they are the foundation of your fence. If posts are not well set, the entire fence will be less durable or even unsafe because wires will sag or sections of pole or board panels will get pushed over. Wood posts should be treated with a preservative and set at least 2.5 to 3 feet (0.7–0.9 m) deep, well tamped so they won't give. Posts for any kind of wire fence must be braced to keep the wire tight, and the wire itself (or netting) must be stretched tightly when

Pipe fence with strands of cable

installed. There must be a very solid brace at each corner or end, as at a gate, and every so often along a lengthy stretch of fencing; exact spacing will depend on the terrain and the pressure the fence will get. If a fence is not braced properly, the pull of the wire will loosen some of the posts or pull the corner posts up out of the ground.

Wire fences of any kind should have braces at every end or corner, to keep the wire taut.

Gates

Gates should be as durable and safe as the fence to which they are attached. A horse may spend a lot of time waiting by a gate or leaning on it because he knows it is the place to go in and out. The gate should be at least as tall as the fence and have a horse-proof latch. A chain with a snap will work if horses fiddle with a latch. Because an impatient horse may paw at a gate, never use wire or netting for a gate.

A gate gets more wear than the fence, so it should be sturdy (but easy to open and close), well hung, and well braced so it won't sag. The post on which the gate is hung should be stout enough to hold its weight, set deep enough to never move or lean. The longer and heavier the gate, the more likely it may pull the post out of line unless the post is well set and well braced. The gate should be smooth so a horse pressing against it won't be injured, and there should not be much gap between gate and post so a horse cannot get his head or foot caught if he tries to wedge through or paw at it.

Gates should be wide enough (at least 8 feet [2.4 m]; 12 feet [3.7 m] is better) for horses to go through easily without running into the gate or post. Gates between pastures should be situated in good locations — not in a corner or next to an obstacle — and wide enough that horses do not crowd one another and cause injury while passing through. Don't leave gates open between pastures or pens; many horses have been injured by running a shoulder or stifle into a gate or gatepost.

A gate in an electric fence should have an insulated handle so you can unhook it without getting shocked. Design the gate so the handle disconnects from the side toward the charger so the gate wire is "dead" when unhooked. Then it won't shock you or a horse when open, nor spark and snap (or start a fire in dry grass or weeds) if you put it on the ground or throw it over a fence, out of the way, when open. You can bypass the gate by using tall posts on both sides to run the electric wire over the top.

For safety, electric wire can be routed over the gate, using tall posts.

Pasture

Good pasture is the best feed for mares and foals and growing young horses. Mature idle horses may not need all the nutrients green pasture provides and often get too fat (unless their time on pasture is limited), but having pasture gives you some options for feeding and for allowing horses room to exercise.

The size of pasture needed to support a horse varies greatly depending on climate (a wet climate grows more grass than a dry one, unless you provide extra water with irrigation), soil fertility, and types of grasses growing there. It takes a lot more dry-land acreage to support a horse than well-watered green pasture.

A pasture should be large enough for the number of horses intended to graze it and should carry several types of palatable grasses. A mix of clover or some other legume with grass makes good pasture; legumes are high in protein. A pasture with too much clover can be detrimental to horses — if they eat too much, they may founder. Some types of clover, such as alsike clover, may also contain dicoumarin, which is toxic.

If you plan to have pasture supply most or all of your horses' feed, assess soil fertility and grass types. Your county extension agent can help with this. Then you'll know if the pasture can provide proper nutrients for horses or if it needs fertilizer, renovations, or reseeding to better types of forage plants. The county extension agent can advise you on the best type of fertilizer for your soil or plants, as well as what kind of plants do well in your climate and soil type that are nutritious for horses.

Disadvantages of a Small Pasture

In a small pasture, horses may beat out the grass in certain areas near fences or watering places, under shade trees, and so on. They'll overgraze their favorite types of grass, leaving others. Horses also don't like to graze grass that grows where they have defecated.

Weed Control and Pasture Management

Check pastures for weeds. Some are harmless (or nutritious, like dandelions and lamb's-quarter), others take up space that could be better used by good forage plants, and others are poisonous to horses. Certain poisonous plants are common to the Southwest, the East, the plains states or mountain West, or coastal areas, so check with your county extension agent or veterinarian for what to look for in your area. (Some of the most common poisonous plants are listed in chapter 11.)

Assess the weeds you have. Large patches of thistles or other undesirable plants should be eliminated. Patches of foxtail or cheatgrass (downy brome) are undesirable because the sharp awns of their seeds can become embedded in a horse's mouth and cause sores or abscesses.

The best defense against undesirable plants is a good stand of vigorous forage plants to keep weeds from becoming established on bare or disturbed ground. Patches of weeds can be controlled by spraying (check with your county extension agent for proper types of herbicides and how to use them) and seeding the area back to grasses, or by mowing diligently. If you mow weed patches before they mature, you keep them from going to seed and spreading. You also reduce their competition with grass plants, which come in thicker once given a chance. If weeds regrow you may have to mow several times during the growing season, but eventually this will weaken and eliminate them.

To keep pasture healthy and prevent it from going to weeds, don't have more horses than it can support, and don't leave them too long on it. Horses are selective grazers that eat favorite plants down to the roots, grazing them over and over, letting others go to waste. They like new green regrowth and keep eating the short grasses, leaving mature plants untouched. In wet seasons they tromp part of the pasture into mud and eventually destroy the plants.

For best pasture health, let it grow 4 or 5 inches (10.2 or 12.7 cm) tall before you turn horses onto it, and then remove them after they've eaten their favorite plants down to about a 2-inch (5.1 cm) stubble. If you leave them too long they'll graze the best grasses down to the dirt or even paw them up to eat the roots. The best solution is to divide your pasture into several sections so you can rotate the grazing, moving horses to new pasture periodically so the grazed area can regrow. If you don't have this option, lock the horses in a pen part-time; limit their time on pasture to what the plants can withstand. If they graze down some areas and leave others, mow the pasture periodically to clip off tall mature plants. When the plants regrow they'll be more palatable and the horses may eat them.

Water

Horses need water available at all times and should be provided with fresh, clean water frequently. The average horse drinks 8 to 12 gallons (30.3–45.4 L) a day and even more if he is eating dry feed like hay, if the weather is hot, or if he's working and sweating. A horse drinks less than 10 gallons (38 L) a day if the weather is cold or if he is eating lush green pasture that has high moisture content.

Water can be provided by a stream or pond, water tank, or tubs. A paddock or stall should have a tub, bucket, or some other type of waterer. For filling tubs or buckets, a water faucet or hydrant in the barn or barnyard is handy for hooking up a garden hose. All water containers should be checked daily and cleaned often to make sure they are free of debris and feed particles, which can ferment and lower the water quality.

Automatic waterers are nice, but they have disadvantages. They can malfunction and flood a stall or pen, or they can quit working. Many people neglect to check them daily, and a horse may go without water for too long if they stop working. It is also impossible to know how much water a horse is drinking when he uses an automatic waterer. A sick horse may stop drinking, but you won't necessarily see this early warning signal. If a horse goes without water for very long, he may suffer impaction. You need to know your horse's drinking habits and be aware of any change.

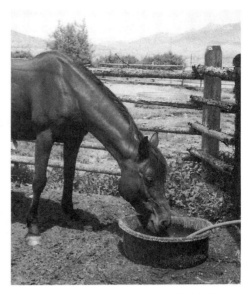

Provide water in a rubber tub that can easily be rinsed out daily and filled with a garden hose.

Cold-Weather Precautions

If you live in a cold climate, be sure to bury water lines to hydrants well below the frost line so they never freeze in winter. This usually means at least 3 to 5 feet (0.9–1.5 m) under the ground. Water lines under driveways or areas where there is heavy vehicle or horse traffic have to be even deeper because this type of impact drives frost down; the ground freezes deeper and stays frozen longer in spring.

In cold weather, drain water hoses completely after each use. Water buckets, tubs, and troughs should be

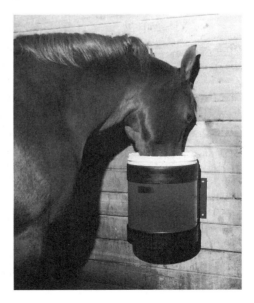

Insulated water containers work like a thermos bottle. The water is kept from freezing, without the risks of using a bucket heater. (Courtesy of Country Manufacturing)

insulated. Buckets with insulation, acting like a thermos bottle would, can keep water from freezing for
many hours; consider using these in stalls or paddocks
when watering single horses in small containers (a
small amount of water freezes more quickly than a
large volume in tank or trough). If water is warmer,
horses drink more in cold weather and are less at risk
for impaction. Electric water warmers can be used in
buckets or tanks but are dangerous if they malfunction. Remember that all electrical equipment is a
potential fire hazard in a barn.

*Freezeproof,
self-draining
water hydrant*

Even with insulation and depending on the container's size, you may have to break ice out of a tub,
bucket, or tank daily. Rubber tubs and buckets are
better than plastic ones, which crack at low temperatures. Plastic tubs are also
not as durable if you have to pound on them to get the ice out.

Water hydrants in a cold climate should be freezeproof and self-draining
(i.e., the water drains out of the upright portion, going deep into the ground,
each time you turn the hydrant off, so there is no water left in the upright pipe to
freeze). They should be located where horses don't have access to them — outside the pen or pasture, or in a barnyard area well out of the way. Otherwise a
horse may be injured by running into the hydrant, or a playful horse may nibble
the handle and accidentally turn on the hydrant.

Make Sure All Facilities Are Hazard Free

Before putting horses into a new pen or pasture, check it carefully for anything
that might cause injury. Make a habit of looking closely at all facilities every
time you enter a pen, pasture, or barn stall. Maintaining a watchful eye can
prevent serious problems.

*Hazardous fence:
Note the sagging
wire and netting
that could catch a
horse's foot.*

Common Hazards and Solutions

Hazard	Examples	Danger
Sharp projections	Exposed nails, hooks of wire, projections from broken hardware or fencing	Can cause eye injuries or tear an eyelid if a horse rubs on them or bumps them
Fences in disrepair	Sagging fence wires	Strangulation; cuts on legs if pawed or on head and neck if horse reaches through
Halter left on		A horse may injure himself catching a foot in it or hooking the halter on a projection of stall hardware or fence; a horse rubbing on a fence or sticking his head through poles or around a post may catch the halter on a nail or staple, stub of wood or a bolt — and pull back and hang himself
Unfilled holes in pastures		If a horse steps in a hole, he can pull a tendon or break a leg
Wood posts sawed off at ground level that have rotted		If a horse steps in it, he can pull a tendon or break a leg
Wood or metal posts broken off above ground level		May puncture a foot or cause lower leg injuries
Badger holes, ground squirrel holes, or other burrows made by rodents		If a horse steps in a hole, he can pull a tendon or break a leg
Sharp edges	Feeders, waterers, water tanks, mangers	Horse may suffer cuts and wounds
Protruding corners or edges	Feed mangers, gates, metal sheds, roofs	Horse may suffer injury or bruising
Narrow spaces	Doorways; gateways; space between feeders, waterers, or wall or fence	Horse might get caught
Low overhangs	Roofs, guy wires from power pole or telephone pole	Horse may run into it
Junk, old machinery, garbage	Old batteries Sharp edges, rusty bolts	Contain poisonous substances such as lead Can cause injury or snag the horse
Road borders	Broken beer bottles, cans, paper, other litter	Cans and broken glass can cause injury
Baling wire or twine		Young horses may chew or eat them

SOLUTION	PREVENTION
Remove or cut off flush with surrounding surface	Use bolts with rounded ends when building fences, gates, or stalls, or cut them off flush with surrounding surfaces; remove broken or unused gate hinges
Repair or replace	Maintain fences at all times
	Never leave a halter on a horse at pasture or a horse in stall
	Fill all holes
	Remove and fill hole
	Remove and fill hole
	Fill all holes
Repair or cover sharp edges	
Repair or cover or buffer sharp corners and edges	
Rehang gates, fill spaces, redo gateposts	
Make it more visible by installing post and horizontal pole	Install post and horizontal pole to shield horses
Remove or fence off junk pile	
Remove or fence off the old machinery	
Patrol and pick up litter regularly	
Make a habit of never leaving twines or wire where horses can get to them; dispose of these hazards in a safe place	

Two

FEEDING AND NUTRITION

Feeding is probably the most important single aspect in caring for a horse. What he eats — quality and quantity of feed — determines to a large extent how well he can perform (or grow) and how healthy he'll be. If he's still young, the feed he eats will make a difference in how well he grows and develops.

A horse inherits its potential for height and size, form and appearance, speed, strength, endurance, and agility. But how fully potential is realized depends greatly on the feed he gets. A horse with good breeding and much potential may develop and perform poorly if fed inadequately. He may have potential for size and strength but end up stunted due to improper or inadequate feed. He may develop joint and bone problems and be a cripple if overfed while young, causing him to grow too fast and carry too much weight on immature bones and joints. Even a sound horse with great ability may perform poorly on insufficient or improper feeds, lack energy, or suffer problems such as founder or muscles tying up. The conscientious horseman must make sure the diet is balanced and provides the proper amounts of what the horse needs.

The Well-Fed Horse

The horse is unique among domestic animals because he is used for athletic ability rather than for production of meat, milk, or wool. Whereas a beef animal is fed to gain the maximum body weight in a short time, the horse should be fed for energy and longevity instead — never to be fat. If he is fat his health, endurance, longevity, breeding ability, and athletic usefulness will be impaired.

Digestion in the Horse

The horse is a roughage eater like a cow, but unlike the cow he has a simple stomach and only chews food once. His relatively small stomach holds only 2 to 5 gallons (7.6–19 L), compared to the cow's rumen capacity of 15 to 30 gallons (57–114 L). The horse eats smaller amounts but more often — continually throughout the day and night if on pasture, stopping only to rest, switch flies, or nap before grazing again. The cow eats more at one time, more quickly, then lies down and rechews it at her leisure.

The horse has a small, simple stomach and depends on his well-developed cecum for fermentation and breakdown of roughages. The cow has four stomachs and breaks down roughages in her large rumen instead of in her not-so-well-developed cecum.

Ruminants such as cows, sheep, goats, and deer process and digest roughage in the largest stomach, the *rumen*, where micro-organisms help break down cellulose and other indigestible material into usable nutrients. In contrast, horses (and rabbits, kangaroos, and other herbivores that do not have a rumen) break down roughage in the *cecum*. This "fermentation vat" does much the same job as a cow's rumen. Because it is located past the stomach and small intestine, the horse cannot process extremely coarse feeds, such as straw, as easily as the cow. She rechews her food and breaks it down by fermentation in the rumen early in the digestive process, whereas the horse accomplishes the final stages of digestion in the 4-foot-long (1.2 m) cecum and 12-foot-long (3.7 m) colon. Together the cecum and colon have a combined capacity of more than 25 gallons (95 L). They are designed to hold, process, and digest the cellulose and other fibrous parts of the horse's diet that are broken down by bacteria, protozoa, and yeasts.

Cellulose and Body Heat

Cellulose, the fibrous part of roughage, is broken down into glucose and other nutrients, but no animal makes an enzyme that can do this. All grass-eating animals rely on micro-organisms to ferment and digest cellulose, changing it to energy-producing fatty acids. The chemical change that takes place during this process also produces heat energy and can help keep a horse warm in cold weather. When temperatures drop, a horse can get more good from extra hay than from extra grain because digestion of roughage creates more body heat to keep him warm.

Importance of Proper Feed

The horse's "fermentation" vat works well under normal conditions when he's eating pasture or good hay, but sometimes problems can develop. Impaction of the cecum can occur if a horse is fed finely chopped hay or lawn clippings. In nature the horse never ate grain or alfalfa; his digestion was not designed to handle these feeds. Therefore, these "unnatural" feeds may cause trouble. Abrupt changes in feed can upset the population of micro-organisms (with some dying off and producing toxins), leading to colic or laminitis due to a change in the rate of feed breakdown and creation of excessive gas or toxins.

The horse depends on microbes and fermentation for digestion of feed and absorption of nutrients and vitamins. The microbes are his allies — essential to his health — but if he is fed improperly they become his deadly enemy. Understanding the horse's digestive process enables us to feed him wisely.

The Digestive Tract

The horse's digestive tract is a long tube: esophagus, stomach, small intestine, cecum, large colon, small colon, and rectum. Because the stomach is small, a hay-fed horse should ideally be fed several times a day instead of only once. The daily ration should be divided in two (preferably three) parts and fed at intervals, the largest portion in evening so the horse can eat it at his leisure during the night. An option for broodmares, as well as inactive horses during winter, is to put out large (half-ton or larger) bales so they can eat as they wish.

Vomiting

Unlike a human being, the horse cannot readily vomit because of the way the back of his mouth is constructed (he cannot breathe through his mouth very well for the same reason). If he tries to regurgitate, feed is forced out through his nose; it cannot be ejected easily through the mouth. Material coming back up the throat may go into the windpipe, causing choking or pneumonia if food particles get into the lungs. A horse's stomach may rupture from the pressure of trying to vomit. Because he cannot unload his stomach of bad feed, the horse should always be fed the best quality feeds. Spoiled or moldy feeds, or toxic plants that cause digestive upsets, can be disastrous.

The stomach initiates the digestive process, breaking down feed a little and starting the digestion of proteins. The small intestine continues digestion of proteins and also breaks down starches, sugars, and to some extent the fats. The pancreas and liver supply enzymes to the intestines so that the digestible parts of the feed can be broken down and absorbed. The horse has no gall bladder (in other animals it releases bile into the intestine for breaking down fats); instead, bile is secreted continuously in the small intestine, but not in large enough amounts to thoroughly digest the fats. Fats are digested when they reach the large intestine.

Many horsemen used to think watering a horse during or directly after feeding would wash food — especially grain — out of the stomach and into the intestine before it was digested, causing inadequate nutrition. But under natural conditions a horse drinks water periodically as he eats. When eating hay, he needs adequate saliva and water to moisten the dry feed. The horse should have free access to water.

The large intestine (especially the cecum) digests roughage more thoroughly; here the fibers are broken down by microbes that change the cellulose to energy-producing fatty acids. They also manufacture amino acids — from which the horse gets usable proteins — and B vitamins, which are essential to good health.

Digestion is completed in the large colon, after which the excess water is absorbed from residue in the small colon, forming balls of feces. The horse normally passes 30 to 50 pounds (13.6–22.7 kg) of manure daily.

Basic Nutritional Requirements

The horse needs carbohydrates, fats, protein, vitamins, and minerals — all in proper amounts. In many cases, too much can be as harmful as too little. On a natural diet of grasses grown in fertile soil, the horse gets all his nutrient requirements. But sometimes feeds are grown in geographic areas with mineral deficiencies (check with your county extension agent or veterinarian for information on your area), and sometimes horses eat unnatural diets (grain, and hay or pellets instead of grass). Many horses are fed supplements as well. Because a well-balanced diet is so important, the responsible horseman must have an understanding of all the horse's basic nutritional requirements.

Carbohydrates

The major source of energy for the horse, carbohydrates include sugars, starches, and cellulose. Carbohydrates are derived from cellulose as it is broken down and digested. Oats, barley, corn, and other concentrates may be as much as 60 percent sugar and starch. Cellulose, found in all roughages, is less easily digested and must be eaten in larger amounts to produce the same amounts of energy.

Fats

Fats are needed only in small amounts, and natural feeds do not contain much fat. Small amounts of fat, however, do aid the horse in absorption of fat-soluble vitamins and furnish fatty acids necessary to sustain healthy cell membranes. The only fatty acid the horse cannot manufacture for himself (linoleic acid) is found in the vegetable fats of roughage and grain.

Some hard-working horses such as endurance animals are given fats as a supplement to be metabolized as a source of energy. Adding corn oil or lard to a hard-working horse's diet can increase the energy content of his ration, as fats contain more than twice as much energy as an equal weight of carbohydrate (such as grain) and can reduce the amount of grain needed. But fat as a supplement should be fed carefully, starting with a few tablespoons and increasing gradually to a cup poured over the grain; don't feed too much, for it may have an adverse effect on absorption and utilization of certain vitamins and minerals. Recent research has shown that flaxseed oil (as found in the oil-cake some horsemen use as a protein supplement) may be healthier because it contains the proper essential fatty acids needed by the horse.

Does Fat Contribute to a Shiny Coat?

The average horse doesn't need fat added to his diet, although some horsemen like to use vegetable oils and other fats to give a shiny coat. Genetics plays a greater role than diet in determining whether a horse has a shiny coat. This explains why one horse may have a glossier coat than another when both are equally fit and healthy and eating the same feeds.

Protein

Protein (made up of amino acids) is necessary for growth — building and repairing of muscles and other body tissues. Protein helps build bones, blood, skin, hair, hoofs, and muscle. A deficiency can impair a horse's growth if he is young, interfere with hoof development, and cause a poor hair coat. Legume hay is a good source of protein, as are soybean, linseed (flaxseed), and cottonseed meals.

Horses need extra protein while they are growing. For example, a mare needs additional amounts during the last third of pregnancy — when the fetus is making most of its growth — and while she is nursing a foal. A little extra protein in a horse's diet in spring can help him shed his winter coat (as new hair comes in) quickly and easily. Excess protein is metabolized as energy; it also produces heat when digested, so a high-protein feed is a good addition to diet in cold weather but a poor choice during hot weather when a horse is working hard.

This old horse is slow to shed in spring. A little extra protein in the diet can help a horse shed more easily; the new hair will grow in more quickly.

Vitamins

Vitamins are organic compounds that are required in small amounts for normal body function. They regulate chemical reactions, keeping the body operating smoothly. There are two kinds of vitamins — fat-soluble ones (A, D, E, and K) and water-soluble ones (B vitamins and C). All green forages are rich in most of the vitamins the horse requires. A horse on good pasture should need no additional vitamins for maintenance; pasture provides adequate body requirements in a natural and inexpensive form.

If you have no pasture, limited pasture, or poor pasture (poor soil, poor types of grasses, or mostly weeds), the horse needs good quality hay. Carefully analyze your situation — pastures, feeds, your horse's condition and use — and then decide if he needs vitamin supplementation. You may want to consult your veterinarian. Vitamins can be costly if the animal doesn't really need them.

Vitamin A

Vitamin A is important to reproduction, eye health, and hoof growth. A deficiency can cause abortion, birth of weak or poorly developed foals, or vulnerability to digestive upsets, colic, and respiratory ailments (cough, colds, or runny nose).

Vitamin A is made by the horse from carotene in various plants. Green pasture and green, leafy alfalfa hay are good sources of carotene. Grass hay is not as high in this vitamin, especially if cut when mature. Vitamin A is fat soluble and can be stored in the body. Horses on pasture during summer usually store enough to last through winter. If your horse is never on pasture or alfalfa and your grass hay is not very green, or if the horse is under special stress during winter (e.g., a pregnant mare), consider giving vitamin A when supplementing the diet.

Vitamin D

Vitamin D plays an important role in the conversion of calcium and phosphorus into healthy bones. A deficiency can cause rickets in growing horses and osteomalacia (soft bones) in mature horses.

Vitamin D is produced in the skin if a horse is out in the sun. Sun-dried hay is also a good source. Under normal conditions a horse should never be short on this vitamin, but if he is kept stabled all the time he may need supplementation.

Vitamin E

Vitamin E is necessary for the development and function of muscle, as well as for healthy reproduction. It is very important to the immune system and for broodmares, young horses, and horses under heavy use or stress. It is important in neurologic function and can help horses with certain diseases such as equine protozoal myeloencephalitis (EPM), which affects the spinal cord. This vitamin is usually very adequate in natural feeds, but during periods of stress horses may need extra, especially if natural feeds are not available.

Vitamin K

Vitamin K is essential to blood clotting and for activating certain proteins in the horse's body. Vitamin K is made in the horse's intestines.

What's the Proper Amount of Vitamins?

All vitamins are required only in very small amounts, and the fat-soluble ones (A, D, E, and K) are stored in body tissues. Feeding excessive amounts of fat-soluble vitamins can be dangerous. For example, too much vitamin D can cause bruises to calcify, and too much vitamin A can cause bone breakdown. Green alfalfa may contain up to seventy times the amount of vitamin A needed by a horse, but this is not a problem unless he is receiving additional sources of vitamin A as well, such as feed supplements that bring the total up to toxic levels. The maximum daily requirement of vitamin A for a lactating mare is 27,500 International Units (IU). The maximum daily requirement for a mature, non-reproductive horse is 12,500 IU. Always check with your veterinarian before giving vitamin supplements.

B Vitamins

B vitamins — including thiamine, riboflavin, niacin, pyridoxine, pantothenic acid, biotin, choline, folic acid, and B12 — are created by bacteria in the horse's cecum and large colon. The B vitamins are important for energy metabolism and the health of red blood cells.

Vitamin C

Vitamin C is also synthesized by the horse. The horse does not need supplementation with B vitamins or vitamin C unless he is ill, anemic, and unable to produce them. Vitamin C is an antioxidant that protects fats, proteins, and cell membranes from damage. Vitamin C also helps in the formation of bones and teeth.

Minerals

Minerals are inorganic elements (i.e., not produced by living things), that are required in small amounts for good health and normal body function. Under natural conditions the horse gets essential minerals by eating plants that store them from soil. The major essential minerals are calcium, phosphorus, magnesium, sodium, chlorine, and potassium. Others needed in minute (trace) amounts are iodine, cobalt, copper, iron, zinc, manganese, and selenium. Supplements are necessary in regions where certain minerals are lacking in the soil. Overdose can be toxic, so check with your veterinarian before giving mineral supplements.

Know Your Soil

If your horses on pasture eat hay and grasses from your property at any time of year, have your soil tested by your county extension agent to determine what supplements may be needed in the diet.

Iodized salt should be given in geographic regions that are low in iodine. Lack of iodine in a pregnant mare's diet can result in stillbirths, weak foals, or foals born with contracted tendons. Other trace minerals, such as iron or selenium, are also lacking in some areas. Ask your veterinarian or county extension agent about possible deficiencies in your area.

The minerals needed in greatest amounts are calcium and phosphorus. The trace minerals (needed in tiny amounts) that are most commonly supplemented are iodine, selenium, and iron. Salt also has an important role in the horse's diet.

Calcium and Phosphorus

Calcium and phosphorus are important for strong bones and teeth. A lack of either mineral can impair growth or lead to unsoundness. Forages like pasture and hay are usually high in calcium, and grain is high in phosphorus. A mature animal needs a little more calcium than phosphorus (the usual ratio in pasture grass), whereas a growing horse needs the same amount of each. Many breeders give their young horses a little grain because it is high in phosphorus.

Vitamin D is essential in the proper relationship of these minerals. If there is not enough of it the calcium cannot be absorbed, throwing off the calcium-phosphorus balance. Too much grain can also upset the balance, by providing too much phosphorus and tying up the calcium in a compound that is not absorbed by the body. Feeding more calcium does not solve the problem; it is not properly absorbed. Overfeeding a young horse on grain can result in soft bones, crooked legs, and unsoundness. No horse should be overfed, especially on grain, or over-supplemented with minerals. Salt is the only exception; it can be fed free choice.

Iodine

Iodine helps control body metabolism and also is important for growth. The tiny amount of iodine required by a horse (about 1 milligram daily) can be supplied by iodized salt or mineral salt if your region is deficient. But the horse may get an overdose if you're also giving trace mineral salt, plus a grain mix that contains iodine, plus a vitamin-mineral or "shiny hair coat" supplement. Kelp products contain a lot of iodine. Check all your feeds. (See "Know Your Soil" box for more on iodine.)

Selenium

Selenium is essential for proper nutrition and health. Deficiency can lead to muscle disease and weakness, especially in foals, and can cause some horses' muscles to tie up. A combination of vitamin E and selenium is often prescribed for horses that are prone to this problem. Like vitamin E, selenium is essential to immune function. Horses occasionally get too much selenium (which is toxic) when eating certain plants that store it from the soil (e.g., locoweed, almost all vetches, goldenweed, saltbush, snakeweed, woody asters, and some of the other asters).

Iron

This trace mineral is often over-supplemented. Iron is crucial to the formation of red blood cells and hemoglobin, which transports oxygen. If a horse is eating hay, he is probably getting enough iron. If you are also feeding him a commercially prepared grain mix that has an iron supplement, he may get too much. Toxicity from misuse of this supplement is more harmful than iron deficiency. Horsemen sometimes give iron supplements thinking it will increase endurance or athletic performance, but this is dangerous. Overdose can be fatal in mature horses, and foals given supplemental iron at birth may die.

Salt

Sodium chloride constitutes ordinary salt — a combination of two important minerals needed by the body. Grass and other roughages do not contain salt; the horse must get it from some other source, and it should always be available — in blocks or loose. Some feeds have salt added, but it may not be enough if the horse is sweating or lactating. Mare's milk contains 160 to 365 parts per million of sodium and 300 to 640 parts per million of chlorine — the chemical components of salt. You should always provide free access to salt even if a horse is receiving a feed containing salt.

If the horse likes to lick his salt block for something to do and eats more than he needs, the excess is flushed out of his body with urine and does no harm, as long as he always has plenty of water to drink. Salt is the only mineral that can be safely fed free choice. By contrast, many minerals and trace minerals are only needed in very small amounts, and overdoses can be quite harmful.

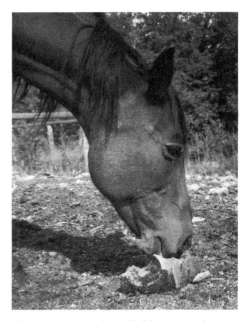

Always make salt available to your horse.

Providing Salt

It is impossible to meet a horse's salt needs by feeding a specific amount per horse or a fixed percentage in a feed ration, because horses vary widely in individual salt needs and consumption. The safest way to provide salt is to give free access at all times, along with plenty of water, to balance the horse's own salt and water intake.

Under normal conditions, mature horses doing minimal work eat about half a pound (0.2 kg) of salt per week, varying with the salt content of prepared feeds being given and how much the horse sweats. Salt can be provided in granule form — in a salt box — or in blocks. Blocks work well if they are not too hard. Rock salt is too hard, causing sore tongues if horses lick them a lot and inadequate salt intake because horses don't eat enough.

If there are many horses in a pasture, provide more than one source of salt so timid horses can get their chance at it. If there's heavy rainfall or the pasture is irrigated, put salt in an area where it will stay dry or in a covered salt box. The "roof" over the salt should be high enough for the horse to reach in and eat the salt. Once dissolved by water, the salt is wasted and also kills the grass around it.

This is one reason it's best to offer a horse plain salt rather than mineralized salt. A horse that sweats a lot in hot weather will need to replenish his salt loss by eating more salt, and if it contains trace minerals he may eat more of them than is safe.

The sodium and chlorine in salt are the main ingredients of the fluid circulating within the the horse's body. They help keep the proper water balance in the body and aid in its release through the pores as sweat for cooling during hot weather and periods of exertion. The more a horse exercises, the more salt is lost through sweat and the more he needs to replace it.

Roughages (Forages)

The primary forages fed to horses are pasture and hay. Silage is sometimes used but is more risky. Good corn or grass silage can substitute for one-third to one-half the hay ration, but it should be introduced into the diet gradually and never be the only roughage fed. Because it has high moisture content and weighs more than hay of equivalent nutrient value, it usually takes about 3 pounds (1.4 kg) of silage to equal the food value of 1 pound (0.5 kg) of hay. Silage that is spoiled, moldy, or frozen should never be fed; it can cause digestive upsets or poisoning. Horses are much more susceptible to poisoning from spoiled silage than are cattle or sheep.

Straw should never be used as horse feed. If cut as hay (while still green and growing, before grain heads are mature), oat hay is palatable and can be used as part of a ration, but mature straw is too coarse. It has little food value and is inefficiently digested. Grains other than oats or beardless barley have rough beards and seed awns and if cut for hay are not suitable for horses because they irritate mouth and throat. Take care in using oat hay; it can cause nitrate poisoning when grown under certain conditions, such as a period of drought followed by rain. It should be checked for nitrates.

Pasture

Pasture is the most ideal feed for horses; it is usually inexpensive and nutritious, and it constitutes the best environment for a horse because he gets sunshine and exercise. The greatest single factor in determining quality of pasture grass is stage of maturity. The older and drier the grass, the less palatable it is and the less food value it has. Mature grasses are more coarse and not as easily digested as young growing plants.

Some hay grasses (e.g., timothy, smooth brome, intermediate wheatgrass) are good horse feed when pastured, especially if plants are young. Kentucky bluegrass is good pasture if mixed with other palatable grasses or a suitable legume. Any pasture or dry-land grasses (except downy brome, often called cheatgrass) are preferable to swamp grasses; the latter are usually coarse. Pastures should have adequate drainage to prevent swampy soil. Permanently damp soil grows inferior types of grasses and serves as a breeding ground for horse flies.

Pasture is the ideal feed for horses.

Hay

Hay is usually the major part of a horse's diet if he's not on pasture. Hay must be of good quality to supply essential nutrition. It should be cut before fully mature (before bloom stage or heading out of grass seeds), while the grasses are still young, tender, and growing. Good hay is green in color (not yellow or brown), leafy, with small, fine stems. Coarse, stemmy hay is hard to digest. Clean, bright, sweet-smelling hay is most nutritious. Dry, dusty hay not only is unpalatable but also can cause respiratory problems. Moldy hay may cause colic.

Legume Hay

Alfalfa, clover, or other legumes should be fed sparingly because horses like them and tend to overeat. Mixed hay, containing both grasses and legumes, is good for horses; the legume adds extra protein. Alfalfa hay should have moderate stems (first cutting). Horses don't need fine-stem dairy quality hay; it is not only too rich but does not have sufficient roughage for proper digestion. Make sure alfalfa is not too mature and stemmy. If it has bloomed (look for purple flowers), it is overly mature.

If you are feeding straight alfalfa, don't feed more than 1 pound (0.5 kg) per 100 pounds (45.4 kg) of body weight; fill in with another roughage if possible.

Too much rich alfalfa can cause digestive upsets, loose bowels, or metabolism problems in some horses when worked hard. If you have no grass hay, select alfalfa that is not too rich and leafy so the horse will have adequate roughage.

Avoid Mold and Fungi

If they are well cured — dry, not damp and moldy; but not so dry that they are sun-bleached, losing vitamins and food value — both grass and legume hay make good horse feed. Grass hay is less likely to be dusty or moldy than alfalfa and other legumes. The richer hay is a more ideal environment for molds and fungi if it is damp. Never feed moldy hay; it can poison a horse, cause colic, or bring on abortion in pregnant mares. Discard all moldy and dusty hay.

If it's just a little dusty, shake it to remove as much dust as possible, then sprinkle with water to settle the rest. Sprinkled hay should be eaten right away; if it sits wet a long time it may begin to mold. Extremely dry hay is usually dusty because the leaves have shattered. It doesn't hurt the horse to *eat* dusty hay; harm is done when the dust gets into his lungs. Hay particles cause irritation and coughing and can lead to lung disease.

Feeding Hay

Whether you feed alfalfa, grass hay, or a mix will depend on your horse's needs and the types of hay available in your area. First-cutting alfalfa usually has some grass mixed in and can be a very good hay for horses. Second, third, and later cuttings are usually straight alfalfa and may be too rich for most horses.

Not all hay is the same in weight or nutrients. It may take twice as much poor hay as good hay to keep a horse in condition. Poor hay is never a bargain, for it can lead to digestive problems, nutritional deficiencies, or respiratory trouble.

Weighing Hay

It is easier to estimate the weight of baled hay than loose hay. There can be a lot of difference in size and weight of bales, so weigh a few from each batch you buy to know how much is in a bale. Some are more loosely packed than others and may not weigh much even though they look big. Some hay is light because it is too dry or the bales are too loose, whereas others are heavy because they are moist and moldy. Bales from the top or bottom of the stack may be heavy and spoiled because they have taken on moisture.

Alfalfa hay of the same volume is usually heavier than grass hay. Because it has more nutrients, horses need less total weight per daily feeding than with grass hay, so this means much less volume. Good alfalfa hay is so palatable, nutritious, and easily digested that most idle horses gain too much weight if allowed to eat as much as they want, and some may get indigestion or colic.

Yet if you restrict a horse to the amount of alfalfa needed for maintaining proper weight, without providing any other source of roughage, he may start chewing wood. There is so little volume of alfalfa needed, the horse does not feel full and looks for more to nibble on. Alfalfa can be gobbled up in a hurry, and he's left with nothing to do. This problem can be reduced if you choose alfalfa with enough stem to provide more roughage.

Grass hay takes longer to eat, occupying more of the horse's time, and provides more fill; it is usually higher in fiber than alfalfa hay. Grass hay more adequately satisfies the horse's need for "chew time," and it provides the roughage needed for the digestive tract to function properly.

Changing Hay

If you change hay, especially from a grass to a legume, do it gradually, mixing the hay for several feedings so the horse's digestive tract can become accustomed to the new feed. A sudden change to alfalfa can alter the microbial environment in the cecum. A gradual change, over several days' feedings, is less apt to cause indigestion or colic.

Health Tip

Lawn clippings should not be used as horse feed because the animals may overeat and become impacted. Clippings often spoil and ferment if they are left piled for any length of time; if the horse eats them the consequences can be devastating. If a horse gets access to lawn clippings and develops colic, call your veterinarian.

Concentrates (Grains)

Concentrates (grain and grain by-products) are low in fiber and high in digestible nutrients. The horse needs digestible nutrients but must also have roughage for proper function of the digestive tract. Roughage provides extra bulk — fiber — as in the stemmy parts of hay or hulls on grain. A bulky feed is one with relatively little weight for its volume. The more weight, the more energy available in the feed. Hence the name *concentrates* — grains contain more nutrients per unit of volume than do forages. Grains that are often used for horses are oats, barley, and corn.

Oats

Oats weigh about 32 to 40 pounds (14.5–18.1 kg) to the bushel and are approximately 70 percent total digestible nutrients, with 9.4 percent digestible protein. They contain about 11 percent crude fiber. Oats are a standard horse feed, palatable and nutritious. They are the safest grain to feed, even in large quantities, because they are

Grain provides more energy per pound than hay.

roughly 30 percent hull (outer covering around the kernels), providing some bulk. Shelled corn, oilseed meals, and wheat (heavy concentrations without much bulk) tend to form a doughy mass in the stomach, which can cause colic, whereas oats form a loose mass that can be easily digested. The bulky oat hulls keep the feed moving through the tract at a normal rate.

Eschew Whole Grain

Crimping or rolling oats increases their food value because they can be more easily chewed and digested. In contrast, whole grain doesn't always get chewed — kernels that get caught in the hollows of the teeth are not ground up but swallowed whole and pass through the digestive tract, wasted. If you examine the manure of a horse that has been fed whole grain, you'll see grain kernels that have passed through intact.

Barley

Barley is 78 percent digestible nutrients — about 10 percent digestible protein and only 5.9 percent crude fiber — and a bushel weighs roughly 48 pounds (21.8 kg). Barley hulls, like oat hulls, provide bulk to aid digestion, but not as much. Some oats or wheat bran is often added to barley to increase the bulk. Barley kernels are very hard and are best if rolled or steam rolled. Veterinarians

Feed by Weight, Not Volume

Hay

Not all bales of hay weigh the same. Until you are good at estimating how heavy the "flakes" of a given bale of hay are, weigh them — especially when changing hay — to know how much you are feeding. A 1,000 pound (453.6 kg) idle horse usually needs 25 pounds (11.3 kg) of grass hay daily, given in two or three feedings. If the horse's work increases, the amount of hay should be increased. Some horses get by on less hay; others need more.

Grain

Not all grains weigh the same, so feeding by quarts or gallons can be confusing. Corn weighs more than oats; a quart of corn may give twice as much energy (or potential body fat) as a quart of oats. Weigh your feed, find out how much your scoop or bucketful holds in terms of weight, and recheck it when changing feeds. Making a change in the ration without adjusting for the weight of the new feed may lead to digestive problems or colic. As a general rule, a quart of oats weighs 1 pound (0.5 kg), a quart of barley 1.5 pounds (0.7 kg), and a quart of corn 1.75 pounds (0.8 kg).

Always weigh grain so you know how much you are feeding.

recommend that *all* grains be steam rolled or processed in some manner to kill the organism that causes EPM (see chapter 8). No grains should be ground for horse feed; ground grains are too dusty and fine, tending to dough up in the stomach.

Corn

Corn, with no hulls, is high in energy and low in bulk. Rich in energy, corn should be fed sparingly or it will make a horse too fat. Being lower in protein, if fed to growing horses or pregnant mares corn should be supplemented with legumes or another high-protein feed. Corn weighs about 56 pounds (25.4 kg) to the bushel and is 80 percent digestible nutrients, but only about 6.7 percent digestible protein and 2 percent fiber. It takes 15 percent less corn than oats to keep horses in good condition the ration is balanced with additional protein.

Wheat

Wheat is usually not fed because of cost, but in wheat-growing areas it is sometimes economical. It should be rolled or crushed to make it easier to chew and digest, fed only in moderate amounts — less than 20 percent of the grain ration, and mixed with a bulky grain like oats or bran to avoid colic. Wheat is high in energy and protein. It weighs about 60 pounds (27.2 kg) per bushel and is 80 percent digestible nutrients. It is about 11 percent protein (one of the highest grains in protein content) and only 2.6 percent fiber.

Wheat Bran

Wheat bran, a by-product of milling wheat, is the coarse outer coating of wheat kernels. It can be mixed with concentrates for additional bulk and for a laxative effect. Bran is twice as bulky as oats, poorly digested, and low in energy. Being poorly digested, it increases intestinal fill and amount of manure passed.

Wheat bran can be added to a daily ration (no more than 10 to 15 percent of the ration) or given occasionally as a bran mash. A hot bran mash can be given on a cold winter evening or after a hard day's work. Some horsemen use it as a tonic for a tired or sick horse, or to get more moisture into the bowel, or after foaling to keep mares from becoming constipated. Because of its laxative effect, it is often fed to older horses to prevent impaction due to sluggish bowels. Never feed it dry unless it's less than 10 percent of a grain mix and well mixed in. Dry bran will swell in the stomach, pulling fluid from the contents, causing digestive problems and discomfort.

Recipe for Bran Mash

Make a bran mash by adding hot water to about 2 pounds (0.9 kg) of bran, stirring until it is crumbly but not soupy. Some horsemen like to use boiling water, adding enough to a 2-gallon (7.6 L) bucket to get the consistency of oatmeal, then covering the bucket to let it steam until cooled. Feed the mash to the horse after it has cooled enough to be still warm but not hot.

Wheat bran is high in fiber, protein (about 17 percent), and phosphorus but very low in calcium. It adds bulk to the diet, but its nutrient value is low except for the high protein content.

A drawback to feeding bran (especially on a regular basis) is its calcium-phosphorus ratio — approximately one part calcium to ten parts phosphorus. If fed frequently, it can cause serious imbalance in the diet, interfering with proper absorption of calcium. This can cause damage to growing bones in young horses and deformed bones in mature animals. An extreme imbalance in calcium and phosphorus can cause bones to soften and thicken until the skull and jaws are noticeably deformed, a condition called bighead.

Sorghum

Sorghum grain, also called milo, is a good horse feed in areas where it is grown. This is a hard, round grain that should be rolled or crushed, and because it is lacking in bulk it should be mixed with bran, oats, barley, or some other bulky feed to avoid constipation. Milo weighs roughly 56 pounds (25.4 kg) per bushel and is about 80 percent digestible nutrients and only 2.3 percent crude fiber. It contains approximately 8.5 percent digestible protein.

Molasses

Molasses, a by-product of sugar refining, provides carbohydrates. It is often added to grain to reduce dustiness and to increase the energy content of the ration. Molasses contains about 54 percent digestible nutrients. The proportion of molasses in grain should not be more than 10 percent; too much will give a horse loose bowels. Refined sugars can also be a problem in horses that tie up, making them more likely to suffer this problem because of the way their muscles use this form of energy.

Oilseed Meals

Linseed meal, soybean meal, and cottonseed meal are by-products of extraction of oil from seeds. These meals are high in protein and are often used as supplements, added to a grain ration to increase protein content. Because these are very concentrated feeds, a horse should never be fed more than 1 pound (0.5 kg) per day.

Commercial Grain Mixes

Even though it is generally cheaper to buy grain in bulk (e.g., a ton of oats or barley) or to purchase several types of grain and mix them yourself, many horsemen like the convenience of commercial grain mixes and buy it by the sack. There are many good products available, but be sure to read labels carefully and know what is in the feed and in what quantities. Otherwise you may overfeed some nutrients, especially if you are also giving the horse a special supplement whose ingredients are duplicated in the grain mix.

Pelleted Feeds

Pellets are handy and economical if hay is expensive in your area. They are helpful if a horse has respiratory problems when eating hay. There are three types of pellets — grains, pelleted hay, and complete pellets that contain hay, grain, vitamins, and minerals.

Advantages of "Complete" Pellets

Pelleted feeds are convenient. Rations can be easily measured, contents are uniform, pellets are easy to store and transport, there is no waste, ingredients are nearly all utilized, and little passes on as manure, making for easier cleanup in stalls. Pellet-fed horses are trim (no hay belly), and old horses, individuals with bad teeth, or horses with heaves (allergy-caused respiratory disease) often do better on pellets than on hay.

Disadvantages

The main disadvantage to a diet of complete pellets is lack of bulk. In fact, pellet-fed horses usually chew wood because they crave roughage. Pellets that are a complete feed contain hay, but it's finely ground. The horse eats less total bulk and doesn't feel full. He needs some roughage; bulky fiber stimulates the digestive process and keeps the tract moving at

A horse that doesn't get enough roughage in his diet will chew wood.

normal rate. Most horses chew wood when first put on pellets, then taper off somewhat after they have become adjusted to the new ration and their stomachs have shrunk a bit.

Pellet Warning

If a horse accustomed to a diet of pellets gets access to pasture or hay accidentally, he may be in danger of colic or impaction. His stomach and intestines have shrunk and cannot manage the extra bulk if he overeats on roughages he is not used to. His access to forage should be monitored and limited.

Supplements

A feed supplement is anything fed to horses other than pasture, hay, salt, or water. Technically grain is a supplement, but the term as generally used connotes any additional preparation of vitamins, minerals, proteins, fats, energy, and so on.

Horses have managed for thousands of years simply by eating grass. But in feeding domestic horses we've come to rely on unnatural feeds, partly because horses are kept in unnatural conditions — stalls and pens instead of pasture. Some horses are stressed more than their wild ancestors were. The

Many horses are overfed. A fat horse like this one has less stamina and more health risks than a normal horse.

race horse, for instance, needs readily digestible high-energy feeds to give "fuel" for maximum performance. Many horsemen try to help their horses perform better or grow faster, so they feed more — and use a wide array of supplements.

Many horsemen overfeed, which can be just as detrimental to health and soundness as underfeeding. Fat horses cannot perform well, foals that grow too fast are likely to develop skeletal problems (especially in feet and legs), and some types of supplements are toxic if overdosed. Part of the overfeeding problem involves misuse of supplements. If a horse doesn't need them, they are an unnecessary cost and a potential danger.

Many commercial feeds and all kinds of supplements are available. Quite a few horsemen add supplements to commercial grain mixes to enhance growth or performance, but as a general rule a supplement should be added only when the hay (or hay and grain) diet does not supply adequate levels of essential proteins or other basic nutrients.

Coordinating a Balanced Diet

The balanced diet for a confined horse with no access to pasture should start with good grass hay. Add grain only if necessary, selecting a mix based on nutrient content of available hay and adjusting the feed amount to meet the horse's needs. You can add alfalfa hay or a protein supplement if a horse needs more protein, such as a young horse or nursing mare, and grain if the horse needs more total energy and calories. Beyond that, always be careful when supplementing.

Use Supplements Wisely

Few horses need vitamin or mineral supplementation. Horsemen who feed or inject vitamin or mineral supplements without knowing if they are actually needed can cause serious harm or even kill the horse. Injectable supplements are especially dangerous because they are easily misused. Injectables should never be given unless a veterinarian determines a horse needs them, and feed additives should be given only if an important nutrient is missing in the hay/grain diet. To determine whether a horse needs a supplement, or what kind, consider the nutritional value of the feed being used as well as the requirements for the individual horse. In the case of a health problem, consult a veterinarian.

General Rule for Feeding

The safest feeds are natural roughages — pasture and hay. As a general rule the young, growing horse or hard-working adult can be fed as much good pasture or grass hay as he will eat.

Feed Each Horse as an Individual

The amount of feed or supplement a horse needs depends on his age and activities. Old horses, especially older broodmares with the added stress of pregnancy and lactation, may need supplements to keep from becoming thin or deficient because their digestive systems may not be efficient. The amount or type of feed and supplement that keeps one horse fit, healthy, reproducing, or performing well may not be enough or may be too much for a different individual.

Horses have different needs, even when doing the same work. It's not uncommon to find differences of 20 to 30 percent in the amount of energy needed. For instance, some endurance horses stay fit on pasture or grass hay alone, with no grain. Others need 6 to 12 pounds (2.7–5.4 kg) of grain added to maintain the same body condition and fitness under the same work schedule.

An old horse (right) *may have trouble keeping his weight. This young horse and old horse are on the same pasture, yet one is fat and the other thin.*

Occasionally a horse needs a special feed to complete his diet — such as an individual being fed bleached or poor quality hay or dried-out winter pasture, a young horse or broodmare having trouble meeting nutritional needs on feeds available, a horse recovering from severe parasite infestation or debilitating disease, a horse doing strenuous work, or an old horse having trouble keeping on his weight.

Energy-to-Protein Ratio

Be careful when adding grain — more corn or oats, for instance — to a commercial feed mix, as some horsemen do to try to increase growth or performance. Adding supplements or extra grain can cause an imbalance in the diet, changing mineral ratio (calcium-phosphorus balance) or protein levels. Adding corn to a grain mix to give increased energy may decrease the protein level too much for a young horse. Proper energy-to-protein ratio is important for sound skeletal growth.

The Mature Idle Horse

The idle horse needs less feed than one working daily. He needs only enough nutrition for bodily functions and limited activity. A maintenance ration for the idle horse is about 2.5 percent of body weight in good grass hay. A 1,000 pound (453.6 kg) horse would need about 25 pounds (11.3 kg) of hay daily. This will vary according to the body metabolism of the individual. If hay is of poor quality, feed more pounds of hay or add a small portion of grain. Idle horses often become too fat on good pasture even if they have exercise. An idle horse that is an easy keeper may also grow too fat on a limited amount of good hay and need to have his ration cut.

The Broodmare

The pregnant mare has more nutritional requirements than the idle horse. Most owners feed a pregnant mare more but tend to overestimate her needs while pregnant and underestimate them after she has foaled and is lactating. Pregnancy does increase the feed requirements of a mare, but only significantly during the last 4 months of gestation when the fetus is growing the fastest. Feeding extra as soon as a mare becomes pregnant causes her to become too fat by foaling time, which might cause problems with delivery of her foal.

Nutritional Requirements During Pregnancy

About 80 percent of the fetus's growth takes place during the last 4 months of gestation. The mare's requirements for the first 7 months of pregnancy are not much more than a maintenance ration.

A 1,000 pound (453.6 kg) mare in the last quarter of pregnancy — when her feed requirements exceed maintenance requirements — needs about 7.3 pounds (3.3 kg) of total digestible nutrients (TDN; portion of the ration actually utilized) per day. The same mare will need 14.4 pounds (6.5 kg) of TDN daily when she is nursing her foal. That's nearly twice as much. The lactating mare needs almost three times as much nutrition as she would for body maintenance alone.

The Young Horse

A growing horse requires adequate feed for the growth needs of his body. The young foal is growing the fastest. He gains about 45 percent of his mature body weight by the time he is 6 months old, 58 percent at 1 year of age, and 87 percent by age 2. This varies somewhat with the individual and the breed; some breeds mature more slowly than others.

A foal will grow well on good green pasture if his mother milks well. Some mares need grain in order to milk well; and if a mare is not on green pasture she will need alfalfa hay or some other high-protein feed to gain the protein necessary for lactation. If a mare does not milk well, foals should be started early on grain and a little alfalfa hay.

For proper bone growth and muscle development, the young horse needs adequate protein, vitamins, and minerals. Good pasture (or alfalfa hay) and a grain ration are usually adequate. Young horses should never be overfed. Their immature bones, tendons, ligaments, and joints are more easily injured when they carry excess weight. Also, too much grain or too many vitamins and minerals can result in decreased absorption of calcium (causing soft bones or improper bone growth) or other skeletal problems, leading to crippling and unsoundness. Fast-growing foals fed too much rich feed are especially prone to bone and joint problems. Overfeeding can cause permanent damage.

The Stallion

A stallion should be well fed, for fitness and fertility, but never overfed. Excessive fat is detrimental to fertility, stamina, and sex drive, but so is lack of adequate nutrition. He needs a balanced diet, with the right amount of protein, vitamins, and minerals — essential for production and quality of sperm. A program of good feed and exercise will keep him in fine condition for breeding.

The Working Horse

A horse being ridden or driven needs more feed than the idle horse. As his work increases, he may need grain if he is unable to maintain proper body condition and energy output on pasture or hay alone. Feed requirements vary for individuals, but in general a horse doing light work (2 to 3 hours daily) requires about 50 percent more energy than one doing no work. A horse doing medium work (4 to 5 hours daily) uses about 70 percent more energy than the idle horse. A horse doing heavy work (6 to 8 hours daily) requires even more energy.

A horse doing strenuous work, such as sorting cattle all day, needs more feed than an idle horse or one doing light work.

The energy required depends on the type of work being done. For instance, a horse being ridden for 3 hours at a walk along a trail won't need as much energy as a horse being ridden for 3 hours to move and sort cattle — galloping, stopping, and turning at high speed, continually exerting strenuously.

A horse with efficient metabolism may stay in good condition on pasture or hay, without grain, even when working hard; but many hard-working horses need a little grain to boost energy and keep from losing weight. Find the amount of feed that seems right for each horse. Some are gluttons, some more picky eaters. Some need more roughage, others more concentrate.

The Old Horse

The old horse eats more slowly and has less efficient digestion than a younger one and thus needs a good diet with ample vitamins and minerals. He may become thin on the same feed that used to keep him fat a few years earlier. If he's not on pasture, feed him by himself so he doesn't have to compete with more aggressive eaters.

Feed him frequently in small portions. Several small meals during the day are better than one or two large ones. Keep his diet somewhat laxative, so if his digestion is a little sluggish he won't become constipated or impacted. Green pasture is ideal, provides adequate nutrition, is easy to eat, and is somewhat laxative — and he can nibble on it as often as he likes.

If he can't be on pasture, give him alfalfa along with his grass hay, and add an occasional bran mash if he tends to become constipated. If alfalfa makes him cough, use pellets instead — they are less dusty than hay. Prepared feeds and supplements designed especially for older horses can be beneficial to individuals that lose weight on traditional feeds. Old horses do not absorb vitamins and minerals (especially zinc) as well as younger ones and are more likely to need supplements.

Make sure his feed is always fine and palatable, never coarse, dry, or dusty. Because coarse feeds are hard to chew, he may not clean up his hay. If he has trouble eating hay, check its quality and also have a veterinarian check his teeth; he may need dental care to relieve a painful eating problem. Grass is usually easier to eat than hay, and most older horses maintain good body condition on pasture. But if the horse is retired and on pasture, watch him closely to make sure he doesn't become overfat and founder.

Feeding Tips

◆ Regularity of feeding is important. Abrupt changes in feed can cause digestive problems. Small amounts given regularly are far better than a large amount all at once. This is especially true of grains, because too much can cause a horse to founder. The horse's small stomach can only handle a certain amount at one time.

◆ Never overfeed. A horse that is too fat cannot perform well and will tire easily when you ride him. Fat horses founder more readily than horses in fit condition. Feeding too much grain when a horse is not working can lead to serious problems, such as azoturia (muscles tying up), when you do ride him. To prevent this, eliminate the grain ration on days he is not ridden or exercised.

◆ When changing from one feed to another, do it gradually over a period of several days or a week.

◆ Keep feed boxes and tubs clean so moldy or spoiled feed doesn't accumulate.

◆ If you are putting a horse on pasture after he's been fed hay, let him graze only 30 minutes the first day, 1 hour the second, 2 hours the third (while still giving him hay), until after 4 or 5 days you can leave him out at pasture.

◆ Make sure the horse has plenty of fresh water, salt, and exercise. Exercise is just as important as feed when it comes to developing good feet, legs, and lungs in the growing horse and in keeping the mature horse in good condition.

◆ "Feed the horse to stand the riding, then ride the horse to stand the feed" is an old horsemen's adage that is very appropriate — muscle and fitness are never built by feed alone.

THREE

SEASONAL CARE

Certain aspects of horse care vary by season and by region. The responsible horseman prepares for weather conditions and problems related to climate at all times of year.

Winter Care

Winter care may be minimal in mild climates or more complex in regions with sub-zero temperatures. Care depends on factors such as whether the horse is outdoors or in a barn, and amount of cold weather, snow, rain, and mud. Many horsemen in northern climates ride less or not at all during winter. Even if a horse is on vacation until spring, this doesn't mean he needs less care. In fact, some aspects of care become more important in cold weather to ensure the horse stays healthy and comfortable.

In winter he needs additional feed (to create body heat as well as body maintenance), access to unfrozen water, and some kind of shelter or windbreak — or blanketing if he doesn't have a winter coat of hair. He should be regularly dewormed so he can utilize feed efficiently and won't harbor internal parasites that rob him of nutrients. Shoes should be taken off if he's not being ridden, and feet trimmed periodically so they won't grow too long. Horses on winter pasture need extra feed if snow gets deep and covers the grass. Old and young horses need special care: the horseman should make sure they get adequate nutrition.

Cold weather in itself is not a problem for a horse if he's had a chance to prepare for it gradually by growing winter hair. Wind and wet weather create more hardship than does cold. In windy regions, be sure to provide horses with shelter to protect against windchill that robs them of body heat.

Windchill Index

Knowing the windchill can be useful, whether you are planning to ride (and need to know how warm to dress) or need to know how "cold" it actually is for an outdoor horse — whether he might need shelter or a blanket.

WIND SPEED	TEMPERATURE (FAHRENHEIT)

WIND SPEED	-20	-18	-16	-14	-12	-10	-8	-6	-4	-2	0	2	4	6	8	10	12	14	16	18	20	22	24	26	28	30	32	34	36	38	40	42	44	46	48	50
0	-20	-18	-16	-14	-12	-10	-8	-6	-4	-2	0	2	4	6	8	10	12	14	16	18	20	22	24	26	28	30	32	34	36	38	40	42	44	46	48	50
2	-23	-21	-19	-17	-15	-13	-11	-9	-7	-5	-3	-1	1	3	5	7	9	11	13	15	17	19	21	23	25	27	29	31	33	35	37	39	41	43	45	47
4	-25	-23	-21	-19	-17	-15	-13	-11	-9	-7	-5	-3	-1	1	3	5	7	9	11	13	15	17	19	21	23	25	27	29	31	33	35	37	39	41	43	45
6	-28	-26	-24	-22	-20	-18	-16	-14	-12	-10	-8	-6	-4	-2	0	2	4	6	8	10	12	14	16	18	20	22	24	26	28	30	32	34	36	38	40	42
8	-30	-28	-26	-24	-22	-20	-18	-16	-14	-12	-10	-8	-6	-4	-2	0	2	4	6	8	10	12	14	16	18	20	22	24	26	28	30	32	34	36	38	40
10	-31	-29	-27	-25	-23	-21	-19	-17	-15	-13	-11	-9	-7	-5	-3	-1	1	3	5	7	9	11	13	15	17	19	21	23	25	27	29	31	33	35	37	39
12	-33	-31	-29	-27	-25	-23	-21	-19	-17	-15	-13	-11	-9	-7	-5	-3	-1	1	3	5	7	9	11	13	15	17	19	21	23	25	27	29	31	33	35	37
14	-35	-33	-31	-29	-27	-25	-23	-21	-19	-17	-15	-13	-11	-9	-7	-5	-3	-1	1	3	5	7	9	11	13	15	17	19	21	23	25	27	29	31	33	35
16	-37	-35	-33	-31	-29	-27	-25	-23	-21	-19	-17	-15	-13	-11	-9	-7	-5	-3	-1	1	3	5	7	9	11	13	15	17	19	21	23	25	27	29	31	33
18	-38	-36	-34	-32	-30	-28	-26	-24	-22	-20	-18	-16	-14	-12	-10	-8	-6	-4	-2	0	2	4	6	8	10	12	14	16	18	20	22	24	26	28	30	32
20	-41	-39	-37	-35	-33	-31	-29	-27	-25	-23	-21	-19	-17	-15	-13	-11	-9	-7	-5	-3	-1	1	3	5	7	9	11	13	15	17	19	21	23	25	27	2
22	-43	-41	-39	-37	-35	-33	-31	-29	-27	-25	-23	-21	-19	-17	-15	-13	-11	-9	-7	-5	-3	-1	1	3	5	7	9	11	13	15	17	19	21	23	25	27
24	-46	-44	-42	-40	-38	-36	-34	-32	-30	-28	-26	-24	-22	-20	-18	-16	-14	-12	-10	-8	-6	-4	-2	0	2	4	6	8	10	12	14	16	18	20	22	24
26	-49	-47	-45	-43	-41	-39	-37	-35	-33	-31	-29	-27	-25	-23	-21	-19	-17	-15	-13	-11	-9	-7	-5	-3	-1	1	3	5	7	9	11	13	15	17	19	21
28	-52	-50	-48	-46	-44	-42	-40	-38	-36	-34	-32	-30	-28	-26	-24	-22	-20	-18	-16	-14	-12	-10	-8	-6	-4	-2	0	2	4	6	8	10	12	14	16	18

Note: Thermometers with both Fahrenheit and Celsius scales are available.
To convert Fahrenheit to Celsius, subtract 32 from the Fahrenheit number. Divide the answer by 9. Multiply that answer by 5.

How Horses Keep Warm

Horses have a normal body temperature of about 100°F (38°C). They maintain this temperature in cold weather by means of several mechanisms that include shivering, change in hormone levels, increased body metabolism, increased digestion of fiber, growing long thick hair that can "stand up" on the skin, increased feed consumption, and increased activity. On a frosty morning cold horses often run and buck to warm up.

A well-fed horse can manage temperatures down to -30° or -40°F (-34° or -40°C) if there's no wind and he's not wet. Wind ruffles the hair and eliminates its insulating quality. A wet horse loses body heat up to twenty times faster than a dry horse, because moisture flattens the hair and reduces air spaces between the individual hairs, greatly diminishing the insulating effect.

The horse's best defenses against cold are long hair and a layer of fat beneath the skin to protect against loss of body heat. The insulating quality of winter hair is lost if the horse is wet or covered with mud. Thus he needs a dry, sheltered area during cold or wet weather, as well as regular grooming (especially if he's muddy) to keep his coat from matting.

Shelter

Your horse will prefer the warmth of winter sun to a shady shed. Most horses, if given a choice, stay outdoors even in the coldest weather but use a windbreak or shed to get away from cold wind, blizzards, or rain. They don't like to stand in mud; if horses are in muddy paddocks they need a high, dry spot where they can stand or lie without being in wet, cold mud.

Many horses in cold climates spend winter outside, using canyons and creek bottoms, timber or brushy areas as windbreaks. If a horse is accustomed to being outdoors, don't put him in a barn just because of foul weather. A horse in good condition, given a chance to toughen naturally as winter approaches, is best left outdoors. In fact, horses kept outdoors with access to a shed or windbreak have fewer respiratory problems than do those kept in a poorly ventilated barn that is full of dust from hay and bedding.

Body Condition

Make sure horses go into winter in good flesh. A thin horse won't winter as well as a fat one. Horses should never be too fat but should have a layer of insulation under the skin. Horses in cold climates need to go into winter with extra flesh if wintering outdoors, for energy reserves as well as insulation.

Health Tips for Older Horses

Old horses should not start winter thin. They may have poor teeth, making it hard to eat well and keep weight on. They generally have less resistance to cold and may have arthritic joints that become painful and stiff in cold weather. Unless an old horse is fat and an easy keeper, he should not winter outside. If he has respiratory problems when indoors, he's better off outside with a windbreak shed and blanket. Many older horses have breathing problems in an enclosed barn due to years of stress on air passages and heightened sensitivity to dust and molds. If your horse has this condition, provide dust-free bedding in his outdoor shed or where he likes to sleep. Dry, clean bedding will keep him warmer and also make it easier on his old bones when he gets up and down.

Some horses gain weight in winter when not working; others can't handle the cold and lose weight unless feed is increased. It's sometimes hard to tell how fat a horse is and how much feed he needs, because long winter hair can hide body contours. Check your horse's body condition periodically by feel — running your fingers over various parts of his body to determine how much flesh is under all that hair. Feel along neck, withers, hips, and backbone and over the ribs.

If you can't feel each rib individually he has plenty of fat, maybe too much. If you can feel a layer of soft tissue between ribs and skin but can still feel each rib, he has about the right amount. If you can feel the ribs quite prominently, like running your fingers over a washboard — and there is no soft layer of tissue over them — the horse is too thin.

Check a horse's body condition by feeling the ribs.

Give Roughage to Create Heat Energy

If the weather gets cold or continually wet, increase feed for all horses kept outdoors; they need more total calories to keep warm. If a horse doesn't get enough food to generate heat, he burns fat and muscle tissue to keep warm and loses weight. As temperatures drop you should feed more hay, not more grain. You can even increase the amount of hay to all he will eat. More body heat is created by digestion of roughage than by digestion of grain. Remember that hay is digested in the large intestine and cecum by means of fermentation — which produces heat in the process — whereas grains are digested in the small intestine. A pound of hay generates more heat than a pound of grain.

Digestion of protein also produces heat, so alfalfa hay can help a horse keep warm in winter. Although corn is higher in energy content than hay, it produces less heat when digested. Corn is more useful for weight gain than for heat production. If you are feeding grain in winter, it is often better to use oats than corn. Because of their higher fiber content, oats produce more heat than a similar weight of corn. Two pounds (0.9 kg) of oats and plenty of hay usually produce adequate body heat for keeping warm.

Give a Larger Feeding at Night

Horses should be fed at least twice a day. During cold weather the larger feeding should be in the evening because winter nights are not only colder but longer. In northern regions there may be 16 hours of darkness and only 8 hours of daylight, and if the horse gets his evening feeding before dark it's a long time until breakfast! The evening feeding should include as much hay as the horse will eat.

Don't shortchange your horse on roughage at night or he may start nibbling fence posts, bedding, or anything else within reach — even his pasture-mate's tail or mane — because he is cold and craves more roughage. Having access to plenty of hay at night will keep him busy and warm. If horses are bedded on straw they'll eat the bedding, which can lead to impaction or other digestive problems. It's better to give the horse extra hay so he won't resort to eating the bedding. Alfalfa can be fed during cold weather; horses metabolize the extra protein into heat energy. Take care to avoid dusty or moldy alfalfa, which can lead to coughing, heaves, or colic.

Feeding on Cold Nights

A rule of thumb is to increase the hay ration by 10 percent for every 20°F (10°C) below freezing.

If horses are fed in groups, make sure the feeding area is roomy enough for all to eat without bossy ones chasing timid ones away from the hay. If a horse is a slow eater, keep him separate from the herd so he gets his share. Some horses like to eat their meal periodically through the day (or night), whereas others gobble theirs in an hour or two. If penned together, fast eaters get more than their share and slow eaters are shortchanged. Older horses with arthritis may not move well in cold weather, hanging back from the feeding area to avoid being chased by younger, bossier herdmates.

Water

Horses generally drink less water in cold or wet weather. They don't like ice-cold water, but they still need an adequate and constant supply or they may become impacted. If the water tub freezes and a horse doesn't get enough fluid, he won't eat enough and may lose weight or be unable to maintain body heat.

Don't Depend on Snow

Horses will eat snow, but if being fed hay, pellets, or grain they need more fluid in the digestive tract than snow alone can provide. It takes body heat — and extra calories — to melt and warm the snow in the digestive tract. A horse should never have to depend on snow. He may die of constipation and impaction if he doesn't get enough water to maintain his body functions.

Check manure to see if he is drinking enough. If fecal balls are hard, dry, and small instead of soft and moist, he's not getting enough water. Other signs of dehydration include loss of skin elasticity, failure to clean up hay (it's too difficult to eat much dry feed without some water), or drawing up of the flanks and abdomen.

If weather is cold, check the watering place twice daily and break ice if necessary. If the horse waters at a stream, pond, or ditch he may be reluctant to step on ice to reach a water hole. You may have to spread dirt or gravel over the ice to give him safer footing if he has to drink from a hole in the ice of a ditch or creek.

If watering in a tub or tank, make sure the water doesn't freeze too quickly. If possible, give warm water or use a tank heater. This way the horse will drink more and the water won't freeze as fast. Mature horses need 6 to 10 gallons (22.7–38 L) a day in winter; if a continuous supply is not available, give fresh water (45°–65°F [7.2°–18.3°C]) twice a day, with the largest portion in the morning. Many horses don't drink much at night when temperatures are colder, drinking more during the warmer daylight hours. If the water tub is empty or frozen throughout much of the day, the horses will be seriously shortchanged.

Foot Care and Safe Footing

Shoes should be removed unless pasture is rocky or snow is deep and the horse must paw for grass. In that case the front shoes can be left on but should be reset periodically so hooves don't grow too long. Usually a horse is better off barefoot during the winter, since snow won't ball up under his feet so badly. He'll have better traction on ice and frozen ground — and less strain on feet and legs than when walking on four slippery balls of ice.

Pay attention to ice and slippery footing on paths, in gateways, near watering areas, and so on. Serious injury to feet and legs can occur if a horse slips and falls or spraddles — possibly stretching a tendon or tearing a muscle. You can prevent such accidents by putting gravel, sand, or rock salt on icy places to improve traction.

Blankets

A horse grows thick hair as protection against cold and wind, rain and snow. If he's blanketed to prevent extra hair growth, or clipped for ease in cooling out (so he won't have a long hair coat drenched with sweat after a ride or workout) he'll need protection from winter weather. Don't clip a horse that has to spend time outside unless you are prepared to blanket him.

Beneficial Winter Hair Coat

Long winter hair traps a layer of body heat between skin and the cold air. When weather is cold, tiny muscles in the skin make the hair stand up fluffy, increasing the insulating effect. A winter hair coat is better insulation than most blankets. Adding a heavy blanket or piling on several light ones can actually make the horse colder, flattening down the hair and severely reducing its insulating effect.

Blanketing may be necessary for a clipped horse or one that is moved to a colder climate without having a chance to grow a heavy coat. Even if a horse has a good hair coat, once he becomes so wet and cold that he has to shiver to maintain body temperature he will burn more calories and need extra feed to keep warm — otherwise he'll lose weight. Under these conditions he's better off with a blanket.

What Kind of Blanket?

There are many kinds of blankets; some are warmer and more weather resistant than traditional wool blankets. Some have an inner and outer shell surrounding an insulating filler material. The outer shell (usually made from a durable synthetic fiber) is waterproof and windproof. The filler provides warmth without bulk or heaviness. The inner shell is smooth and nonabrasive so it won't chafe the skin. These blankets are more durable than cotton or wool and light enough to not press down the horse's hair.

Outer shells are usually made of nylon or canvas. Nylon is strong and fairly indestructible; it won't snag, tear, rot, or mildew. Canvas is durable and economical, giving partial protection from wind and rain because of its dense, tight weave, but it isn't truly waterproof. It eventually soaks through in heavy rain. A net cooler (usually cotton) under the blanket can wick away moisture and provide an air space for insulation.

All-cotton canvas is susceptible to mildew and color fading. Some canvas outer shells are treated with a special coating to increase water resistance. Advantages of canvas outer shells are ease of cleaning (they can be brushed off, hosed down, and hung on a fence to dry), durability and toughness, water resistance (fibers swell when wet, making less spaces for moisture to soak through), and ability to breathe. They hold in body heat but wick away moisture such as sweat. Canvas makes a good top layer over other blankets or sheets, and the weight helps hold everything in place. Disadvantages of canvas are its weight, slowness in drying, and tendency to soak up urine and moisture from manure.

Many blankets have a waterproof finish that keeps moisture out and holds heat in. The disadvantages of these coatings are that they may crack in temperatures below -20°F (-28.9°C) and are not good during changeable weather because the seal does not allow heat to escape; a horse may overheat under the blanket.

See pages 68 and 69 for tips on using blankets and proper blanket care.

Tips for Using Blankets

◆ **Keep track of temperatures. Don't load a horse with blankets he
doesn't need or leave him bare when he needs protection.** If he's
shivering, he needs more covering. If he's sweating under the blanket,
he's too warm. You don't want him to be sweating and then chilling at
night when the temperature drops and he's wet under the blanket.

◆ **Don't blanket a horse for turnout in the cold early morning and
leave him blanketed throughout the day, especially if it gets warm
in the afternoon.** Feel his chest or just behind the elbow at his girth;
these are the first places he'll start sweating. It's a good idea to have
a heavy blanket for cold nights and a lighter one for daytime.

*Make sure the blan-
ket is properly fitted
and appropriate for
the horse's needs.*

◆ **Put the blanket on correctly, starting it well forward and then
pulling it back into place so it won't ruffle the hair.**

◆ **No matter what type of blanket you use, make sure it fits the horse.**
A blanket that is too tight may rub and chafe. A high-withered horse
kept blanketed all the time may develop skin irritation and infection
where the blanket rubs the withers. Blanket rubs are a common
wintertime problem and can cause sores. Some horsemen sew a
band of cotton (such as a leg wrap) around the shoulder edge of
the blanket to minimize rubbing.

A blanket that is too large or loose may slip and end up under the
horse's belly. If his legs become entangled in the straps, he may

seriously injure himself while struggling. The blanket should fit properly, with straps well adjusted so there is no slipping.

◆ **Make sure the blanket is designed to fit well and stay in place while still allowing freedom of movement.** This is especially important if your horse is quite active and playful when outdoors with a blanket on. There should be a loop that goes around each hind leg (below the stifle and above the hock) to keep the back end of the blanket in place. These leg straps are easier to keep clean if they are not made of leather. Some blankets have elastic leg straps or straps with elastic portions that hold well yet "give" when the horse moves. Others have replaceable straps or swivel snaps that don't get twisted up. Some horsemen cross the leg straps under the horse when putting on the blanket (running the straps between the hind legs, then crossing them before attaching them at the sides) to keep straps from chafing the legs and also to ensure the blankets stay in place even if one strap comes loose.

Most blankets have a surcingle strap under the belly that helps hold the blanket in place; some have double straps that cross under the belly. Elastic inserts on these straps are nice; the extra give can keep a horse from tearing out of his blanket. Some styles have an option for closed or open front (with fasteners), cut back at the withers or covering withers and shoulders (to provide extra warmth and to keep the blanket from shifting). Other styles have shoulder gussets for more freedom of movement, reducing the rubbing or pulling across the chest and adding "give" at the shoulders.

◆ **Don't use the same blanket on several horses.** This will prevent the spread of skin problems.

◆ **Keep blankets clean.** Wash them at least twice during winter if used often. If a horse constantly gets his blanket muddy or dirty, choose a type that can be washed frequently. Most horse blankets should be washed in cold water with soap and a disinfectant, then very thoroughly rinsed. You don't want any soap left in the blanket or it may irritate the horse's skin. Dry cleaning is not recommended. It won't remove odors, solvents tend to make the waterproof coatings disintegrate, and heat makes the bindings shrink. Choose a good blanket that fits your horse well, keep it clean, and it will last a long time.

Spring Care

Spring brings longer days, warmth, melting snow, and rain instead of blizzards. Frozen ground or deep snow gives way to mud. If spring is wet, horse pens (and places in a pasture where horses congregate) may become a deep sea of mud. If muddy conditions persist for long, they can be a health hazard for your horse.

Beware of Mud

Mud makes bad footing, which is hazardous when horses run and play. They may pull muscles or strain joints and tendons, especially if the horses are out of shape from winter vacation.

Mud Problems and Health Concerns

Mud can make it hard for a horse to keep warm. If he stands in it he will chill more easily in a storm than if he were standing on drier ground. A hair coat matted with mud loses its insulating quality. Mud also creates ideal conditions for problems such as thrush, scratches, rainrot, and other skin diseases.

A muddy paddock provides no place for a horse to lie down for a nap or even for a good roll. Being reluctant to lie down in deep mud, he may spend most of his time standing or try to lie down in a dangerous place just because it is a little less wet, such as on a built-up manure pile next to the fence. If he lies too close to the fence and tries to roll, he may become *cast* (on his back, against the fence, unable to get up).

In wet, muddy weather, pay special attention to your horse's feet and skin. Be aware that moisture makes an ideal environment for bacteria and fungi and that long hair can hide a lot of skin problems. During a wet spring you may encounter rainrot, scratches, lice infestations, or foot problems such as thrush or abscesses. (See chapters 12, 10, and 4, respectively, for more information.)

Grooming a horse daily — keeping him free of mud and preventing the hair coat from becoming matted — and cleaning his feet can give you a good handle on what's happening. In this way you can minimize or prevent certain wet-weather problems. If you catch a skin problem or foot problem early, it will respond much more quickly to treatment.

Mud Management

Mud that freezes on cold nights can make treacherous footing, as the humps, bumps, and ruts in pens and around water tanks or travel areas and gateways become solid. Horses may strain their legs or cut themselves on frozen lumpy ground and ice pockets created by deep hoofprints. The frozen ridges and holes make it difficult to walk, let alone run and frolic — and injuries can easily occur.

The problem of frozen humps on cold mornings can be alleviated by hauling in gravel or coarse sand to build up main traffic ways around the barnyard and to create smoother, drier areas in paddocks where horses can safely walk to water or lie down. Keeping a stockpile of sand or crushed rock for winter and springtime emergencies will help you reduce hazards of ice or mud.

Pens should be situated on high ground, if possible, rather than in low areas that collect rainwater and melting snow. You can resolve most drainage problems in a boggy pen or paddock by hauling in sand or gravel to create high spots. Where there are problems with spring run-off, dig a ditch or make a swale (with tractor and blade) to channel excess water away from or out of pens, paddocks, and riding areas so moisture does not pond up.

Shedding

Hormonal, environmental, and nutritional factors trigger hair growth and shedding. The lengthening days of spring stimulate shedding, whereas the decreasing hours of daylight in fall stimulate growth of winter hair.

Hormones and Shedding

Hormones affect hair coat and shedding. A stallion usually has a sleeker, shinier coat than a gelding, sheds faster in spring, and retains a short sleek coat longer in fall. Thyroid imbalance can affect shedding; some horses with this problem don't shed winter hair until late spring and others don't shed at all — they have to be clipped to prevent discomfort in hot weather. Older horses that don't shed may have Cushing's disease, which mainly affects horses in their late twenties or thirties; a tumor in the pituitary gland interferes with the proper balance of hormones. Mineral deficiencies, especially selenium, can also cause slow shedding.

A healthy horse, well fed and eating a balanced diet, sheds more rapidly in spring than an undernourished horse does. Because hair is made up of protein, the horse must have adequate protein for a healthy coat. A horse infested with worms — especially a young horse — often has a dull hair coat and is slow to shed because these parasites rob him of essential nutrients. Stress, illness, starvation, or other abnormal conditions also make a horse slow to shed.

The key to promoting healthy hair and speeding up shedding is a combination of balanced diet, regular grooming, and massage. Blanketing also stimulates shedding; the sweating process initiates getting rid of the extra hair.

Horses often rub on fences or trees or roll a lot when shedding to work the old hair loose. They also enjoy being groomed and scratched. If the horse exercises and sweats, he's even more itchy and eager to rub off the extra hair. If a horse seems too itchy — rubbing himself to the point of damaging the skin — or has a lot of dandruff or greasy skin, this could be a sign of lice or skin disease. Take a closer look.

If spring is rainy, winter hair may harbor fungus or bacteria. The combination of moisture, dirt, and dead hair can lead to skin problems that require medical attention. At this time of year a daily grooming will speed up shedding and minimize the chances for skin disease. Use a rubber curry comb or shedding blade to remove the dead hair, and do a lot of brisk brushing and massage, which increase circulation and promote skin health by stimulating the oil glands. "Elbow grease" is the best key to a healthy hair coat.

Grooming regularly each day helps a horse shed faster.

Summer Care

Summer brings heat and humidity, flies and other parasites, and the potential for problems caused by dry climate and bright sunlight.

Dry Climate

In dry conditions the hoof dries out and gets harder, but if feet are too dry they become brittle and likely to split and crack. A little moisture in the pasture can help solve this problem. If the horse waters in a creek or pond, stands in water when he drinks, or walks through water, it can keep his feet from becoming too dry. If there is no natural moisture in the pen or paddock, you can create a mud puddle near the water tank. An alternative approach is to use a good hoof dressing. Your veterinarian or farrier can advise you on types of hoof dressing that are best for dry, brittle feet. If feet are brittle, trim and smooth them often (if the horse is barefoot) to prevent splitting.

Bright Sunlight

Horses with white markings or light-colored skin are sometimes bothered by bright sunlight. Dark skin is the best protection against the sun's rays. Pigment is nature's sunscreen. Horses that have light-colored or pink skin are prone to sunburn and peel, which makes them miserable.

Sunburn often occurs on white markings (especially on the nose) or where hair is thin and gives little protection. If a horse sunburns, use human sun lotion containing sun-screening ingredients or zinc oxide to protect the skin from ultraviolet rays. Another way to protect pink skin is to paint those areas with an organic dye such as methylene blue or gentian violet. These dark colors block the sun's rays.

Photosensitization is much more serious than sunburn. It occurs when a horse eats certain plants (e.g., Saint-John's-wort, perennial ryegrass, dried buckwheat, and alsike clover) containing a chemical that causes a toxic reaction in the unpigmented skin; the chemical, or photodynamic agent, is absorbed from the gut into the bloodstream and travels to the skin, where it is activated by sunlight and kills the skin cells (see chapter 12). A horse with this problem should be turned out only at night.

Eye irritation from sunlight often occurs in horses with white markings around the eyes. Bright summer days (and bright winter days when sun reflects off the snow) can make the eyes water. Dark skin absorbs the sun's rays, but light skin reflects it into the eyes. Sunlight can irritate and damage the eye and surrounding tissues. In fact, cancer of the eyelid is common in horses with unpigmented eyelids. If a horse has a white face and is bothered by bright sunlight, apply something dark around the eye such as mascara, the dark purple coloring used in pink-eye medications for cattle (obtain this from your veterinarian), or black theatrical greasepaint used by actors to reduce the glare of stage lights and by athletes to reduce sun glare. The greasepaint is actually a cream that can be easily applied around a horse's eyes, especially if it is warmed for easier spreading. Anything used near the horse's eyes should be nontoxic and nonirritating.

Insects

Biting insects such as flies and mosquitoes are a nuisance to horses in summer; many species cause annoyance and injury. (These are discussed in detail in chapter 10.) Shelters or sheds can help protect your animals from horse flies and deer flies; these flies don't like to go into an enclosed building or shady, dim place. If flies are bad you may want to stable your horses during the day. It also helps if pastures are well drained and not swampy; mosquitoes and horse flies require wet areas to complete their life cycles.

Insecticides for controlling flies on the horse can be applied as dusts or sprays or wiped on. (These are discussed in chapter 10.) Fly masks can give relief from those that chew on the horse's eyelids.

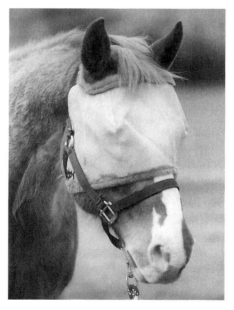

A fly mask can protect the horse's eyes when flies are at their worst.

Grooming Tip

When grooming your horse in summer, take into consideration the fly problem before you cut off his mane or make his tail so short that it doesn't reach his hocks. Mane and tail are the most effective weapons against flies.

Health Concerns

If humidity is high, horses have difficulty cooling themselves by sweating. Sweat does not evaporate when air is already full of moisture, so the horse stays hotter. Temperatures above 80°F (26.7°C; especially if humidity gets above 50 percent) increase the chances for trouble, particularly if a horse is being worked or ridden.

The horse gets rid of extra body heat created by exertion through air exchange in the lungs, through radiation effect from the skin, and through evaporation of sweat. The hotter the air, the less efficient are his cooling efforts by means of air exchange or radiation effect. The horse can efficiently cool himself if the environment around him is cool, but as air temperatures rise and approach that of body temperature he must rely on the evaporation of sweat to dissipate heat. If the air is humid, the sweat does not evaporate very well.

Hyperthermia

Extremely hot weather can cause problems even in a horse that is not exerting. When he has abnormally high body temperature, he suffers from hyperthermia (as opposed to hypothermia, which occurs when body temperature is too low). It's normal for a horse's temperature to rise slightly during a stretch of hot weather and humidity or when he is exerting on a hot day. But if he must continually endure heat and humidity in a shadeless paddock, remain enclosed in a hot trailer, or work hard in the heat, he may develop hyperthermia and heat exhaustion. If the problem is not treated immediately to bring down his temperature, he may go into shock. This type of shock is called heat stroke, and it can be fatal.

Be Careful When Trailering

Heat stroke can occur in horses that are confined in a hot, stuffy stall or trailer with no ventilation. In hot weather, transport horses at night when temperatures are cooler, and never let a trailer sit in hot sun with horses inside. The trailer should spend as little time as possible standing still once horses are in it; the movement of travel creates air flow that prevents heat build-up and helps with evaporation of sweat to cool the horses. Make sure the trailer has adequate ventilation for good air flow. Don't leave horses tied to a trailer for long periods in the hot sun (as at a horse show or trail ride). If they must be tied to the trailer, use the shady side and provide some sort of awning.

Heat Stroke

Heat stroke is more the result of the horse's inability to cool himself than a direct effect of hot sun. Dehydration from sweating can interfere with the body's ability to cool itself. Anything that inhibits the horse's ability to cool himself (e.g., a long hair coat, too much fat, a heavy blanket) can lead to heat stroke.

Signs of heat stroke may begin as restlessness and anxiety, progressing to erratic or irrational behavior, depression or excitability, disorientation, rapid breathing, weakness, and dry skin — the horse stops sweating. Body temperature may rise as high as 106° to 110°F (41.1°–43.3°C). The affected horse won't eat, and in spite of the heat he stops sweating. If you suspect heatstroke, call your veterinarian immediately and start trying to reverse the condition at once while waiting for him or her to arrive. Without prompt treatment, the horse may die. (For instructions on how to take a horse's temperature, see page 127.)

Treatment for mild cases of heatstroke involves reducing the high body temperature as quickly as possible by putting the horse in a shady area where there is a breeze (even if you have to use an electric or manual fan) and spraying him with cold water or using ice packs. Apply cold water or ice packs to the

Heat Stroke Reference Guide

SYMPTOMS	TREATMENT	PREVENTION
• Restlessness • Anxiety • Erratic or irrational behavior • Depression or excitability • Disorientation • Rapid breathing • Weakness • Dry skin (inability to sweat) • Body temperature as high as 106° to 110°F (41.1°–43.3°C) • Lack of interest in eating	• Call veterinarian immediately and then do the following while you wait for him or her to arrive: • Move horse to shady area • Create breeze with fans • Spray body with cold water or apply ice packs, especially to insides of legs and thighs, lower side of neck, and head • Back off on rapid cooling after body temperature comes down to 105°F (40.6°C) to avoid causing a chill • Have veterinarian administer intravenous fluids and medications to control shock	• Provide adequate water • Provide shade • Feed appropriately (avoid large amounts of protein) • Monitor feed and water intake, as well as behavior

head, neck, and areas with large veins, especially the inside of legs and thighs, the lower side of the neck, and the head. When the horse's temperature drops below 105°F (40.6°C), back off a little on the rapid cooling; too drastic a reduction in temperature at this point can chill the horse and make the condition worse.

In serious heat stroke the horse needs intravenous fluids and medications to control shock. If he recovers from the shock he will probably survive, but his ability to endure hot weather or hard work may be permanently impaired. It is better to prevent this problem in the first place.

Insufficient Water Intake

Lack of water during hot weather can cause heat stroke. Problems can arise if an automatic waterer quits working; you might not know it until horses have been without water for a while. It's hard to monitor water intake or tell if the waterer is working unless you are watching when the horse is trying to drink. It's better to use a large tub so you know if a horse is drinking, and how much; you can monitor his intake because you are personally involved in the daily filling of the tub.

If weather is cooler than about 85°F (29.4°C), a 1,000 pound (453.6 kg) horse will drink about 6 to 12 gallons (22.7–45.4 L) of water a day. At higher temperatures he will sweat a little to cool himself (or sweat a lot if exercising); water consumption may increase as much as three times. A lactating mare in hot weather may drink as much as 100 to 150 gallons (380–570 L) per day.

When necessary, cool an overheated horse with water.

Health Tip

Fat horses and heavily muscled ones don't dissipate heat well and should be watched for signs of heat stress. Limit a horse's exercise when temperatures and humidity are dangerously high. A rule of thumb for danger zones: When temperature and humidity are added together and the total exceeds 130 (as when temperature is 80°F [26.7°C] and humidity 50 percent), there is great risk of horses overheating. This is also true whenever the temperature exceeds 100°F (37.8°C), even if humidity is low.

Appropriate Feed

The type of feed a horse eats can make him susceptible to heat stress, because horses generate more body heat when digesting large amounts of protein (grass hay is better than alfalfa when a horse is working hard in hot weather). During hot, humid weather a horse may become impacted if he eats dry hay without access to abundant water. Any horse recovering from heat-related illness should be on a diet of green grass or other feeds with high water content.

Inability to Sweat

Sometimes horses in hot climates (e.g., tropical countries or Gulf Coast states) suffer from *anhidrosis,* which means "without sweat." This problem usually develops during the hottest part of the year and may come on quickly or gradually over several weeks. If a horse has to sweat continually to keep cool during a hot, humid summer (as when confined in a hot, stuffy stall), his sweat glands work overtime and may shut down. The horse is then dry-skinned and his temperature climbs. He may have a few patches of sweat under the mane, behind the ears, under the elbows, and at the flanks, but most of his body stays dry. Prolonged inability to sweat can be fatal.

If discovered early, the condition can be corrected by cooling the horse — bathing with cool water to bring his temperature back down to normal (99°–100.5°F [36.7°–38°C]), and keeping his temperature low enough that he won't need to sweat (about 100°F [37.8°C]). He should be kept out of the sun but not in a hot, humid stall. Use a portable fan to help stabilize his body temperature until his sweating ability recovers.

Adequate Shade

On the hottest days of summer your horse will greatly appreciate shade. A few trees in a pasture can provide him a cool place to fight flies on a hot day. If there is no natural shade, consider building an inexpensive sun roof or run-in shed (see chapter 1).

Horses appreciate shade on a hot summer day.

Good Care Is Essential

Your horse's health is very important, and the seasonal care you give him will make all the difference in keeping him healthy and in good condition. This includes having a yearly vaccination program for infectious diseases (see chapter 7); regular deworming, including treatment for bots after the first hard frosts in fall (see chapter 10); fly control (see chapter 10); and foot care (see chapter 4). The horse also needs regular exercise and a safe environment. In some respects summer can be the easiest time for horses, especially if they can enjoy green grass and nice weather, but in other ways it brings challenges and problems — so good care and management are essential.

FOUR

FOOT CARE

The old saying "No foot, no horse" can be taken literally as far as the working ability and soundness of the horse are concerned. Because the horse is an athlete, his feet and legs are crucial parts of his structure. Care of the feet is essential; neglect can lead to unsoundness and pain.

Handle the Feet Often

If you handle a horse's feet often, he grows accustomed to it as part of his daily routine and becomes more well mannered for trimming and shoeing. Also, you become more apt to notice problems. As you clean his feet daily, you can remove any rocks stuck in the feet and prevent buildup of packed mud or manure that can lead to thrush. You are also likely to notice any heat in the leg or hoof (or swelling of the lower leg) that could be a sign of infection or injury.

How to Pick Up a Horse's Foot

To pick up a front foot, run your hand down the leg and gently squeeze the lower leg just above the fetlock joint — in the indentation in front of the back tendon. This will encourage the horse to pick up his foot. Hold it between your legs for cleaning. To pick up a hind foot, stand close to the horse and run your hand down his leg in the same way, squeezing gently at the fetlock if he does not pick it up readily. Rest the leg across your thigh for cleaning the foot. If the horse does not want to pick up his foot, lean against him a little to encourage him to shift his weight off it to the other three legs. (See chapter 5 for more on safe hoof handling.)

What Makes a Sound Hoof?

Horses are individuals; the feet of each one differ slightly from those of others in terms of shape, hardness, and rate of hoof growth. The hoof is a specialized horny shell that covers sensitive living tissues — bones, blood vessels, and nerves. The outer shell is a unique covering that grows continually to compensate for wear and tear.

The sole should be somewhat concave to allow for expansion when weight is placed on it. The hoof wall is designed to carry most of the weight and the bars serve as a brace to prevent overexpansion or contraction of the foot. The V-shaped frog serves as a cushion in the middle of the foot, helps absorb concussion, and regulates hoof moisture. If any of these outer tissues are abused by injury or excessive trimming, the normal functions and soundness of the hoof are impaired.

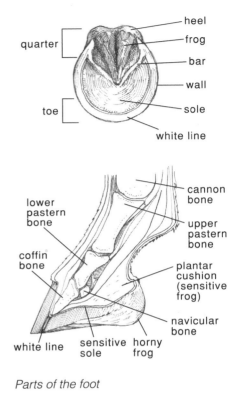

Parts of the foot

Hoof Growth

The hoof wall grows down from the fleshy coronary band that goes around the hair line at the top of the hoof. Any injury to the coronary band affects growth of the hoof. The average rate of hoof growth in a normal horse is about ¼ to ⅜ inch (0.6–0.9 cm) per month. The entire wall is usually replaced by new horn every 8 to 12 months. A horse doing normal activity usually wears his hoofs at the same rate they grow. If the horse is on pasture or in a paddock without much riding, his feet don't wear fast enough to keep up with growth and need periodic trimming. If hoofs are allowed to grow too long, they may split or splay out and hinder his movement. Toes that are too long put strain on the leg tendons. If a horse is shod, his shoes should be reset or replaced every 3 to 10 weeks depending on the rate of hoof growth.

Foot Conformation

How the foot is built plays a role in its ability to hold up. Front hoofs should be larger, rounder, and stronger than hinds because front legs support about two-thirds of the horse's weight. Hoofs should be wide at the heels, not narrow or contracted. The sole should be slightly concave in the front feet and even more so in the hinds. A horse with flat feet is likely to suffer stone bruising or develop navicular disease. (See chapter 9 for more information.)

Ideally the hoof wall should be thick, pliable, and resistant to drying out and should grow at a normal rate. The sole should be thick so it can't be easily bruised, and the bars should be strong and well developed. The frog should be large and healthy and centered in the foot (an off-center frog is an indication of crooked feet and/or legs; the hoofs probably do not wear evenly).

Abnormalities and Malformations

Few horses have perfect feet and legs. When breeding horses, you should select individuals with the best possible conformation because this characteristic is passed on to offspring. When acquiring or keeping a horse with less than ideal feet — a dependable old child's horse or pony, or an animal with malformed feet due to earlier injury, or a horse you bought for its wonderful disposition or exceptional ability in spite of a foot problem — you may decide to overlook the foot faults. This horse may need special foot care or more than routine foot trimming to stay comfortable and reasonably sound.

Contracted Feet

A contracted foot is more narrow than normal, especially at the heels and quarters; the frog becomes small and atrophied, shrinking to such an extent that it no longer has contact with the ground. A horse with contracted heels or feet is likely to go lame due to inability to properly absorb and dissipate concussion — making him susceptible to problems such as navicular disease or concussion-related breakdowns in other structures of foot and leg.

Contracted heels are often due to lack of frog pressure caused by improper shoeing, injury, or unnecessary shoeing — leaving shoes on too long or keeping the horse shod all year round. If shoes are left on too long without trimming the foot and resetting the shoes, the hoof wall grows long and heels become underrun, inhibiting expansion at the heel and quarters.

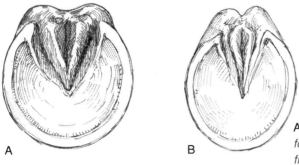

A. *Normal foot.* **B.** *Contracted foot (heels too close together; frog small and atrophied).*

Lameness from any cause may make a horse put less weight on a foot, and over time the lack of frog pressure results in contraction. Once a foot is badly contracted it may take more than a year to become normal again, even with special shoeing.

Contracted feet are more common in front feet than hinds, especially if the condition is due to improper shoeing. Sometimes only one foot is contracted due to injury. It is easy to tell the difference between the normal foot and the contracted one; the normal foot has healthy frog and heels whereas the heels of the contracted foot are too close together.

Foot contraction is often accompanied by dished or concave sole. The foot no longer flattens when weight is placed on it because the heels cannot expand; the sole becomes more concave, arched upward. If contraction becomes severe, the hoof wall may start to press against the coffin bone inside the foot, making the horse lame and unsound — a condition called "hoof bound."

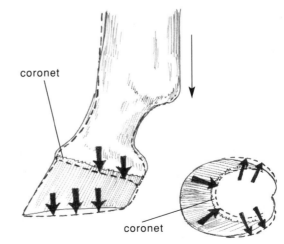

Foot expansion helps counteract concussion; the normal foot has a concave sole that flattens when weight is placed on it, the heels springing wider apart. The coronet narrows and drops backward as the overall height of the foot decreases.

Treatment

Corrective trimming and shoeing can help reverse contraction (you'll want the help of a good farrier), but the primary cause should be identified and corrected — such as lameness, improper shoeing, or dry feet. If feet are hard and dry, efforts should be made to restore proper hoof moisture by using a good hoof dressing daily. A dry foot lacks elasticity for proper foot expansion.

Flat Feet

Flat feet lack natural concavity of sole — an inherited condition. There is not much you can do to correct this, but you can keep the horse from stone bruising by using Easyboots (EasyCare, Inc., Oro Valley, Arizona) over the shoes when riding among rocks or by having your farrier attach hoof pads under the shoes.

Some flat-footed horses get by without hoof pads if the soles are kept toughened so they don't bruise easily. This can be done by applying a little iodine over the sole of the foot, taking care not to spill any on the horse's skin (because it burns) or let it run over the hoof wall (because it dries out the tissues). The iodine can be applied on days when the horse will be traveling on rocks or gravel.

Applying Iodine

An easy way to apply iodine to the sole is with a small syringe. Then you don't get it all over your hands or spill it on the horse. Draw up a little into the syringe and squirt it gently, a little at a time, covering just the sole area.

Club Foot

A club foot is one that has a steep pastern, with hoof and pastern angle of more than 60 degrees. The hoof angle is often steeper than the pastern angle, and the heels grow faster than the toe. If just one foot is this way, it may be due to an old injury (lack of use can cause contracting and shortening of the tendons, making the foot more upright). Sometimes the problem is genetic; in certain family lines several individuals may have one abnormal foot (always the right or the left), possibly due to conformation of that leg and the way the foot and pastern grow. Other horses may inherit the club foot trait in both fronts. This disability makes a horse less agile, with a rough and stumbling gait.

Treatment

Frequent and proper foot trimming can help a horse with a club foot, especially if corrective trimming is begun while he is still young and growing, before the condition gets worse. Letting the hoof grow too long between trimmings can aggravate the problem, causing the horse to have a very long heel and upright foot and pastern.

*Club foot (**A**) compared with normal foot (**B**).*

Basic Foot Care

You may wish to leave all foot trimming to your farrier. However, if your horses have normal feet and legs and do not need corrective trimming or special work, you may wish to learn how to trim. Ask the farrier to give you lessons on trimming so that you can take care of the routine matters. Even if you never trim a foot yourself, be sure to handle and clean all feet regularly. Then you'll know if there are any problems and be able to appropriately schedule trimmings with the farrier.

When cleaning a foot, a hoof pick or the blunt edge of a hoof knife is the best instrument for getting out all dirt or embedded rocks. *Note:* Never use a sharp or pointed object for cleaning a foot. You could injure yourself or the horse if he happened to jerk his foot at the wrong time.

When cleaning a foot, use a hoof pick.

Pulling a Shoe

Once in a while it becomes necessary for a horse owner to pull off a shoe. Even if a farrier does all your trimming and shoeing, you may at some time face a situation when you can't wait for him or her. A shoe may become loose (e.g., hooked on a fence, caught in a deep bog, stepped on by a hind foot and pulled loose, or loosened when traveling through rocks) even if the shoe was well clinched to start with.

In these instances it's best to pull off the shoe. If it is not immediately removed it may injure the horse if hanging loose on one side, catch on something and get the horse into trouble, or cause a corn or bruise because it has slipped out of position and is putting pressure on the sole or bars. If the shoe is accidentally pulled off it may break the hoof wall, taking out a chunk and making it harder for the farrier to reshoe the foot.

Tools for Removing Shoes

It is fairly easy to remove a shoe without breaking the hoof wall if you have a few appropriate tools or use adequate substitutes. Shoeing hammer, clinch cutter, nippers, and rasp make the job easy; but you can use a flat-edged screwdriver in place of a clinch cutter, and a regular carpenter's hammer if necessary. Pulling nippers or hoof cutters are best for pulling a shoe, but if you don't have them (and the shoe is fairly loose) you can use a pair of vice grips or regular pliers to hold on to the shoe, and gain leverage for pulling it.

Unclinching the Nails

The shoe is easiest to remove (and pieces are less likely to break out of the hoof wall as you pull it off) if you first unclinch the nails that are still holding. You can use any kind of hammer to drive the clinch cutter or screwdriver under each nail end to pry up the clinched end and straighten it out. This is easiest while the horse's foot is on the ground. Be careful not to cut into the hoof wall if you are using a screwdriver. Once each nail is unclinched, cut off the straightened nail end with nippers if you have any. If you don't have a clinch cutter or screwdriver, rasp off the clinched nail ends with a rasp or file.

Removing the Shoe

If you don't have a rasp or file, you can still pull the shoe with nippers or vice grips, especially if the shoe is partly loose already. Hold the hoof in regular shoeing position — between your legs for a front foot, across your thigh for a hind. Place the nippers or vice grips between shoe and hoof at the heel (starting on the looser side to make it easier), and use an outward and downward force to pry and loosen the shoe. Work alternately along each branch of the shoe, starting at the heels and moving toward the toe as the shoe comes loose.

If you are unable to undo the clinches you can still remove the shoe in this manner, though it takes a bit more strength and leverage, because you must pull the clinches loose and on through the hoof wall. The nails will straighten out as you pull the shoe; they will come out with it.

If some of the nails are still tight, the hoof wall may break unless you take each nail out as you go. This may happen when you pry on the shoe, because the clinches are still secured in the wall. To get hold of a nail head, you may have to gently pound the shoe back down against the hoof so the loosened nail head protrudes up enough to grasp with hammer claws, nippers, or pliers. Pull it out, then loosen the shoe enough to take out the next nail, alternating down each side of the shoe.

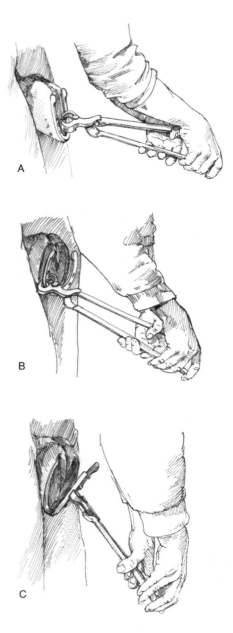

A. *Pull a nail out to help loosen the shoe.*
B. *Work down each branch of the shoe with alternate pulls and take each nail out as you come to it.* C. *Pull the shoe.*

Safety Tip

If a nail breaks off in the hoof wall, grasp it with pliers or nippers and pull it out. Don't leave nails or pieces of nails in the wall sticking out; the horse may hit and cut himself with them.

Thrush

Thrush is hoof rot caused by bacteria (and sometimes fungi) commonly found in barnyards or pastures. Because these organisms thrive in wet, decaying material such as manure or mud, thrush is common in horses that live in muddy pens, wet pasture, or dirty stalls. If a horse's feet are frequently packed with dirt, mud, or manure, the lack of air next to the frog and the constant moisture in the hoof make ideal conditions for the organisms to flourish, rotting the frog tissue.

The telltale foul odor is unmistakable when you clean a horse's foot. There are also black secretions along the edges of the frog, and it may be soft and eaten away at the edges. The early stages of thrush are indicated by just a little dark coloration and grime around the frog, or dark spots along the white line of the sole, plus the bad odor. At this point the problem can be quickly cleared up by keeping the foot cleaner and applying iodine, bleach, or a commercial preparation for thrush *daily* to the affected area to kill the bacteria. Just be careful not to let any of these medications run down the hoof wall to the coronary band or they may burn the skin.

If a horse is kept in a muddy pen, boggy pasture, or dirty stall and his feet are rarely cleaned — never having a chance to dry out — thrush can progress to the point of lameness as the infection penetrates and spreads to sensitive parts of the foot. If thrush is long-standing and deep, the horse will flinch when his feet are cleaned or trimmed because the frog is undermined with infection.

Prevention and Treatment

Prevention is the best "treatment": keep the horse in a clean environment, clean his feet often, and ride or exercise him regularly. If he can get out of the wet paddock and travel on dry ground, his feet will have a chance to dry out and air will get to the bottom of them, inhibiting thrush. No hoof dressing or medication can keep thrush from recurring if the horse is constantly kept in dirty surroundings and his feet are packed with mud and manure most of the time.

Winter Requires Extra Diligence

Winter can be a hard time to prevent thrush if pens and pastures are wet and muddy. It may take diligence — daily foot cleaning and care — to prevent thrush or keep it from getting worse. A bad case that invades the inner tissues of the foot must be treated as a puncture wound would be; seek advice from your veterinarian.

If you detect the beginnings of thrush when you clean a foot, don't ignore it. Clean the foot thoroughly, then swab the affected areas (usually the edges of the frog and any black spots in the sole) with iodine-soaked cotton or squirt on a little iodine with a small syringe. Daily treatment with strong (7 percent tincture) iodine for 3 or 4 days will clear up an early case. Clean feet often and keep track of their condition. Then thrush won't get a head start.

Foot Injuries

The horse's foot is strong and durable, but it can be injured if he steps on a sharp rock or nail. The tough outer covering of hoof wall and sole protect the delicate inner tissues from most types of trauma, but sometimes an exceptionally deep penetration, severe blow, or constant pressure causes bruising of the underlying tissues.

Corns and Sole Bruises

Corns are bruised areas on the sole, usually involving the tissues at the angle where the hoof wall meets the bars. A corn appears as a slightly reddened area and is especially noticeable when feet are trimmed and the sole pared down. The red area may be hot and tender in the early stages of the bruise. Corns are most common in front feet because they bear more weight and are more subject to bruising than are hind feet. Horses with flat feet are more likely to get a corn or sole bruise from stepping on sharp rocks or gravel.

Corns are most often caused by improper shoeing (ill-fitting shoes that are too narrow) or by leaving shoes on too long. The hoof wall begins to grow down around the outside of the shoe at the heel area, and the shoe puts pressure on the sole at the angle between hoof wall and bars, bruising the sole. Trimming a horse's heels too low may also cause corns as a result of increased pressure at the angle of the wall and bar. Corns in this area are rare in a horse that goes barefoot, but he may suffer sole bruising (a similar sore spot) from stepping on rocks. Horses with thin soles or chronic laminitis (founder) with dropped soles are very susceptible to sole bruising. Bruises in the sole generally occur in the toe or quarter area.

Definitions

- ◆ A **dry corn** is bruised tissue in which there is red staining due to bleeding inside the sole.
- ◆ A **moist corn** is caused by severe injury to the sole, resulting in seeping fluid beneath the injured horny tissue.
- ◆ A **suppurating corn** is one that has become infected and abscessed.

Locating the Problem

Corns or sole bruises can make the horse lame. If the problem is a corn, the horse will favor the heel and put more weight on the toe. If the problem is a bruise in the toe area, he will try to land on his heel to keep the weight off his toe. Your veterinarian or farrier may use a hoof tester to locate the corn or bruise. The horse will flinch when the sore area is pressed. If the sole at that spot is pared down with a hoof knife, a reddish (or bluish if an abscess is developing) discoloration may be seen in the area where the horse shows pain.

Hoof Testers

A hoof tester looks like big hoof nippers, with curved parts large enough to fit around the entire hoof. The hoof tester puts pressure (by pinching) on the sole to check for soreness. When you put pressure on the sore spot, the horse will flinch. Your veterinarian or farrier can show you how to use a hoof tester.

Use a hoof tester to pinpoint the area of soreness or to locate a corn or sole bruise.

Treatment

When corns are caused by improper shoeing, removal of shoes may be all that is necessary for healing. The horse should not be ridden or reshod until the corn and lameness have disappeared. To make sure shoes do not cause corns, check that the heels of each shoe extend well back and cover the hoof wall completely at the heel and quarters to allow room for hoof expansion. The hoof wall in this area must rest on the shoe. If the hoof extends beyond the shoe when the foot expands, the shoe is not large enough.

If a horse develops an abscess from a corn or sole bruise, your veterinarian will pare down the sole at that area until the infection in the sensitive tissues begins to drain. Then soak the foot daily, packing and protecting it between soakings, as you would for a puncture wound.

Puncture Wounds in the Foot

Puncture wounds in the body are always serious because they can lead to tetanus if a horse has not been vaccinated, but a puncture in the foot can be very bad because it may damage the inner tissues and cause any one of several serious problems — inflammation or fracture of the coffin bone, decay of the bone, or decay of the digital cushion. A puncture in the middle third of the frog may puncture the navicular bursa.

A puncture in the bottom of the foot is not always easy to locate if the foreign object (e.g., nail, stick) is no longer embedded. A puncture in the frog can be difficult to find if the spongy tissue has closed up again after the object makes its hole.

Treatment

If you suspect a puncture (or the object is still embedded), consult your veterinarian. You may not realize there is a problem until an abscess develops within the foot and the horse goes lame — from infection putting pressure on surrounding tissues. It's best to start treatment immediately after the puncture occurs.

Hoof Abscess

An abscess may be caused by a puncture wound, infected stone bruise, misplaced horseshoe nail that "quicked" the horse (i.e., entered the sensitive inner tissues), or a case of deep, neglected thrush that penetrates into sensitive inner tissues. Sometimes a hoof crack becomes deep enough to allow entry of bacteria that start an infection.

Treatment

The area should be opened to drain, soaked, cleaned with disinfectant, and the hole plugged with a wick that will allow drainage. A puncture wound in the sole should be opened so there is at least a ¼-inch (0.6 cm) hole into the infected tissue, with the walls of the drainage hole widening toward the ground surface of the sole so it won't become obstructed. When the wound is in the frog, the frog should be trimmed away at the site of the puncture until adequate drainage for the abscess is established.

Open up a sole abscess or puncture wound with a hoof knife to establish drainage for the infection.

Part of the treatment for a puncture wound or foot abscess consists of soaking the foot. After the veterinarian has dealt with the original condition — opening the area for drainage and flushing it out — and given the horse a tetanus shot and possibly antibiotics, he may tell you to soak the foot daily for several days to draw out infection and promote rapid healing. Soaking the foot in a warm water and salt solution not only pulls out any remaining infection but also makes the sore foot feel better.

Soaking a Foot. If a horse has never had a foot soaked, get him used to the idea carefully and gradually so it will be a good experience — so he will cooperate rather than resist. Put his foot gently into an empty rubber tub. Once he stands there with his foot in the tub, wash the foot thoroughly with water — scrubbing with a rag, if necessary, to remove mud or dirt — and replace it in a clean, well-rinsed tub. Carefully add a little warm water, slightly warmer than the horse's body temperature (100°F, or 37.8°C). After he accepts it and is relaxed, you can add several tablespoons of Epsom salts (magnesium sulfate) to the water. Don't add the salt to the water until you're sure he's comfortable about the soaking, in case he moves around and spills the first batch.

After he is standing with his foot in the warm salt solution, periodically add hotter water until the water in the tub is quite warm but not burning hot. Keep the water as warm as the horse will tolerate. It's easiest with two people — one to keep the horse from spilling the tub water if he picks up his foot (hold the foot and guide it gently back down into the water so he won't put it on the ground and get it dirty), and one to manage the teakettle and refills. Usually 20 to 30 minutes of soaking per day is adequate. Most infections clear up after 3 or 4 days of this.

Tub or Bucket?

A rubber tub or bucket is best — if you don't have a special soaking boot — because it won't injure or startle the horse as much as a rigid (noisy) metal one. If the injury is low (e.g., at the coronary band or heel) or on the bottom of the foot, a rubber tub works fine. If the area to be soaked is higher (e.g., at the fetlock joint), the tub may not be deep enough to cover the area in warm water. In this case you'll need a rubber bucket.

Protecting a Drainage Hole. If there is a drainage hole in the sole or on the bottom of the foot, protect it between soakings by wrapping and bandaging the foot to keep out mud and dirt. An easy way to bandage is to wind Vet-wrap (stretchy material that sticks to itself) around the foot to cover the bottom; then use duct tape from mid-hoof on up to the pastern, to hold everything in place; and put the bandaged foot into an Easyboot or some other type of protective boot so the bandage stays clean and dry. If the pen or pasture is wet or muddy, use a waterproof boot to keep the foot and bandage clean between soakings.

An Easyboot (EasyCare, Inc., Oro Valley, Arizona) helps keep the foot clean and dry.

Coronary Band Injuries

Sometimes a horse suffers injury to the coronary band at the top of the hoof by running into something, stepping on himself with a shod hoof, or having his foot stepped on by another horse. A deep wound at the coronary band that results in scar tissue may disrupt the hoof growing cells and make a weak spot in the hoof as it grows — an area that may easily crack or disrupt hoof growth completely.

Treatment

A serious or deep injury at the coronary band requires immediate veterinary attention so it can heal properly and prevent proud flesh (excessive granulation tissue) (see chapter 16).

Hoof Cracks

Occasionally a crack develops in the hoof wall from splitting and chipping, concussion, or injury. Barefoot horses often develop chips and cracks, but these are usually not serious unless neglected. In a long, untrimmed foot, cracks can become worse quickly, traveling up into sensitive tissues. These can be hard to correct. A deep crack may make the horse lame.

Brittle Feet

Brittle feet — due to dry conditions, poor nutrition, or genetics — make a hoof very susceptible to cracks. Brittleness is more common in white hoofs than dark ones, but actual hoof strength depends on the individual horse. Some have tough, resilient feet regardless of hoof color. If a horse has brittle feet that crack easily, a good hoof dressing can help, as can certain feed supplements. Consult your veterinarian.

A serious hoof crack should receive immediate attention to halt its progress and protect the underlying tissues from infection. If the crack reaches sensitive tissues the horse may need a tetanus shot, antibiotics, and foot soakings. If a horse ever develops lameness in conjunction with a hoof crack, consult your veterinarian.

Toe cracks, quarter cracks, and heel cracks start in the hoof wall at ground surface, usually from a chip or split in an over-long hoof, and travel upward. Sometimes a crack starts in the heel or quarter area and travels horizontally around the foot due to weakness in the wall at that area (from a blow or injury, as when a horse strikes the hoof against a rock). Sometimes a crack starts at the coronary band. This type of crack travels downward because of weakness in the hoof wall at that area.

A barefoot horse with a serious crack must be trimmed often to relieve pressure on the crack and keep it from splitting farther. It is difficult to grow out a bad crack on a barefoot horse because it's impossible to take all pressure off the crack and keep it from spreading. The horse may have to be shod to grow out the crack.

Treatment

When trimming a foot with a crack, the farrier will cut away the hoof wall at the crack so it does not bear weight (which would expand the crack). He may also try to stop the progress of the crack by rasping a small notch across its highest point. This helps stop the splitting, for it sends the force along the notched groove instead of on up the hoof wall if weight is placed on the crack.

Halting the progress of a crack by shoeing with clips (to keep the hoof from expanding at that point and spreading open the crack) and by notching the top of the crack with a bar and triangle design sends the force of impact along the notched grooves instead of up the hoof wall.

Corrective Shoes. A crack that originates in the coronary band, due to an old injury that affects horn growth, may be a more persistent problem. In this case corrective shoeing may be necessary for the rest of the horse's life. Regular use of a hoof dressing, lanolin ointment (such as Bag Balm or Corona), or olive oil may keep the coronary band and hoof wall more pliable in that area and less apt to crack.

A shoe can help take the stress off the hoof wall and keep a crack from splitting farther, enabling it to grow out. A clip on each side of the crack helps keep the hoof from expanding there when weight is placed on the foot.

Fast-Drying Glue or Plastic. Some cracks can be repaired with a strong, fast-drying glue or plastic that fills the hole and holds the cracked area together so it can't keep expanding. This protects sensitive tissues from contamination and infection if the crack is large or deep. Your farrier can help you with this. The foot will still need to be trimmed often or the shoe reset — until the crack grows out — so the glue or filler material may have to be reapplied several times.

Diagnosing Lameness

Pain is nature's way of keeping an animal off an injured leg or foot. If a horse limps, try to determine the cause. A sore front leg (or foot) will cause him to bob his head more than normal as he hurries to get off the painful limb and takes more weight on the good leg. A sore hind leg or foot will cause a hitch in his stride as he puts less weight on the bad one and hits the ground harder with the good one.

To tell which leg is lame, watch his head and hips. The horse's head goes higher when he steps on the bad front leg and drops lower as he takes more weight on the good front leg. His hip stays higher on the lame side if he's got a bad hind leg and drops lower on the good side that is taking more weight.

Check his feet and legs to see what the problem might be. It may be as simple as a rock caught in his foot, or it may be a stone bruise or leg wound. If there is infection in a leg, there will be heat and swelling. A pulled tendon or injured joint will also cause heat and swelling. If you cannot determine the exact problem or how to treat it, call your veterinarian.

FIVE

HORSE HANDLING

The horse is an intelligent, sensitive animal with great athletic ability. Part of our fascination with him comes from the realization that he's a wild animal that has been domesticated — we work so closely with him that there's practically no limit to what a good horseman and well-trained horse can accomplish as a team. To develop that unity, or to handle any horse safely, we must understand how horses "think" and react. The horse owner needs to be mentally in tune with the horse.

Horse Sense

Horses have better memories than humans beings, but they don't "think" the way we do, and often the horseman makes the mistake of assuming they do. The horse is a prey animal. Most of his natural reactions are geared toward flight unless he's at ease with the situation at hand. When he feels alarmed he becomes nervous and ready to flee. He has a strong herd instinct; he's happiest and feels most secure with other horses.

Horse and human can develop a strong loyalty bond — the horse transferring his needs for safety and comfort to the person he trusts.

Horses are most at ease when their human handlers act like the dominant herd member (boss horse) without being too aggressive or submissive. As a herd animal, the horse is accustomed to being bossy or bossed. It comes naturally to him to submit to a more dominant individual, so this is the role the horse owner must take when handling the horse. The extremes in horse handling methods are force and bribery. Force may make an aggressive horse resentful, always fighting back, or a timid horse more afraid. Bribery will quickly spoil a horse; he soon learns that he (not you) is in control. The most effective handling methods fall in between, with the horseman taking an assertive leadership role in tune with that particular horse's ability and nature.

Understanding the Horse's Temperament

The horse in nature is a member of a group. Most horses become frantic, lonely, or neurotic if there are no herdmates around. The lone horse is in a situation contrary to his natural instincts; his attitude is healthier if he has some company — even a goat or a cow or some other animal if there are no other horses nearby. Some adjust more readily to an isolated existence than others.

There is a direct relationship between temperament (emotional characteristics — whether a horse is bold, calm, easygoing, timid, or insecure) and ease of handling. A horse who is readily upset or distracted can be more challenging to handle than a calm one.

When handling horses it's important to work with their temperament, making sure bold ones don't try to dominate you, lazy ones don't cheat and try to avoid doing what they are supposed to do, and timid ones don't become more afraid and skittish. Use good judgment when catching, leading, handling feet, and so on, and your horse handling will go a lot more smoothly.

Levels of Communication

You are always sending signals to the horse — via your emotions as well as actions and body language — whether you realize it or not. Every time you handle him you are training (or untraining), making things better or worse. Everything you do with the horse should be with thought as to how he will respond. Ask yourself whether you are aiding or defeating your goal of training him to be well mannered and easy to handle with every action you take.

Cultivating a Good Relationship

Cultivate the horse's respect and cooperation so he knows what to expect from you. If you use consistency and quiet confidence in all your actions, transmitting a message of benevolent dominance, the horse feels more secure and trusting than he would if handled by a nervous, inconsistent, timid, or angry person. Your attitudes and feelings are crucial when handling any horse, contributing to success or failure.

Working with a horse is much easier if you are tuned in to what he is doing and thinking. He should also be comfortable with signals he picks up from you. Remember that you control and communicate not only by means of physical cues and actions but also with your mind. This type of control over a horse comes with familiarity (knowing the horse well) and your understanding of one another, gained through working together — the horse knowing what you want and being conditioned (through proper handling and training) to obey you. He senses your relaxed but dominant attitude rather than fear, anger, or nervousness. He respects you as the dominant one (boss horse); it's part of the relationship you develop with him as you work together and handle him daily. He can read what you are thinking by your body language and the way you act.

Relax

Horses can read a person well. They sense changes in physical or emotional tension that tell them a person is frightened, nervous, or angry. Indeed, they can sense our mood and feelings no matter how much we try to disguise them. Animals have the ability to communicate with each other through a combination of body language and empathy and they are able to apply this to human beings to the extent that they can read us like a book.

You must truly be at ease in your relationship. Even if you go through the motions of being the dominant member of the team in your words and actions, if in your mind you are afraid of the horse he can sense it. Your fear will overshadow your actions and will make him more nervous or more aggressive.

A combination of body language and confident attitude gives you dominance over the horse and puts you in the position of boss and teacher, enabling the horse also to be at ease in the relationship — not feeling the need to aggressively disobey you through lack of respect or being unduly afraid of you. If you have confident dominance over him, the horse will respect and trust you, accepting you as boss. You will rarely have to resort to a loud voice, punishment, or any physical contest of strength while handling him.

Be Relaxed

The best way to get a nervous horse to relax and cooperate is to be relaxed yourself, both mentally and physically. If you are tense and nervous the horse becomes agitated also, and since he thinks that whatever is alarming you is also potentially dangerous to him, he may become very upset.

Use Appropriate Tone of Voice

The sound and tone of your voice are important. An approving voice can be a powerful reward or positive reinforcement for good behavior. A disapproving tone is often punishment enough for bad behavior; the horse realizes you are displeased with him. Horses don't have to fear us to submit to our wishes; punishment and reward by use of body language, voice, and attitude are enough when there is good communication in your working relationship.

Punishing Bad Behavior

If a physical reprimand is needed, it should be instant and appropriate. The horse may need just a jerk on the halter rope or a tap with a whip (to remind an overeager individual not to drag you along too fast, or to punish an act of bad manners). One short application is enough. Most horses understand one swat — after all, in the herd situation one bite or a single kick is adequate to keep a member in line. They don't understand beatings or continual punishment.

Avoid Excessive Punishment

Excessive punishment makes a horse lose respect for the handler, replacing trust with fear or strong resentment. Continually pecking at a horse can thoroughly confuse him. He will quit trying to do the right thing or become afraid of everything you do.

Any punishment should be of short duration — immediate — and forceful enough to fit the circumstance. The psychological aspect is more effective than physical discomfort or pain; punishment should be adequate to have proper effect on that particular horse. For some, vocal disapproval is enough. For others, especially bold or highly aggressive individuals that test your authority, a sharp spanking may be necessary. If the horse is testing your dominance, as he might do in a herd situation trying to work up to being boss horse himself, you must remind him that you are still on top of the pecking order.

Use Caution

The horse is a large, strong animal. If he becomes upset or frightened, or moves suddenly, the person handling him may be injured if he or she is in the wrong place at the wrong time. Most accidents can be prevented if you make a practice of handling yourself and the horse in such a way that there's less chance of unexpected trouble. If you always have "safety first" in mind — thinking ahead and being prepared for anything the horse might do — you are less apt to get hurt.

Safety Tip

Never take any horse for granted. Even the most dependable one might move suddenly if startled and hurt you unintentionally if you are in the way. Have safety-conscious working habits even around horses you know and trust; it makes for fewer stepped-on toes, bumped heads, or more serious mishaps. Always be aware of which way the horse might move, and be prepared to move with him.

An important factor in minimizing accidents is having a good working manner and quiet confidence — being firm but gentle. Then the horse is less apt to try you out (if he's bold) or be afraid and flighty (if he's timid). You want him to relax and trust you. A nervous horse that is uneasy about the way you handle him is more likely to become unmanageable than the one that feels secure about what you are doing with him.

Handling horses safely is a matter of awareness (being tuned in to what the horse is doing and thinking, being prepared for what he might do next) and body position. You dominate a horse through confident attitude and body position. This type of mental control comes with understanding one another, the horse knowing what you want and being conditioned to obey — knowing "Whoa" means stop and stand still; accepting that he must accept restraint by the halter, that he must behave when you pick up his foot.

The horse is stronger than you are, but through training and your confident attitude he accepts your dominance. If, however, he becomes momentarily frightened or upset, or worried about the pain you or the veterinarian may cause when he must be vaccinated or treated, he may forget his manners. At those times you must be able to calm and restrain him and keep from being hurt by him, even though you are not as strong as he is. You can use body position and leverage (and physical contact with the horse) to your best advantage, to keep him under control and avoid being kicked, bumped, or injured.

Always keep in mind that the horse's instinct is to protect himself from danger — to run away, strike, or kick to defend himself. The veterinarian's needle, medication spray, or any procedure the horse isn't used to may provoke a defensive response that is dangerous to people handling him. Be prepared for evasive or defensive actions, handling the horse in a way to minimize or prevent these actions — or be in a position where his actions won't make physical contact with you or inflict harm.

Approaching a Horse Safely

When you catch and halter a horse — whether in paddock, pasture, or stall — speak to him as you approach; make sure he knows you are there. Even a gentle horse may kick if you startle him while he's napping. Talk to him before you touch him. Don't approach from directly behind him (in his blind spot) but at an angle, so he can see you better. Speak softly and move slowly.

Holding a Horse

At times you must hold a horse for the farrier or the veterinarian for treating wounds, vaccinating, and the like. The person holding the horse can make a difference in whether the procedure is successful, difficult, or dangerous. Keeping the horse from moving at the wrong time, keeping him calm or preventing him from kicking, can be crucial.

Use a properly fitted halter. With some horses a bridle may be necessary for more control. Some need a chain over the nose to keep them restrained (or even a twitch or lip chain — see subsequent discussion) but if a chain is used it should not cause pain or annoyance. Otherwise, the horse may become more unmanageable, fighting the restraint as well as the procedure being done. If you use a chain, never allow your fingers to get accidentally wrapped in it.

Choose a safe place to hold the horse. Sometimes an open area is best, with no obstacles to bump if he moves around. If he must be still, hold him next to a stall wall or solid fence, never next to a wire fence. If he usually tries to rush backward when confronted with something he doesn't like, plan in advance to back him against a fence or wall so he can't use this tactic. For most procedures, if there is a solid fence or wall on one side and the handler and person working on him are both on the other side, the horse will stand still.

Don't Stand in Front

When holding a horse, don't stand in front of him. If you're at his shoulder you'll be out of the way if he lunges forward or strikes out with a front foot. Stand facing his shoulder with one hand on the halter and the other on his neck or withers, with feet positioned so you can move when he moves and you won't be stepped on or jumped into. Body contact is crucial. If you are braced against him you can move with him rather than being bumped. Holding the halter, facing his shoulder, you can read his expression and intentions.

With your other hand *rub his neck or withers to help calm and distract him.* Often a soothing voice, soft whistling, and rhythmic rubbing will help keep his mind off the procedure and he'll tolerate the situation instead of fighting it.

The person holding the horse for medical treatment should stand on the same side as the person treating the horse. Stand at the horse's shoulder, with one hand on the halter and the other against the neck or withers, ready to move with the horse if he moves. With his front leg turned up, he is unable to kick on that side.

When holding a horse for the farrier, keep the horse calm and still so he won't move at the wrong time.

Prevent Kicking

Maintain proper contact with the horse's head to gain leverage over his whole body and his movements. This is important for the person at the halter, who should be tuned in to the horse's state of mind and possible reactions to prevent movement at a crucial time or keep the horse from kicking. Raising the head or pulling it toward you at the right moment can make it awkward for the horse to kick the person who is working on him. If the halter is too loose, hold it snugly under the horse's chin (keeping the loose part together) so your hand has contact with his jaw while holding the halter. The more contact you have with him, the better control you have and the less chance there is for him to bump you.

Lead Safely

When leading walk beside the horse's left shoulder, holding halter rope or lead shank a few inches from the halter or grasping the halter itself if more precise control is needed. This will afford the best control over his movements. Always have a rope or lead shank attached to the halter; never try to lead a horse with just a halter. The rope should be at least 6 feet (1.8 m) long, preferably 8 (2.4 m). If you use a lead rope the horse is less able to pull away, hurt your arm, or dislocate your shoulder should he jump, rear, or turn quickly. With a rope attached you can play him like a fisherman plays a fish; he cannot jerk you as hard as if you were hanging on to the halter alone. Never wrap the end of the rope around your hand. Keep extra length in neat loops, not a coil that might get caught around your hand or arm if he bolts.

When leading a bridled horse with loop reins, take the reins down over the head. If the reins are still up on the neck and he balks, backs up, or bolts, you have no control and only make the problem worse, because one side will be pulling at his mouth.

Proper position when leading a horse: Walk beside his left shoulder, and hold the lead rope close to the halter with your right hand, with the extra length of rope looped (not coiled) in your left hand.

Don't Walk in Front

Don't walk in front of a horse you are leading. If he spooks and leaps forward, he'll bump into you or step on your heels. You have no control of his movements if you are in front of him. When you walk beside his shoulder you move with him and have control over his head and all his actions. You can keep him going at the speed you wish and halt him when necessary. If he tries to go too fast or bolt, you can use body leverage to halt or slow him by leaning into his shoulder and pulling his head around so he has to circle. When you have this kind of contact and body leverage you can be as strong as he is, making him pivot around you so he can't bolt forward. Having your arm or body against his shoulder gives you more control than just holding him by the halter.

Use body leverage to your advantage when leading an overeager horse. Keep your hand close to the halter and your arm braced against his neck when necessary. With your body against his, you are strong enough to spin him around and make him circle if he tries to bolt.

Tying a Horse

When tying, always keep safety in mind; an improperly tied horse can be potentially dangerous. *Use strong halters and ropes, and keep them in good repair.* Nylon web halters should be at least three-ply with sturdy hardware. Remember that a halter rope is only as strong as its snap; it's often best to have a rope permanently braided onto the halter ring or tied with a secure knot rather than depending on a snap to hold if a horse sets back. If you don't leave halters on horses (and you shouldn't — it's a dangerous practice because a horse may catch the halter on something in his pen or pasture), you don't need a snap-on lead rope and can keep your ropes securely and permanently fastened to the halter.

Always tie to something solid. Careless tying, use of a flimsy rope or halter, and tying with bridle reins or to an insecure object all invite disaster. If a horse sets back (as even a well-trained one will do if suddenly startled) and breaks the rope or bridle reins, he could fall over backward, pull the top pole or board off a fence, or bolt down the road dragging whatever he was tied to. If this happens he may seriously injure himself, people, or other horses. It may also be such a traumatic experience that he can never be dependably tied up again.

Never tie to a wire fence. The horse could get a foot caught if he paws. Tie to the post itself or to a pole or board only if it is nailed to the other side of the post and cannot be pulled off. When tying, don't leave so much length of rope that the horse might get a foot over it. *Tie short, but with enough rope to give some freedom of head and neck movement, and level with his head or higher*. Then if he pulls back it won't damage his neck muscles as readily as if he were tied too low.

Properly tied, the horse has some freedom of head and neck but never can get his foot over the rope. The rope should be tied level with his head or higher.

When tying to a tree, make sure the rope won't slip down the trunk. Choose a tree with no branch ends that might injure his head or eyes. When tying to a horse trailer, be sure the trailer is attached to the pulling vehicle or securely blocked so the wheels cannot move. When tying near other horses, keep the horses far enough apart so they cannot kick at each other. Never tie with bridle reins; this is a sure way to end up with broken reins or an injured mouth.

When tying a horse to a tree, tie high, make sure the rope cannot slip down the trunk, and be sure there are no branches to create a hazard for his eyes.

Use a Safe Knot

Always use a knot that will securely hold yet be easy to untie in a hurry. The "manger tie" is a good quick-release knot that can be undone with just a pull on the loose end. If a horse nibbles at the rope or tries to untie himself, you can put the loose end through the loop so he can't undo the knot.

Handling the Horse's Feet

When handling a foot, the more body contact you have with the horse, the better. Not only are you braced against him for holding the leg, but you can also sense his mood (tense or relaxed) and anticipate any movement he might make.

Safety Tips

Use basic rules of safety when working around a horse for any purpose — proper positioning and body contact will work to your advantage. You are less apt to be bumped, kicked, or stomped if you're close to the horse and braced against him in such a way that you keep him from moving improperly. If he does move, you'll be moving with him.

When working around front feet, take care never to be right in front of a leg as you bend over to examine it or apply medication or hoof dressing while the foot is on the ground. The horse can unintentionally bump you just by picking up his foot; his knee can hit you in the face. Always watch out for the knee.

When working around hindquarters, be aware that maintaining close contact is always safer than being a short distance away where you'd get the full force of a kick. Close contact can prevent a kick if you are braced against the leg. It also allows you to move with the horse and not be hurt.

Picking Up a Foot

To pick up a front foot, stand at the shoulder, facing to the rear, and slide your hand down the leg. Press your fingers just above the horse's fetlock joint. If he doesn't immediately pick up the foot, press a little harder in the area just in front of the back tendon — the indentation between tendon and cannon bone. If necessary, lean into his shoulder at the same time to encourage him to shift weight off that leg and pick up the foot. Pinching just in front of the back tendon generally makes even the stubborn horse pick up his foot and is easier than trying to wrestle with him.

Run a hand down the back of the front leg and prepare to pick up the foot. Press your shoulder against the horse's to encourage him to shift his weight off that foot. Squeeze gently at the fetlock joint if he doesn't pick the foot up for you.

To pick up a hind foot, lean against the horse to help him shift his weight. Run a hand down the leg, pinching the back tendon gently if necessary, then pick up the foot.

To pick up a hind foot, stand next to his hindquarters facing to the rear and run one hand down the leg while you lean against the horse with your other hand to encourage him to shift his weight and pick up that foot. If he does not want to pick it up you can pinch the back tendon gently.

Once you've picked up the hind foot, prepare to rest it against your thigh for examination or cleaning.

If the horse tries to jerk the foot away while you are cleaning or trimming it, you can thwart this by bending the fetlock joint, firmly pulling the toe up. He cannot kick at you or take the foot away as readily with the joint flexed. This added leverage works in your favor.

Holding a Foot

When holding a front foot to clean, trim, or doctor, be sure to keep your upper arm or shoulder in contact with the horse's side. This affords more leverage if you have to lean against him and hold tightly to the foot to keep him from pulling it away. When holding a hind foot, try to keep almost the entire hind leg in contact with your own leg. If the horse tries to jerk his hind foot away, firmly bend the foot by pulling the toe up toward you. This makes it harder for him to pull it away and kick you.

Holding Up a Front Foot

Sometimes a horse fidgets or kicks so much that certain tasks become awkward or impossible to complete. In many instances, holding up a front leg can minimize these actions and immobilize him. The safest way to do this is with another person holding the horse. On the side you don't want the horse's hind foot to kick from, pick up the front leg. Hold up the leg with one hand, with maximum contact and leverage, and brace yourself against the horse's shoulder with the other. Be prepared to move with him in case he tries to lunge and rear. Don't be in a position where you would get bumped or hit by the leg, and be ready for any movement the horse might make.

When holding a front foot, keep your upper arm and shoulder in contact with the horse.

Establishing Body Contact

Body contact with a horse not only gives better leverage for holding or working on him but also enables you to tell whether he is relaxed or tensing up for sudden action. Also, close contact and confidence on your part — being relaxed and matter-of-fact in what you are doing — help the horse to be more relaxed and secure. Because he knows where you are and what you are doing, he is less apt to be jumpy or try to kick. This is especially important when giving injections or treating a wound. Sudden contact may startle or alarm him, as when a needle is jammed in with no prior warning or medication suddenly sprayed on. If you are touching him in that area and reassuring him, and he's aware of and tolerating your presence, you can usually accomplish the procedure without much reaction on his part.

The Importance of Touch

Safety around horses is a matter of knowing the individual horse well and being alert to his potential reactions — intuitively and automatically keeping yourself in a position relative to him that puts you at least risk. This is usually a position very close to the horse, touching him and transmitting your confidence to him through that touch.

Using Restraints Properly

There are several ways to restrain a horse if he does not want to submit to a particular procedure. Extreme measures involve the use of ropes (scotch hobbles or sidelining — tying the hind leg to the front leg — so a horse cannot kick) or tranquilizers. The average horseman should not try these; improper use of ropes and hobbles can put you and the horse at risk for injury, and improper use of tranquilizers can be hazardous (these should be given only by a veterinarian). It's better to rely on other methods if possible. Should these fail, seek the help of a professional horseman or a veterinarian.

Cross Ties

Cross ties are two ropes with snaps that secure the horse in the middle of an open space; the ropes are snapped to each side of the halter. This approach is often used in a box stall or barn aisle. If the horse is accustomed to being cross-tied, this is often adequate restraint. He is held from both directions but with open space on each side for you to work in — he can't crash you or himself into a fence or wall. Yet, the restraint of a rope on both sides of his head limits his movement.

Cross ties often help to keep a horse from moving around while you are working with him.

Chain over the Nose or through the Mouth

For some horses, a chain over the nose gives better restraint than a halter. Whereas they root and pull and misbehave with just a halter, the chain over the nose gives more pressure when you need to remind horses to stand still. In some cases the chain works better through the mouth (like a bit) rather than over the nose. The halter rope or lead strap has a short length of chain at the snap end to put over the nose or through the mouth if needed. When using a chain, be careful not to get your fingers wrapped in it.

A chain over the nose gives better restraint than just a halter if you need to remind a horse to stand still or to gain more control when leading him.

Lip Chain

Some people prefer to put the chain under the upper lip and over the gum. This works well for some horses but not others. Because the gum is tender, the lip chain is effective in making a fractious horse behave and stand. If properly used, without jerking it too hard, it does not hurt the horse. However, it can be a severe restraint that injures the gum and upper lip if the horse rears and plunges.

Twitch

The traditional twitch is made from a length of wooden handle 15 to 30 inches (38.1–76.2 cm) long, with a loop of rope or chain attached to one end. Newer versions that are made of metal clamp to the upper lip and the halter. This enables the horse handler to have both hands free to work on the horse after adjusting the twitch if there is no one to help. The disadvantage of this "one man" design is that if the clamp twitch loosens and the lip

A traditional twitch has a wood handle and a loop of chain that is twisted around the horse's upper lip.

A metal twitch like this one clamps onto the horse's upper lip and is snapped to the halter.

To put a twitch on a horse, place your hand through the loop and take hold of the horse's upper lip, then slide the chain or thong over your hand and around the lip. Next twist the handle until the loop tightens around the lip.

comes free, the horse is no longer under control and is left with the tool dangling from his halter. If it's too tight and he breaks away, he may start running blindly to try to escape the clamp stuck on his upper lip.

Safety Tip

The safest way to twitch a horse is with two people — one to handle the twitch and keep the horse under control, and one to do the treatment. Both people should stand on the same side of the horse so they each know what the other is doing and can be prepared for the horse's reactions. If he lunges, jumps around, or tries to break away, he'll go in the direction away from both people and no one will be stepped on.

Not All Horses Can Be Twitched

Some horses will not tolerate a twitch. Others, such as foals or young horses, should not be twitched to begin with. A youngster that has not had a lot of handling may react with fright and resent any further handling of the muzzle.

The best candidates for twitching are emotionally stable adult horses that are accustomed to human contact — these animals are basically trusting and tractable and don't mind having a person handle the upper lip.

Why a Twitch Works

Twitching increases the horse's ability to handle pain by stimulating receptors in the skin that activate a pain-decreasing, morphine-like substance that causes heart rate to decrease and sensations of pain to be reduced. These substances, called endorphins, also make the horse calmer. After application of the twitch he becomes quieter and appears somewhat sedated, with drooping eyelids. He stands as if in a trance, with greater tolerance for pain and discomfort. A lip chain works in the same way.

A twitch relaxes a horse quickly and easily (and only temporarily, with no after-effects). As soon as you release the twitch or lip chain, he's back to normal. There's no waiting for the effects of a sedative or tranquilizer to wear off.

Hand Twitching

A horse that tolerates a twitch can often be "hand twitched" by gently grasping his nose and squeezing. This can make him stand still and calm for a few minutes, rather than fighting the procedure being done to him.

If done correctly, twitching is not resented by most horses and won't make them headshy. The twitch should never be painfully tight, just enough to put a little pressure on the upper lip. When the procedure is finished you should release the twitch slowly and gently massage the upper lip. Then there are no bad memories for next time.

Stableizer

A relatively new tool for restraining nervous, fractious horses, the Stableizer (Wheeler Enterprises, Inc., Ellendale, Minnesota) is a modern adaptation of the Indian war bridle but is more effective, humane, and versatile, often working better than a twitch, lip chain, or tranquilizer. One person can put the device on a horse, and it stays in place. The Stableizer relaxes and sedates the horse, yet the horse can still move and be led while wearing it, which is not the case with a twitch.

Tightening the device activates pressure points behind the ears, releasing endorphins that tend to block out pain and make the horse feel calm and relaxed. The pressure point beneath the upper lip blocks release of adrenaline, relaxing the horse even more. The portion of the device that goes under the lip is covered with plastic tubing, so it never cuts into the gum and is more humane than a lip chain.

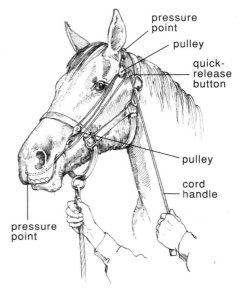

The Stableizer (Wheeler Enterprises, Inc., Ellendale, Minnesota) is a humane restraint that slips over the horse's head, applying pressure behind the ears and under the top lip next to the gum. When tightened by pulling the cord handle, it stimulates pressure points behind the ears and beneath the lip, releasing painkilling endorphins and blocking adrenaline, causing the horse to relax and feel good. The Stableizer can be set at any pressure with the quick-release button.

When Physical Restraints Won't Work

In extreme cases, when a horse is already in pain from serious injury or colic or in an unstable emotional state, a chemical restraint such as a sedative or tranquilizer given by your veterinarian may be necessary to safely handle the horse for examination or treatment. Twitch, lip chain, Stableizer, or hobbles are not adequate in these instances.

Dealing with Bad Manners

Some horses develop bad habits if they are allowed to get away with misbehaving, and they may become dangerous to work with. A young horse just learning how to interact with people should never be allowed to continue bad behavior or he'll grow up spoiled and unruly, and an older spoiled horse should be reprimanded and retrained or he will think he can always do as he pleases. Some horses have bad habits due to improper handling in their formative years or a bossy attitude that was not curtailed by a previous owner. As the old saying goes, "There are no problem horses; just problem people." A horse's habits and manners are a product of his reaction to people and the way he's been handled.

Pushy horses readily take advantage of a timid or soft-hearted person who allows them to do as they please instead of handling them with consistent firmness. Bold horses spoil easily if they feel they can have their own way in the relationship. Regardless of the horse's temperament, aggressive behavior such as biting, kicking, rearing, walking on you instead of respecting your space, or rushing backward is not only annoying but dangerous.

The Biter

If you have a horse that bites, try to figure out why he bites. His biting may be a defensive action if he thinks you are a threat or if you cause him pain. He may also bite in playfulness, since this is how he interacts with other herd members. Alternatively, the biting may be an aggressive action if the horse is trying to establish dominance, not being content to accept you as boss. Within the herd, the horse uses his teeth to reprimand other horses, establish dominance, and attack intruders. Some nervous horses bite when frustrated or bored; confinement may lead to irritability, and they take out their frustrations on the person who comes to handle them.

Prevention

Halt nipping so it won't become a more serious problem. With a young playful horse the solution is to teach him to respect you; make it clear that you are the dominant one who must be obeyed. The biter needs firm, consistent discipline every time he tries, just as he would get in a herd situation from a more dominant horse.

Be constantly alert to his intentions so you can prevent them or reprimand him instantly. All disciplinary action, especially for biting, should be given as the act is happening or about to happen, not afterward. Otherwise it may become a game for him — trying to sneak in a nip, then jerking his head away before you can bump him. If you bump him on the nose after he nips, he just jerks his head up and may become headshy or play games to see who is quicker.

Punishment

If you can punish him *before* he nips, he'll lose interest in this game. Position yourself so you can always prevent a nip by meeting him halfway. When grooming, bending over to clean a foot, saddling, and so on have some body part or tool ready to bump him in the nose whenever he turns toward you to nip. Meet him with your hand or elbow or grooming tool. When you bump him in the nose, do it in a nonchalant manner so he thinks he's bumping himself.

Don't jab at him, reprimand him with your voice, make a fuss, or get angry. Stay casual so he thinks he's doing it to himself. If you remain alert and catch him every time he tries to bite, this "self-punishment" will discourage him and he'll quit; he'll soon tire of initiating his own discomfort.

If a horse is a persistent biter, it may take more than bumping him in the nose with your elbow to discourage him. Use a nail or hoof pick — something he'll pay attention to. You already know when he's most apt to nip at you, so for his lesson you can set him up, putting yourself in whatever position inspires him to take a bite. Then you can catch him at it. Don't jab; let him hit himself on the nose with your nail or hoof pick. Just be ready for him. After a few instances of self-punishment he'll stop.

The Kicker

Many kickers, especially those who kick when startled or upset, can be trained to stop this bad habit. A horse that kicks viciously with intent to hurt needs handling by a professional trainer (unless you decide to sell), but most horses that kick on occasion are just nervous or sensitive individuals who have not been fully acclimated to handling. They can usually learn to tolerate the situations that previously caused them to kick.

A kick toward a human being is generally an expression of nervous fear or self-defense — the horse is protecting himself from a perceived threat. Horses in groups also use kicking to keep subordinates in line or to express annoyance. Often when a dominant horse merely threatens to kick, the subordinates get out of his way.

Warning Signals

Most horses give warning signals before actually kicking — laying ears back, rolling eyes, presenting the rump, clamping the tail, lifting a hind leg threateningly, or backing toward the intruder. When handling horses, be alert to these signs. A startled horse may kick explosively without warning, but many times a kick can be avoided if you are alert to his mood and can see a kick in the making — and prevent it or move out of the way.

Working Close to the Horse

When handling any horse, kicker or not, approach his head and work back to his hindquarters if you need to handle that end, touching his body all the way. Always work close to the horse, especially around his hindquarters, while touching him. A kick is less apt to hurt you if you're close; he can only push you or bump you with his leg. If you are a little distance away you're more apt to get the full force of the kick, which can be powerful enough to cause serious injury.

Watch his ears and tail, not his feet, for signs of mood. Stand on the opposite side from the leg you're working on when applying medication, fly spray, or anything else that might prompt a kick. When handling a young horse, get him used to having hind legs handled early on so he won't be as defensive about his hindquarters. This makes future procedures easier.

Prevention

A horse that tends to kick must be trained to stand still and trust his handler, learning that kicking doesn't work and that what he earlier perceived as threats are not so terrible. He must become less touchy about having his hind legs brushed or bumped, so he's less apt to kick when startled.

Stand at the horse's shoulder and position him against a solid fence or wall so he cannot move away from you. Then use a folded feed sack or old saddle blanket (something soft, with a familiar smell he won't be afraid of) to rub along his back and over his hindquarters. Talk to him, praise him when he stays calm, and scold him if he tries to move or kick — but don't yell at him.

One way to teach the horse to let you touch his hind legs — while staying out of kicking range — is to use a small "pole" 4 to 5 feet (1.2–1.5 m) long, like a broom handle. Poling was used by old-time harness horsemen for getting "green" horses accustomed to the feel of shafts and harness before hitching them to a wagon the first time. Let the horse smell the broom handle. Then rub and touch him with it, over the sides and rump and along the flanks. Don't raise it high or make any threatening movements. When he is relaxed about it, slide it down around his hind legs, then between his hind legs. It may take several sessions to convince him that having his hindquarters and legs touched is no big deal.

Work with the horse held by a helper, not tied. Being tied by the head makes him feel more trapped. If a horse is tied, the post or fence may be a danger for you if he pulls back and then lunges forward, smashing you against it. You do not want him setting back when tied; you don't want him to become a halter-puller.

The Halter-Puller

A horse that sets back when tied is a danger to himself and to people around him. If he rushes backward while you're still tying him, you could lose a finger or hand. The best cure is to tie the horse often (as part of his daily routine, leaving him tied for an hour or two until he accepts the restraint), to something safe and solid, with a rope and halter he cannot break.

Preventing Injury to the Horse

To avoid injury to neck muscles, and to teach the horse he cannot pull free, it helps to tie him to an inner tube securely fastened to a post; this has some give when he pulls back. Another good method is to use a body rope (around his girth) so much of the strain comes on his whole body instead of just on head and neck. Put the rope around his girth and tie it under his belly, then run the free end between his front legs and up through the halter ring. Tie with that to a solid fence post that will hold him even if he sets back with all his strength. Always tie level with his head or a little higher; tying too low puts too much strain on his head and neck when he pulls back.

Tying the Young Horse

A young horse being trained to tie will generally only pull back a few times; once he's tried it and discovered he cannot pull free, he will respect the restraint. But some headstrong individuals or confirmed halter-pullers need daily tying with the body rope until they learn the rope is stronger than they are and resign themselves to standing patiently.

The Hard-to-Catch Horse

Solving this problem requires time and patience. Many horses never improve because their frustrated owners continue to chase and corner them, instead of beginning again with this aspect of handling.

One thing that helps is a reward system. Catch the horse and then give a treat as you groom and saddle him — a carrot or a few bites of grain, or some nibbles of green grass if the horse is on a diet of hay and grass is considered a treat. Don't take the grain with you to pen or pasture. Some horses play games and try to get a bite or two yet still keep their freedom. The horse has to learn to go by the rules and be willingly caught — then he'll get his treat.

The Timid Young Horse

The young, inexperienced horse may be hard to catch if he is timid or afraid, but this can be resolved with more handling and patience; once you have gained his trust, he'll allow himself to be caught. Weaning time is a good opportunity to gain his trust because he is missing his mother and must rely on you for food and companionship. Even the most independent and stubborn youngster will eventually come around to seeing things your way if you catch and handle him every day, making his experiences pleasant.

The Old Spoiled Horse

The confirmed, spoiled horse can be a challenge because in order to persuade him that being caught is pleasant (offer a bite of grain or some other reward), you first have to catch him. If you must run him around the pasture each time, you've defeated your purpose. You have to outsmart him, and he must decide he wants to be caught.

Using a Small Pen

It usually helps to keep the horse in a small pasture or pen for a while until he gets used to being regularly and easily caught. Horses that run from habit rather than fear sometimes don't bother to run if there's no room. Putting a horse by himself can also help. Horses in a group may be hard to catch if one or two are elusive or habitual runners. Keep the problem horse by himself and catch him several times a day until he has been rehabilitated.

Catch him to take him to water, feed him, or give him a bite of grain or some other treat — so he realizes he is totally dependent on you and that catching is associated with good things. Catch him occasionally just to brush him and turn him loose again. When he finds that catching is the most pleasant event in his day, he'll look forward to it instead of avoiding it. Feel your way with each problem horse, and find a way to work around his attitude and overcome it. The key to retraining the hard-to-catch horse is to spend enough time convincing him that being caught is just part of his daily routine and a pleasant part, at that.

Reforming the Mannerless Horse

Some horses are frustrating to handle. They fidget when you try to groom them, step on your toes, bump into you, root and tug at the halter, drag you along when you lead them, or have no respect for your personal space. Often these problems develop when you treat the horse as a pet or lack consistency in handling. The horse thinks of himself as the boss.

Good manners are crucial to control of the horse and to your safety. A horse that is disrespectful or inattentive can be very dangerous — whether he stomps on your toe or takes a parting shot at you with a hind foot as you turn him loose in his paddock.

Establish Ground Rules

Establish consistent rules for the horse so he can understand — in no uncertain terms — the allowable limits of behavior. He must learn that human beings have special status. You not only give the orders but must be respected at all times. He must understand that you are not to be touched without permission (never bumped — even by accident — bitten, kicked, or tail-swatted). If you spoil a horse by allowing him to nuzzle, rub, play, or push, he will consider you an equal, a buddy to roughhouse with; and this can lead to dangerous consequences because human beings are much more fragile than his herdmates.

Enforce the Pecking Order

Social ranking is an important aspect of herd life: who eats first, who leads, and who follows. Pecking order is a major element in horses' interactions with each other and with people. Much of their behavior is based on dominance and submission, and the horse must figure out how human beings fit into that picture — who is higher on the pecking order and who must submit to whom. If a horse is successful at dominating people, he becomes aggressive and does what he pleases unless he comes up against a person who can set him straight.

Maintain a Respectful Distance

One of the basic rules of behavior the horse must learn is to keep a respectful distance. Whenever he presses his limits or violates rules, don't ignore it even if it seems like an accident. Through your own body language let him know you have a personal space that he must not violate.

Even the most gentle horse can be dangerous if he doesn't respect your personal space. When handling any horse, make sure he knows where his place is. Allow him to be comfortable in it, but always make him uncomfortable if he pushes into your space. Let him run into your hand, the rope or lead strap, a stick or a whip; create an imaginary space around you (about 10 to 12 inches [25.4–30.5 cm] or so) that he is not allowed to penetrate.

Discourage Rubbing

Don't let a horse rub on you, look for treats, or nibble your hand or clothing. Be especially firm and consistent with young horses still learning how to relate to people and experimenting with dominance. After all, little nippers grow up to be big biters. If you rub on a horse (some people like to rub the forehead), he will rub you back. By doing this you set yourself up to be challenged as to who will be dominant in the relationship.

Avoiding Bad Manners While Leading

Some horses root and pull, kick at the person leading them, or wander about, paying little attention to the handler. Some become rooters because of the improper way they've been led. Most horses resent being pulled. They don't like being dragged along, and they don't like being held back if they want to go faster than you do. If you must continually restrain the overeager horse by pulling back on his halter, he will lug and root even more. It takes two to have a tug of war, and if you pull on the horse you invite retaliatory behavior. Soon this becomes a habit.

When you apply pressure on the halter to slow him down it should be intermittent, not continuous. A led horse should have a loose rope so he can accompany you in a relaxed manner. If you have to slow him down, give well-timed short tugs rather than a continuous hard pull. He will learn to respond to pressure only if it is intermittent.

The Overly Aggressive Horse

For the horse that wants to drag you around much faster than you want to go, or one that ignores your signals to halt, you may have to use a nose chain when leading. It is used over the horse's nose in much the same way a choke chain works on a dog. It doesn't inhibit movement unless it is engaged by the horse (by going too fast) or the handler (by asking the horse to stop with a short tug). For proper use the chain must be attached to the halter correctly. Wrapping it around the nose band decreases its effectiveness.

When holding the lead line, with your right hand grasp the end of the chain where it meets the line, having the horse on a relaxed lead as you walk beside or just ahead of his shoulder at arm's length from him. There's no set rule for position because the size and neck length of each horse vary. Be somewhere behind his head and in front of his shoulder. If you're ahead of his cheek you'll be dragging him; if you are behind the shoulder he'll be dragging you.

If he tries to bolt you'll be able to hold the line adequately if you have made a knot just in back of the place where the chain fastens into the nylon or lead rope. This makes the lead line jerk against the chain over the horse's nose with the same effect of a choke chain on a dog. If the chain is properly adjusted on the halter, it presses into the bridge of his nose when he lugs into the halter or bolts forward, and it releases and loosens automatically when he stops. You can also engage the chain with one firm downward jerk if the horse misbehaves while being led. After the jerk, you should immediately give slack to release chain pressure. This pattern of jerk and release is an effective form of discipline — much better than a futile tug of war.

The Body Basher

Some horses fidget around, stepping on you, bumping you, or even trying to smash you against the stall wall or fence when you try to groom them, saddle them, or do anything else while they are tied. Distinguish among behavior triggered by discomfort, fear of abuse (from a prior bad experience), and deliberate bad manners. A horse that is jittery about some aspect of grooming or saddling needs patient, gentle handling to get him over his fears, but a horse that is trying to bully you needs firm and immediate discipline.

Usually when a horse is trying to get the best of you he doesn't show any indication of fear or alarm but is quite calm and purposeful — or even angrily aggressive if he's used to bullying people and getting his own way — and is waiting to see what you do about it. A deliberate show of bad manners should be punished with an immediate and firm verbal reprimand or an appropriate smack on the offending part of his body, delivered without anger. Use something other than your hand, such as a brush, hoof pick, or whatever you are holding; using your hand will probably hurt you more than the horse. If a horse steps on your foot, try stepping on his with your boot heel. After making your point go on about your business so the horse knows he hasn't gotten the best of you, must behave as you continue to groom or saddle him, and cannot get away with invading your personal space.

Correction

Horses that habitually bump into a person must learn to stay back or move back on command. Using the blunt end of a whip, you can tap the horse on the chest to make him back up or move over, rewarding him with praise when he responds properly. Anytime a horse steps uninvited into your space, react immediately to get him back where he belongs. It's not enough to command him to halt; he must back up or move over. This makes him realize that he must submit to you and that you are in control, not him.

For the aggressive horse that tries to mash you into the wall, be prepared with a short, stout stick. Hold it so he runs into the stick instead of you. After a few instances of self-punishment, most horses learn it's not pleasant moving into your space.

The Turnout Terrorist

Some horses have bad manners when you take them out of a stall, put them at pasture, or turn them loose after working with them. A horseman may unwittingly foster dangerous habits by letting the horse have his way. Often the person is unaware of his own negative influence, which may be mere inconsistency, timidity, lack of discipline, or inappropriate discipline.

If a horse takes off the instant he is unhaltered at pasture (perhaps bucking and kicking), the handler may try to quickly unsnap the lead shank while hurrying through the gate, hoping the horse will continue on through and go a few steps more before starting obnoxious and dangerous behavior. This timid, undisciplined turnout approach leads to loss of control of the horse and even more trouble.

Prevention and Correction

To help prevent shenanigans when handling high-spirited or aggressive horses, think ahead and stay in control. Always use a halter and lead shank when taking a horse to and from pasture or paddock, no matter how gentle or well trained the horse or how short the distance. Lead only one horse at a time, pay attention, and convey a calm and confident attitude.

To thwart or prevent the "charge" at turnout time, make the horse stop and relax before you let him go. Loop the lead rope around his neck before unsnapping him or taking off the halter, so he can't go rushing off the instant he thinks he's free. Keep him guessing. Make him realize you are still in control — that you are the one who will leave him, not the other way around. When he is calm and relaxed you can quietly walk away. With persistence, you can correct the horse that tries to immediately gallop off.

Promoting Ease of Handling and Trust

Every time you interact with a horse you are making him easier or more difficult to handle, depending on how you do it. If you use firm tact, consistency, and appropriate discipline, working on "manners" as part of the horse's daily routine, you can usually nip bad actions in the bud before they become bad habits. Your horse will be well mannered and pleasant to work with, not a danger to you or others. He will respect you and your personal space and accept you as the dominant force in his life, someone he can feel at ease with and trust.

PART II

Health Care

Six

Reading the Signs of Health or Sickness

Your horse's health is your most important concern. All aspects of care and management are aimed toward keeping him healthy and sound. Give him the best of feed and water to meet his individual needs, vaccinate and deworm to keep him free of infectious diseases and internal parasites, and protect him from pests such as flies and mosquitoes, which not only annoy him but also spread certain diseases. Monitor him closely to make sure he is healthy, comfortable, and happy.

A conscientious owner is in tune with each horse, knowing how he is feeling by reading the signs of health or sickness and recognizing what is normal or abnormal for that individual. Always check vital signs (temperature, pulse, and respiration) if you suspect a problem. If you know what to look for, you can interpret the clues that the horse gives you and detect early signs of illness.

Vital Signs

A horse's temperature, pulse, and respiration (TPR) can give a good indication of whether he is healthy or sick. Your first impression of him as you approach (when you observe him standing in paddock, stall, or pasture) tells a lot — whether he is perky or dull — but a quick check of his vital signs can confirm or lay to rest your suspicions. Because fever and pain elevate temperature, breathing, and heart rate, a quick check of these can tell you if he might be suffering from illness, colic, or some other problem.

Normal Vital Signs for a Horse at Rest

Temperature: 99°–100.5°F (37.2°–38°C)
Pulse rate: 36–40 beats per minute
Respiration rate: 8–10 breaths per minute

Normal Rate

Check each horse when at rest and not exerting. Expect the rates to vary somewhat by horse. Numbers given for "normal" rates are averages, such as a pulse of 36 or a temperature of 100.5°F (38°C). If you take each horse's pulse, temperature, and respiration rates a few times at rest, it is good practice in case you have to check him someday in an emergency situation. Being already familiar with the procedure, the horse will be less likely to resist or try to kick you.

Temperature

Normal body temperature for a horse ranges from 99° to 100.5°F (37.2°–38°C) on average. It is generally lowest in the cool of early morning and a little higher in the evening after he's been active, especially if the day has been warm. A horse might have a temperature of 98°F (36.7°C) on a cold December morning and 10l.8°F (38.8°C) on a hot August afternoon.

For taking temperature any rectal thermometer will do, but an animal thermometer is best; it's sturdy and has a ring in the end for a string. Never put a thermometer into a horse's rectum without a string attached or it may get lost in the rectum. If you don't have an animal thermometer, securely attach a string to the end of a human rectal thermometer with masking tape or duct tape.

Know Your Horses' Vital Signs

Check every horse now and then to determine the normal rates for each individual. If you know what a certain horse's normal temperature, pulse, and respiration (TPR) are at rest, you have a better idea whether they're abnormal when you suspect he's sick. In consulting with your veterinarian about a problem, you can mention the horse's vital signs. This information, coupled with your description of other symptoms, helps the veterinarian determine the seriousness of the problem.

Safety Tip

When taking the rectal temperature of a horse that's not used to the procedure, you can prevent trouble if you stay relaxed and work close. Stand directly beside his hindquarter with your body against his hip and stifle, left arm over his rump. With the left hand, gently rub his tail area until he relaxes. Most horses like their tails rubbed and will be cooperative if you don't rush them.

After the horse relaxes he'll raise his tail a little instead of clamping it down, and you can slide a lubricated thermometer into the rectum. If your body is pressed against his hindquarter and stifle it's hard for him to kick forward, and you are safely out of the way if he kicks backward. Most won't kick, however, if you get them relaxed before inserting the thermometer.

When taking a horse's temperature, stand close to the horse with your body against his stifle and hip so he cannot kick you.

Have the horse tied, or have someone hold him. Shake down the thermometer to below 96°F (35.6°C) so you can get an accurate reading. Have the end well lubricated with petroleum jelly or your own saliva (a little spit works fine) so it will slip in easily and not cause discomfort. A dry thermometer is not as easy to insert, and the horse may object. When you are ready to insert it, stand close to the horse and gently rub alongside the top of his tail. Don't abruptly pick up his tail because he may clamp it down and become unhappy about what you are doing. If you gently rub the sides of his tail he'll relax and raise it a little on his own. Most horses enjoy the rubbing because this is a hard place to reach an itch.

Once the tail is raised it's easy to slip the thermometer into the rectum, aiming it slightly upward and rotating it a little as you go. The twisting motion helps the thermometer go in more easily. If the horse tries to clamp his tail down, gently hold the tail to one side. If you insert the thermometer properly, without poking the sides of the rectum or causing any discomfort, he won't fret about it. Even a nervous individual will become accustomed to it after a few practice sessions and submit to it willingly because he likes to have his tail rubbed. Insert the thermometer all the way in, leaving only the very end (with string attached) visible.

For accurate reading, the thermometer must rest against the rectal wall. If it's stuck in a fecal ball the reading will be low, which can be deceptive when you're trying to determine if the horse has a fever. If the thermometer doesn't go in easily or gets stuck in manure (which will make it hard

Aim the lubricated thermometer slightly upward, and rotate it as you gently push it into the rectum. This technique allows it to slide in more easily.

to push in), take it out and try again. Leave it in for a full minute or more. The most accurate reading will be obtained if you leave it in for 3 minutes.

Pulse

The pulse, or heart rate, of a healthy horse is on average 36 to 40 beats per minute. Every horse's normal is different. One may have a normal resting pulse as low as 28, and another's normal may be 44. Consider anything at 48 or lower to be normal. Foals are a little higher. Athletic, fit horses tend to have lower resting pulses than horses who have never been in top physical shape.

Take your horse's pulse when he is quiet and resting to find out his normal rate. It's easiest with a stethoscope (placed at the heart girth behind the left elbow) but not difficult with your hand if you know where to feel. A good place is along the lower jaw where the big artery runs across under the bone. Move your fingers along the bottom of the jawbone until you find the artery. It feels like a small, firm cord. Press lightly on it with your finger to feel the pulsing blood.

Shortcut to Taking the Pulse

An easy way to determine pulse is to count for 15 seconds if you have a watch with second hand, then multiply by 4. This is a quick and handy approach, especially if the horse doesn't want to stand still for a whole minute. If you wish, you can do it a couple of times (15 seconds each) to double-check your figure.

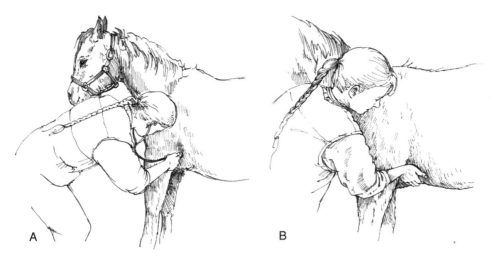

*Heart rate can be checked with a stethoscope positioned at the heart girth behind the elbow on the horse's left side (**A**), and with your hand at the girth area (**B**).*

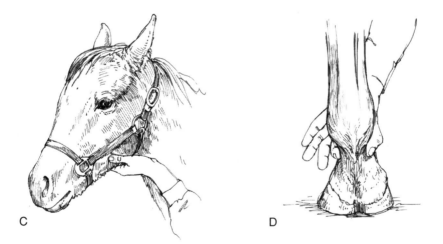

*Heart rate, or pulse, can also be checked at the big artery under the lower jaw (**C**), and at the fetlock joint, by pressing a finger against the artery that runs just under the joint at either side of the foot (**D**) .*

You can also check pulse directly over the heart, just inside the left elbow, by feeling with the flat of your hand against the ribcage. The actual heartbeat is two beats in one — very easy to distinguish if you are listening with a stethoscope. Count each *lub-dub* as one beat. Be aware that pulse can become elevated with exertion, excitement, nervousness, pain, fever, or other serious problems.

Respiration

Respiration is easy to determine — just watch the movements of the horse's nostrils or flanks. Count as he inhales or exhales, but not both. Just as in taking the pulse, it is often easiest to count for 15 seconds, and then multiply by 4. Normal respiration rate in the average horse at rest is about 8 to 10 breaths per minute, if weather is cool and he's not excited.

Body Language

The person who works closely with horses usually develops an intuitive feel for their well-being. An important part of this skill is being able to sense when a horse is sick or uncomfortable by reading subtle signs from body language.

Posture

The way a horse stands or moves can tell the observant horseman at a glance whether or not all is well. General bearing, leg position, and head, neck, and tail carriage (e.g., droopy instead of jaunty) are good indicators of something amiss. If he's standing with head down and looking dull, rather than bright and alert, you should be immediately suspicious. Remember that when a horse is not feeling well or is in pain he becomes less perky than usual, less aware of what's going on around him. He is tuned inward instead, focused on his own misery. It would be wise to take his temperature; dullness may indicate a fever.

If You Aren't Sure, Check TPR

A horse won't necessarily have rapid respiration with a fever. He may just seem a bit dull. If he isn't quite himself, take his temperature. While you're at it, check his pulse. An elevated pulse can be a sign of pain. If you decide the horse's condition warrants a call to the veterinarian, mention these vital signs so he or she can decide whether a thorough examination and diagnosis are necessary.

Mild Abdominal Pain

Every horseman should know about signs of colic (obvious abdominal pain; the term *colic* simply means abdominal pain) — pawing, rolling, sweating, elevated heart rate. Mild abdominal pain may be harder to detect unless you know your horse well and can tell when he isn't quite himself. A horse with mild abdominal pain may be just a little dull, slightly restless, or off feed. He may get up and down more than usual or spend too much time lying down. He may lie with his head tucked around toward his belly or flank.

He may also stand in a corner away from other horses or display slightly abnormal posture. He may stretch or stand with his head up high and front legs forward, with hind feet farther back than normal — trying to ease discomfort or gas pains in his abdomen. If he acts strange

A horse with mild abdominal pain may lie with his head tucked around, or he may keep looking at or nosing his belly.

or looks dull, check vital signs and abdominal sounds. Use a stethoscope if you have one, or your ear pressed to his side, to listen for gut sounds (i.e., rumbles and gurgles). Constant rumbling may mean an overactive gastrointestinal tract; absence of gut sounds (ominous silence, except for the gentle "whish" of lung sounds as the horse breathes) could mean a blockage or shutdown gut. In the latter case, call your veterinarian. Keep the horse under observation. If he has mild colic it could worsen, depending on what is causing it.

When checking gut sounds with a stethoscope, listen for rumbles and gurgles.

A horse with front legs forward and hind legs back (standing in a stretched position) may have gas pains.

Back Humped Up

A horse standing with back humped up and stomach muscle tense is showing signs of severe body pain — injured back or ribs, chest pain, peritonitis (inflammation and infection in the abdominal cavity), or serious digestive tract pain. He may be reluctant to move because any movement hurts. He needs immediate medical attention.

Position of Legs

How the horse is standing — position of legs and overall body stance — can give clues to other problems as well.

Pointing

Standing with one front leg in front of the other (pointing) usually means he's trying to relieve pain in that leg by not bearing much weight on it. He may have bruised the heel or sustained a more serious injury — to the deep flexor tendon, flexor muscles, or related ligaments. Pointing can also be indicative of navicular disease (see chapter 9). Or the horse may be trying to ease the discomfort of a strained biceps muscle, bruised shoulder, or injured elbow. Check his leg for heat and swelling, and check his heel for soreness (see chapter 4). If this does not reveal the cause, have your veterinarian do a more complete examination.

Standing Three-Legged

You should suspect more serious injury when the horse is standing three-legged, not putting any weight at all on a leg and hobbling if he has to move. If this continues for more than a few minutes, get professional help to diagnose

Standing with one front foot forward (pointing) may mean that the horse's leg is sore and he doesn't want to put weight on it.

and treat the problem. The horse may have a foot abscess, strained or sprained fetlock joint, nerve or muscle injury, or even a fracture. Close examination of foot and leg gives clues. Check for heat and swelling; exaggerated, unusually strong pulse (felt in the lower leg when feeling fetlock joint area and pastern for heat); punctures or embedded foreign objects in the hoof; or hoof tenderness (check with a hoof tester; see chapter 4).

However, if the pain is temporary and the horse soon walks normally again after a few minutes he probably just banged the leg during playful antics, was kicked by a pasturemate, or landed the wrong way while running and bucking but did not suffer serious damage. In this case there is no need to call the veterinarian, but it's a good idea to keep an eye on the horse for a day or two to make sure he's all right.

Hind Legs

Hind legs are not injured as frequently as fronts because they carry less weight, but accidents sometimes happen. Standing three-legged, resting a hind foot, is usually nothing to worry about but might indicate pain and trouble. If the horse's stance seems abnormal, check the foot and leg more closely. Make him walk a bit to see if he favors the leg while moving.

If he's standing with one hip dropped and hind leg dangling, or if fetlock joint and pastern are knuckled under (the front of the pastern resting on the ground with the bottom of the hoof pointed up behind him), suspect serious injury. A few horses rest a hind foot this way, but if yours generally does not you'd better check it out. Odd hind leg stance could mean a broken leg or a ruptured ligament or muscle high in the leg. If he's unwilling to move or his leg will bear no weight, do not try to make him move because this could cause the problem to get worse. Seek prompt veterinary help.

Rigid Hind Legs

Another instance in which a horse may be reluctant to move is when major muscles, especially in the hindquarters, "tie up." Due to painful muscle cramping he stands with hind legs rigid, often behind their normal position, and paws the ground with his front feet. This reflects his frustration at not being able to move without discomfort. He will probably break out in a sweat, especially around the flanks.

If a horse paws with his front foot, it may be an indication of colic or of muscles tying up.

Muscles tying up generally occurs during exercise, early in the workout when a horse is being ridden, or right after a ride. The problem can occur any time a horse overexerts. You might find him standing this way after he's been running or bucking in mud or snow, or racing around with pasturemates. Most cases involve horses that are fed grain, the energy of which is stored in the muscles. When the horse exerts strenuously, the muscles produce lactic acid faster than the bloodstream can clear it away, creating this painful condition, with muscles cramping and knotting up.

Is It Colic or Muscles Tying Up?

The sweating, pawing, pain, and anxiety from muscles tying up might be mistaken for colic, but you can tell the difference by noting the horse's stance, actions, and reluctance to budge from the spot. The horse with tied-up muscles does not want to move, and his gait is stiff if he does, especially in the hind legs. Moving will make the problem worse; don't try to make him walk. Your veterinarian can give medication to relieve the condition.

Legs Bunched Together

If a horse is standing with front legs too far back and hind legs more forward than usual (front and hind end bunched together), it is generally a sign that his body hurts. He may be trying to relieve pain in his back or discomfort in his chest. A dull horse standing "camped under" like this could be seriously ill.

Standing with legs bunched together is a sign of pain.

If he seems normal in other respects — alert, eating, drinking, with usual attitude about everything — he may have a simple strain or bruise. If he is dull, sweating, trembling, unwilling to move, wobbly, staggering, or in any way acting strangely, call your veterinarian. The horse may have a seriously injured back, severe intestinal problem, or respiratory condition such as pleuritis — inflammation of the chest lining and lung covering accompanied by fever, fluids in the chest cavity, and painful breathing.

A horse with most of his weight shifted onto his hindquarters (with hind legs forward) indicates that he doesn't want to put weight on his front legs. This could mean neck pain or laminitis.

Weight Shifted Back onto Hind Legs

A horse standing with weight shifted onto his hind legs is showing that bearing weight on his front legs is painful. This could indicate founder (laminitis; see chapter 9) or neck pain. Founder should be your first suspicion, so check his feet for heat or a pounding pulse. See if the horse will move. If he tries to carry most of his weight on his hind feet and is very reluctant to move or turn to the side, you are probably dealing with founder (this causes terrible pain in the front feet — sometimes in all four feet). Call your veterinarian.

Be Observant

Be alert to any changes in normal behavior. "Acting funny" is often the first hint that something is seriously wrong. If the horse is showing unusual alarm, alertness, agitation, or personality change, or if he is dull and preoccupied, droopy, unresponsive, disoriented, or abnormally aggressive, take heed. Changes in behavior and posture may signal an insignificant or temporary problem, but they might be the first signs of serious injury or illness such as rabies, encephalomyelitis (sleeping sickness), or tetanus (all discussed in chapter 8). If a horse is sweating, trembling, straining to urinate or defecate, drooling, grunting, wheezing, panting, or showing other obvious and alarming changes, call your veterinarian. Even subtle changes may be nothing serious but should be carefully checked out, because they could be early warning signs of a major problem.

Obvious Signs of Trouble

- Rolling or pawing
- Sweating when at rest (unless weather is very hot)
- Having rapid respiration when at rest
- Not eating at mealtime
- Not drinking the usual amount of water
- Spending too much time lying down
- Lying in an abnormal position
- Displaying unusual posture or stretching (legs bunched under the body or stretched out), or trying to carry most of the weight on hind legs
- Refusing to put weight on a foot
- Being reluctant to move when asked
- Deviating from usual personality or behavior

SEVEN

DISEASE PREVENTION

Most diseases can be prevented through good care and vaccination. Many of the most deadly infectious diseases — tetanus, influenza, strangles, and sleeping sickness — can be prevented with annual vaccinations. You should definitely protect your animals.

Not every horse needs to be vaccinated for every preventable disease because some are a problem in certain regions but not others. For example, Venezuelan equine encephalomyelitis (VEE) occasionally threatens horses in the Southwest when this disease heads north from Central America, but it is not a concern in the rest of the United States. Likewise, Potomac horse fever is a threat in some regions, primarily where there are rivers and streams, but not in others. The same is true of rabies and equine viral arteritis (EVA; of special concern if you are breeding horses). Ask your local veterinarian which diseases to vaccinate against in your horse's annual schedule of shots.

Some diseases are only spread by horse-to-horse contact. If any of your horses go to trail rides, shows, or to another horse farm to be bred — or if other horses come to your place for any reason, your neighbors have horses, or you ride with other people — you should vaccinate all your horses against the common infectious diseases, especially influenza, rhinopneumonitis, and strangles. If you have a backyard horse that never comes in contact with other horses, or your ranch horses never leave your property and no other horses come in, you may not need to vaccinate for all diseases. If there's any chance of contact, though, play it safe.

What Is Disease?

Disease is defined as an unhealthy condition of the body. There are several types of disease, primarily infectious and non-infectious. An infectious disease is caused by a pathogen, a tiny organism that invades the body and stays there. Most of these invaders are microscopic. Infectious disease can be caused by bacteria (as in the case of strangles), viruses (as in the case of influenza), protozoa (one-celled organisms), fungi, or parasitic worms. Infectious diseases of horses are discussed in chapter 8.

Non-infectious diseases are caused by conditions within the body itself — such as cancer, respiratory disease due to allergy, hormonal imbalances, genetic disorders, and metabolic or stress-related problems that throw the horse's body chemistry out of balance. Laminitis and heaves are examples of non-infectious diseases (see chapter 9).

Definitions

- **Antigens** are foreign substances, such as toxins, that stimulate the body to produce antibodies. A vaccine contains a modified antigen that will trigger an immune response in the body without causing the disease.
- **Immunity** is the body's ability to fight off disease-causing organisms. A healthy horse with strong immunity is less likely to become sick than one with poor immunity.
- **Pathogens** are microbes that cause disease. Illness due to an infectious disease occurs when the body is overwhelmed by an invading pathogen.

How the Immune System Works

The immune system is complex; it involves generalized mechanisms for fighting disease as well as the ability to develop immunity to specific pathogens. When a virus or bacterium enters the body, it causes damage by multiplying and creating toxins, poisons produced by living organisms. The presence of the pathogen or its toxins stimulates the body to create an antibody, which reacts with the invader and neutralizes it.

Antibodies and Vaccines

Antibodies are blood proteins that are produced in response to a foreign substance (toxin) in an effort to neutralize its harmful effect. Produced by one type of white blood cell, antibodies are carried throughout the body in the bloodstream. If an animal already has antibodies against a specific disease organism, whenever that organism invades the body again the white blood cells and their antibodies converge on the site to kill the invader. Ability to recognize and respond to foreign material is the body's immune response; this is the basis for vaccination.

Vaccination stimulates the production of antibodies by serving as the antigen. When a vaccine (a foreign substance) is injected into an animal, the body reacts by producing immune substances — antibodies — to fight the foreign invader. The end result is protection for the body (immunity) against subsequent exposure to that disease. If the horse eventually comes into contact with the actual infection, the antibody is already in the bloodstream, able to inactivate the pathogen.

Vaccines are often created from killed viruses, modified live viruses, or other inactivated disease organisms that still retain their characteristic so the body's immune system can recognize the invader and create antibodies to fight the real thing. For example, antibodies produced in the horse's bloodstream by vaccination against influenza lie in wait, searching for the antigen (the flu virus) that originally stimulated their production. If the horse is exposed to influenza — through contact with a sick horse or its cough droplets — the white blood cells "programmed" by the vaccine to make a particular antibody will quickly respond and produce enough to inactivate the invaders. As a result the horse will not get sick.

How Vaccination Works

Vaccines are small doses of disease — not enough to cause the disease, but enough to stimulate the production of antibodies so the body can recognize and fight a similar kind of invasion (the real disease) if exposed to it.

Developing Immunity

The vaccination process exposes the injected horse to specific antigens. In order to develop immunity his body must respond by producing antibodies and other immune substances. This takes about 7 to 10 days in a healthy horse if the immune system is intact and fully functional. If the immune system doesn't work properly, the horse does not develop strong immunity. Most horses respond to vaccination successfully, but some have only a partial or poor immune response and may not be protected against the disease despite having been vaccinated.

Many factors influence the development of immune response — or lack of it. A stressed, undernourished, or heavily parasitized animal may not make a good response to vaccination. Stress factors typically include pregnancy, fatigue, and extremes of weather.

Vaccination Failure

Sometimes vaccination fails because the horse is already incubating the disease at the time of vaccination (he has been exposed and is coming down with the disease). In foals, vaccination often fails because of interference from maternal antibodies in colostrum (the mare's first milk). If foals are vaccinated they usually need a booster shot a few weeks or months later, or several boosters, to gain full immunity. There are different times to start vaccinating for different diseases and starting too early (especially for influenza) may hinder a foal's future immune response. Ask your veterinarian about this.

Another cause of vaccination failure is the mishandling of vaccines. The most common example is improper storage. Most vaccines must be refrigerated; if left too long unrefrigerated, they may not be effective. In addition, sunlight can be detrimental to any live or modified live virus vaccines. Using a vaccine at improper dosage or beyond the expiration date can also contribute to inadequate immune response.

Health Tip

The best way to develop good immunity is to vaccinate only healthy, unstressed animals, using appropriate sterile equipment — new syringes and needles (the disposable kind) — and properly stored vaccine at recommended injection sites and dosage.

Stress

Stress can be defined as anything that poses a threat to the physical or mental well-being of the animal. There are many kinds of stress — some temporary, others more prolonged and harmful. Among the factors that contribute to it are unaccustomed noise, upsetting changes in management that lead to emotional insecurity, crowding, temperature extremes, lack of water, toxic reactions, allergic reactions, pain, fear, anxiety, strenuous exercise, and exhaustion.

The way the body deals with stress can compromise the efficiency of the immune system to ward off invading pathogens. If the stress is severe or prolonged, it can lead to illness. It can hinder the development of immunity when a horse is vaccinated, and it can cause a previously vaccinated horse to lose his immunity.

The respiratory system is especially vulnerable to the effects of lowered resistance due to stress. This is partly because harmful organisms are a natural part of the horse's environment, almost always present in the air he breathes. Pathogens are constantly being drawn in, so the immune system must be continually active to protect the respiratory system and keep the pathogens under control. This is one reason horses tend to more readily develop flu, rhinopneumonitis, "colds," and other respiratory problems when hauled to shows, competitions, races, and the like. The stress of transport, unfamiliar surroundings, and strange horses weakens the respiratory system's resistance; moreover, the horse is more apt to come into contact with infectious agents in places where horses are brought together from many areas. Stress and contact with other horses give infections a prime opportunity. Stress at weaning and during long periods of extreme weather also give rise to respiratory problems and illness.

Disease Prevention

You can't just give horses their annual vaccinations and expect them to remain disease free. Vaccinations should always be given when the horse is healthy and not stressed, well ahead of stressful situations like weaning, training, conditioning, or transporting, so the horse can build peak immunity. Some types of vaccinations should be given more than once a year if the horse's immunities may be compromised. Flu and rhinopneumonitis are two diseases that may require more than one annual booster, as will equine encephalomyelitis if mosquito season lasts a long time in your area. Discuss a vaccination schedule with your veterinarian.

There are vaccines for most of the common diseases that affect horses. Develop your horse's vaccination schedule according to the diseases prevalent in your area and how you use your horse. Some vaccinations should be given even if your horse never comes in contact with other horses. For example, tetanus and equine encephalomyelitis are a risk in almost all locales; tetanus is caused by a bacterium that lives in the soil and encephalomyelitis is spread from birds to horses via mosquitoes. Both diseases are deadly, and vaccine is inexpensive. Don't risk losing your horse.

Sample Vaccination Schedule for an Adult Horse Needing Only Booster Shots

SEASON	VACCINE
Spring (before mosquito season)	• EEE, WEE, influenza, and tetanus (available in one injection), *or* EEE, WEE, and tetanus
	• *Optional* (if horse will be in contact with strange horses): Rhinopneumonitis (modified live-virus vaccine), strangles
	• *Recommended in some locales:* Potomac horse fever, rabies
Fall (if mosquito season is long)	EEE, WEE

EEE = Eastern equine encephalomyelitis; WEE = Western equine encephalomyelitis.

Note: Young horses being vaccinated for the first time will need a series of injections, several weeks apart, for all vaccines.

Pregnant mares may need several vaccinations during gestation to protect the growing fetus from certain diseases and to maximize antibodies in the colostrum.

Important: Discuss *each* horse's vaccination needs with your veterinarian. What you ultimately give will depend greatly on the individual animal, its use, and your location.

Tetanus Shots

It's a good idea to give a pregnant mare tetanus toxoid so there will be antibodies in her colostrum (first milk) to give the foal temporary immunity. The foal itself should have a series of tetanus shots (how early these should begin will depend on whether the mare was vaccinated), and from yearling age on each horse should have an annual booster. Even then, if a horse suffers a serious wound or deep puncture and has not had an annual booster in the 90 days preceding the injury, he should have another booster to make sure protection is adequate.

Equine Encephalomyelitis (Sleeping Sickness)

In most areas of North America bird (and sometimes reptile) populations serve as reservoirs of infection for equine encephalomyelitis. After biting an infected bird or reptile, mosquitoes carry the disease to horses and, sometimes, to human beings. Migratory birds and mosquitoes can transport the disease a long way, especially in wet years with storms that carry the insects long distances. Vaccinate in spring before mosquito season begins. Immunity from vaccine lasts only about 4 months, so revaccinate later in the summer if mosquito season lasts a long time.

Other Inoculations

Your vet may recommend vaccination against equine influenza, rhino-pneumonitis, strangles, Potomac horse fever, equine viral arteritis, botulism, rabies and leptospirosis, depending on your location. These diseases are discussed in chapter 8.

Vaccinating Your Horse

You may choose to have the veterinarian vaccinate your horses, or you may prefer to do it yourself. If you vaccinate your own horses, take proper care of vaccines — read directions for storage and handling, and make sure all vaccines are fresh and potent, keeping none beyond their expiration dates. Also, learn how to give injections properly. Be aware of the possibility for reactions (some of which can be very serious) and what to do if you encounter this problem.

If you have never given an injection, have the veterinarian show you how. Observing the technique of an experienced person will help you learn to do it with less discomfort to the horse. A properly inserted needle causes little pain; the horse may not even notice.

Health Tip

When giving an injection, make sure the site is very clean, free of mud and dirt. It should also be dry. Contamination is likely to be taken into the flesh with the needle if hair and skin are wet. Disinfecting the area with alcohol is not necessary if the hair and skin are clean. If you do swab the spot with alcohol, wait for it to dry before inserting the needle.

Intramuscular Shots

Most injections are given intramuscularly — deep into muscle tissue. These are the easiest to give. Relatively few shots are given to horses subcutaneously — under the skin. Some medications are given intravenously (e.g., sedatives and pain killers), but these are generally administered by a veterinarian. Do not attempt to give an intravenous injection unless your veterinarian has instructed you on how to do it properly.

Choosing the Proper Needle and Syringe Size

Most vaccines are given in small doses; use a small syringe for accuracy. Antibiotics and some other medications are often given in larger amounts, and for these you'll need a larger syringe. It's best, however, not to inject more than 10 mL in any one site. A large injection can be split into two doses and injected in two places for better absorption and less muscle soreness afterward.

For intramuscular injections, an 18-gauge needle 1.5 inches (3.8 cm) long is about right. A needle of larger gauge is too big — more likely to cause discomfort as well as leakage of injected material because it makes such a big hole in the skin. A needle of smaller diameter goes in nicely, but it's more difficult to push liquid through it; it takes longer, and there's more chance of having syringe and needle "blow apart" (losing the vaccine or antibiotic) because of the extra pressure needed to push the liquid in. A needle longer than 1.5 inches (3.8 cm) is apt to penetrate too deeply and hit a bone or artery, and a needle shorter than that will not deposit the injection deep enough into the muscle for optimum absorption.

Sharp and Sterile Is Best

Use a sharp, new needle — a separate sterile needle for each horse. The sharper the needle, the more easily it goes in with the least pain to the horse.

Filling a Syringe

Some vaccines come in convenient individual doses, already in small syringes. But often you have to fill a syringe from a bottle of vaccine or antibiotic. After inserting the sterile needle through the rubber top, tip the bottle up and slowly pull back the plunger of the syringe to create a vacuum in the syringe. It will be filled by liquid from the bottle.

If you have a large bottle of vaccine and several horses to vaccinate, or a large bottle of antibiotic and are giving a large-dose injection, it will be harder to fill the syringe because of the vacuum created in the bottle when you withdraw some of the fluid. In these instances, fill the syringe with air by pulling back the plunger, and then insert the air into the bottle. The fluid will come out more easily this way. Hold the bottle at an angle so the needle injects air into the air space above the liquid. Injecting air into the liquid

A   B

A. *To fill a syringe, insert the needle through the rubber top of the bottle, then tip the bottle upside down so the needle tip is in the liquid.* **B.** *With the needle tip inserted into the bottle, slowly pull back the plunger on the syringe to create a vacuum, allowing fluid to flow into the syringe.*

itself would create many tiny bubbles and make it harder to fill the syringe to accurate dosage.

Don't Have Air in the Syringe

Try not to get any air into the syringe when filling it. If this happens, hold it vertical (with air pocket at the top) and gently push the plunger until the air is all expelled, making sure the proper dosage of liquid remains in the syringe. Squirt the air back into the vaccine or antibiotic bottle so no fluid gets wasted if you accidentally push too hard.

Before Giving the Injection

If it will be a moment before you give the injection, put the protective cover back on the needle so there is no chance of getting it dirty or contaminated if you set the syringe down for some reason. If the contents of the syringe are cold, hold the syringe in your hand for a moment to warm it before giving the injection to the horse. The exception would be a modified live virus vaccine such as rhinopneumonitis. Once you rehydrate the rhinopneumonitis vaccine (by injecting a vial of sterile water into the dehydrated vaccine powder), give the injection as soon as possible — before the fragile contents are inactivated by sunlight or heat.

Injection Sites

Intramuscular injections should go into thick muscle. The best site is the buttocks. For shots of 2 mL or less, the side of the neck (a short distance above the shoulder) or the breast (chest) on either side of the breastbone can be used.

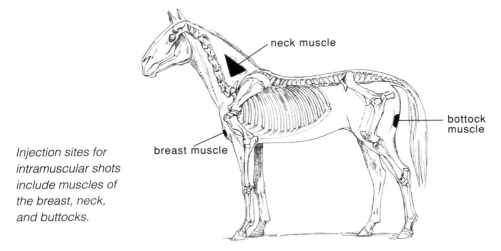

neck muscle

bottock muscle

breast muscle

Injection sites for intramuscular shots include muscles of the breast, neck, and buttocks.

Horsemen traditionally used the large rump muscle, but it's not recommended because it runs horizontally. The thick muscle is one of the best sites for absorbing a large shot, but if an injection should cause an abscess there it would be difficult to establish good drainage. A properly administered injection carries little risk of contamination and infection, but there is always the possibility so it's better not to take the chance. The neck, buttock, and breast muscle have the advantage of being more vertical, and an infection in that area would be able to drain.

Proper Use and Storage of Vaccine and Other Injectables

Note the expiration date on every bottle of vaccine or antibiotic, and don't buy any that is within a month or less of the expiration date; *never* buy any that's already expired. Read labels. Make sure a product that's supposed to be kept cool has been properly refrigerated. Try to buy vaccine in a bottle size that you can use up at one time. Many antibiotics keep fairly well once opened, but a bottle of vaccine should be used right away. Don't save fractional contents. Follow label directions for storage, use, and administration.

Swellings

Sometimes an injection site swells up for a day or two. Large shots given in a small muscle are more likely to cause swelling than if given in a larger muscle. Vaccines are more apt to cause swelling than antibiotics. The amount of swelling from a vaccination depends on the type of shot and the individual horse's reaction to it. Some horses are more sensitive to certain vaccinations than others. In most cases the swelling is merely unsightly and might make the horse stiff and sore for a few days, but usually the site returns to normal within 24 to 48 hours. Sometimes mild exercise after a vaccination helps prevent or resorb tissue swelling. If a horse swells alarmingly (more than 2 or 3 inches [5.1–7.6 cm] in diameter), it generally means he is quite sensitive to the substance being injected. You should discuss this with your veterinarian.

The Neck

A safe and convenient place to give a small injection is the side of the neck, in the triangular area above the shoulder and below the crest. The horse's neck bones are low here; if you aim for this triangle you will be in thick muscle. The advantage to giving injections in neck or breast is that you are working at the front of the horse, where it is easier to control him and avoid being kicked. Don't put large shots into the neck because tissue swelling may make the neck sore; the horse may have trouble raising and lowering his head to eat and drink.

The Breast

For a horse that fidgets or resents shots, the best place is the muscle on either side of the breastbone. The horse is less sensitive here and less apt to feel the needle. You can usually put it in without him noticing, especially if someone else holds him and gives him a bite of green grass or grain at the time you are doing it. For a large injection, however, this muscle is too small for proper absorption, more likely to cause tissue swelling. If giving a shot larger than 2 mL it's best to split it and put half in each breast muscle, or inject it in the buttocks.

1 cc = 1 mL

Syringe size is measured in cc (cubic centimeters); 1 cc is the same as 1 milliliter (mL).

The Buttocks

Large injections can be put into the horse's buttock muscle. This is a good location unless a horse is inclined to kick. When using this site, stand close to the horse.

Giving the Injection

When giving the shot, have someone hold the horse for you if he is a nervous individual. Some old dependables can be tied, but if there's any chance the horse will react violently it's better to have him held. If shots are given properly the horse won't even flinch. Teamwork helps. It's not hard to give shots by yourself to a horse that knows and trusts you, but for a nervous or suspicious horse it helps to have someone at his head holding him still and talking to him, distracting him so he won't move at the wrong time.

Make sure the injection site is very clean, then detach the needle from the filled syringe and, holding the needle in your hand, gently thump the area several times with that hand, using the side of your fist. Thumping tends to desensitize the area and mask the prick of the needle. If you try to jab it in with no prior preparation, the horse will jump every time. Thump quickly, only two or three times, and immediately put in the needle.

Some horses get nervous about the thumping because it's been done before and they know what's coming. For these individuals it's better to just press the area firmly for a moment with your thumb or the side of your hand; this also desensitizes the skin. When giving an injection in the neck, you can desensitize the area by giving the skin a little pinch before inserting the needle.

Putting in the Needle

After thumping, pressing, or pinching, stick the needle in with one swift motion. Most beginners make the mistake of not pushing forcefully enough to send it all the way in. After a few tries you will know how to do it. Remember that the faster the needle goes in, the less it will hurt.

Always take the needle off the syringe and put it in first. Then if the horse flinches or moves around you can wait until he settles down again before attempting to give the shot. Also, you can tell if you've hit a vein. If the inserted needle begins to ooze blood, take it out and start over. Never inject an intramuscular shot into a blood vessel.

Double-Checking

Once you attach the syringe to the inserted needle, double-check to make sure you haven't hit a blood vessel. Pull the plunger back just a little to create a vacuum. If blood appears in the syringe, remove it and start over. If no blood appears, go ahead and give the injection. (A small amount of blood in the syringe is nothing to worry about when making the injection.) Push the plunger in steadily but not too quickly; if you create too much pressure for the needle it will come apart off the syringe.

If you are using a screw-on needle and syringe, make sure it is adequately screwed together before pushing the plunger. Otherwise it will come apart and spill the injection. If you are using a slip-on needle and syringe, make sure it's fully on; hold the syringe and needle firmly at their juncture with your other hand as you depress the plunger, just to make sure they do not come apart as you give the injection.

Removing the Syringe After Giving the Injection

Withdraw the syringe and needle together in one easy movement, pulling straight out. Then rub or press the injection site with your hand or finger. This not only keeps it desensitized for a moment but also prevents the fluid from leaking out before the hole constricts. If an injection starts to ooze out, press the site until it stops.

A

Before giving an injection, desensitize the target area by thumping it a few times with the side of the hand holding the needle.

B

Then thrust the needle into the muscle.

C

Next attach the syringe to the needle.

D

Check to make sure you have not hit a blood vessel by pulling back slightly on the plunger to see if blood appears in the syringe.

E

If no blood appears, give the injection. If blood appears, withdraw the syringe and needle and start over.

F

After withdrawing the needle, rub or press the injection site to keep fluid from leaking out.

Reactions to Injections

Vaccinations offer tremendous protection against some of the most serious diseases of horses, but a few individuals react adversely to even the safest vaccines or to routine shots of antibiotics or other injections. Every horse owner should be aware of this and be prepared to act fast should a serious reaction ever occur. Even if your veterinarian gives the injection, the reaction may occur after he or she leaves and it will be up to you to treat the horse; in a very serious reaction the horse might be dead by the time the veterinarian is able to return.

Despite the potential for adverse reactions, vaccination is very important in controlling the spread of disease and protecting horses from debilitating or fatal illness. If a horse is sensitive to a vaccine such as tetanus or sleeping sickness and you don't want to leave him unprotected, have the veterinarian try half doses or a half dose split into two injection sites. This will depend on the vaccine and the type of reaction your horse has. In some instances you should never use that vaccine again. The veterinarian can advise you on this.

All injections should be given with caution, proper technique, and sterile equipment and at recommended sites. Only healthy horses should be vaccinated. Be aware that a horse coming down with strangles and given a strangles vaccine will have a serious reaction (see chapter 8). Even with proper procedure, problems sometimes occur — so be prepared.

Allergic Reactions

There are two kinds of allergic reactions, though in a sense the difference is one of degree. A *mild* allergic reaction may manifest in excessive swelling of local tissue or hives all over the body. A *severe* allergic reaction may cause congestion of the lungs and shortness of breath, like an attack of asthma. Most reactions can be alleviated by giving antihistamines, but in extremely severe cases the horse may go into shock and die unless promptly treated to reverse the shock.

Some horses react adversely to certain injectables or to the substance (carrier) a vaccine is mixed with. If your horse has a history of such problems, you may have to change vaccines or even eliminate that kind from his annual vaccination program.

Anaphylactic Shock

Shock is a manifestation of hypersensitivity that affects the horse's entire body rather than just the local tissues. Anaphylaxis, a life-threatening reaction to an antigen, usually occurs after the second exposure to a protein substance entering the bloodstream — a substance the horse is already sensitive to (e.g., a particular disease organism or vaccine). Histamine is liberated in the affected tissues because of the reaction between the circulating antigen from the vaccine and the horse's own antibodies, which developed in his bloodstream from the previous vaccination or exposure to the disease. Histamine acts to dilate the capillaries, which start leaking fluid, producing a drop in blood pressure and causing smooth muscle, including the bronchial tissue in the lungs, to contract, making it harder to breathe.

Symptoms

Horses going into anaphylactic shock may react in various ways. Skin, lungs, feet, or blood circulation may be affected. The horse may have trouble breathing and suffocate. He may develop acute laminitis. He may collapse and die before treatment can be given (this sometimes occurs if an injection is put directly into a blood vessel instead of muscle).

Usually the more serious the reaction, the sooner it occurs after the injection causing it. For example, horses that are allergic to certain drugs such as penicillin may collapse after injection of such a drug.

Treatment

Anaphylactic shock is usually reversible if treated in time. Steroids (dexamethasone) and epinephrine (adrenaline) reverse it if given immediately. Keep these drugs on hand in your barn or tack room for an emergency, especially if you live some distance from a veterinarian. Antihistamine works fine to reverse a mild allergic reaction but is inadequate if a horse goes into shock. In serious cases the horse can die in a matter of minutes, so delay in treatment can be fatal. The time it takes for a veterinarian to get to your place (or come back after giving your horses their vaccinations) may be too long, so keep these drugs on hand and know how to administer them. Discuss proper procedure and dosage with your veterinarian.

It's wise to observe horses for at least 30 minutes to an hour after vaccination, so keep them close by in a pen rather than turning them right back out to pasture.

Guidelines to Prevent and Minimize Disease

1. Vaccinate all horses against prevalent diseases in your area.
2. Keep horses in a healthy, safe, stress-free environment.
3. Keep feed and water supplies clean and uncontaminated.
4. Do not let your horse drink from a water tank used by other horses at a horse show, fair grounds, or trail ride.
5. Minimize exposure to strange horses.
6. Isolate any new horse in a separate pen for 2 weeks with no direct contact with your other horses until you are sure the new horse is healthy.
7. Quarantine any horse that becomes sick with a contagious disease (e.g., strangles, flu, or rhinopneumonitis) and take precautions not to spread infection to your healthy horses.

Note: Specific precautions for certain diseases are given in chapter 8.

EIGHT

INFECTIOUS DISEASES OF HORSES

There are many pathogens that cause disease, including bacteria, viruses, fungi, protozoa, and tiny parasitic worms. All are opportunists that invade the body and multiply, causing illness. Vaccines are available to protect horses against many of the most dangerous diseases and some of these (specifically, tetanus and equine encephalomyelitis) should be included in nearly every horse's annual vaccination program. The need for other vaccinations will depend on your locale and circumstances. Consult your veterinarian.

The following diseases are discussed in order of seriousness and likelihood of being encountered. Some diseases near the end of the list may be deadly but are not a consistent problem in all regions.

Tetanus

Tetanus is a highly fatal disease of all animals, but horses and human beings are especially vulnerable. This disease is caused by the toxin produced by *Clostridium tetani*, a spore-forming bacterium present in the digestive tract of many animals and in soils rich in humus and animal manure. Infections usually occur from wound contamination; deep puncture wounds in the feet are the most common portal of entry for the bacteria in horses. Clean wounds rarely result in tetanus; those that contain foreign matter such as soil (especially in an enclosed puncture wound) are most dangerous.

The incubation period for tetanus is usually 1 to 3 weeks for a horse, with some cases occurring even later. After the toxin reaches the central nervous system by traveling along nerve trunks, muscle spasms begin. Mild stimulations to the nerve endings cause exaggerated responses by the animal. Death is usually by suffocation because the muscles used in breathing become paralyzed.

Tetanus May Be Dormant

Tetanus spores may lie dormant in body tissues after the wound that introduced them heals; disease may set in only when conditions turn favorable. The bacteria produce their deadly toxin if there is a lowering of local oxygen level in the tissues as is the case with a wound that damages the tissues, disrupting the blood supply. This may happen immediately after the bacteria enter if there's moderate tissue damage, or it may be delayed until another injury at the same site causes additional tissue damage. The original wound might be completely healed by then.

Symptoms

Early symptoms are muscle stiffness and tremor, followed by chronic muscle spasms. Facial and chewing muscles are affected, making it hard for the horse to open his mouth (hence the term *lockjaw*). The third eyelid membrane protrudes and the hind legs become stiff, making the horse's gait unsteady. He holds his tail out stiffly.

Other symptoms are an anxious and alert expression with ears up, eyes wide, nostrils flaring. The horse may eat and drink normally in early stages, but as the disease progresses chewing becomes difficult and he can't swallow. Saliva may drool from his mouth. Treatment for tetanus is often unsuccessful; it is far better to prevent it.

A horse with tetanus is characterized by a "sawhorse" appearance, with stiff head and neck, third eyelid protruding, flaring nostrils, inability to chew ("lockjaw"), and hypersensitivity to sounds, hence the cotton in his ears.

Prevention

The best protection is vaccination with toxoid, starting the day the foal is born if his dam was not vaccinated during pregnancy. If she was vaccinated, producing antibodies in her colostrum to give the foal a temporary immunity, the foal can be given his first shot of toxoid at 3 to 4 months of age. The usual procedure is to give two shots 3 to 4 weeks apart, then a yearly booster. Antitoxin is sometimes given if a wound occurs when a horse has not been vaccinated, but toxoid is sufficient if given within 24 hours of injury, because horses build immunity quickly. Antitoxin should be used if a wound is not discovered promptly. It should also be given to foals with dams of uncertain vaccination history (the mare might not have had a toxoid shot) or with uncertain colostrum intake at birth. Antitoxin gives quick protection but doesn't last more than a few weeks. Yearly vaccination is best.

Definitions

◆ **Toxoid** consists of inactivated toxins used to immunize against tetanus, stimulating the body to build its own immunity.
◆ **Antitoxin** is a purified antiserum obtained from horses that have been immunized against tetanus. This product gives a passive (temporary) immunity, similar to that gained from the antibodies in colostrum.

Equine Encephalomyelitis

Called "brain fever" and "sleeping sickness," equine encephalomyelitis is often fatal. It's caused by a virus that is spread to horses from birds via mosquitoes. Cases occur when horse owners neglect to protect horses by vaccination. There are three kinds of encephalomyelitis in North America — Eastern, Western, and Venezuelan. The Venezuelan strain (VEE) crossed into the United States from Mexico in 1971 after moving north from South America. U.S. horses had no immunity against this strain, and thousands died in Texas before a suitable vaccine was developed and the northward spread was halted.

Eastern and Western strains have been in the United States for a long time. Horses vaccinated against one type are not immune to the other. The two types now overlap in territory; most horse owners use a combination vaccine that immunizes against both strains.

Symptoms

After being bitten by a virus-carrying mosquito, a horse usually shows symptoms within 3 weeks, often in 4 to 6 days. Early signs are fever, depression, and loss of appetite. The next stage involves invasion of the nervous system, with destruction of nerve tissue and brain swelling, but not in all cases. If the virus doesn't reach the nervous system the horse recovers, developing antibodies that protect him for about 2 years.

If the virus attacks the nervous system the horse becomes deranged by the time the fever reaches its peak. He acts restless, excited, and seemingly blind. He may grind his teeth, walk in circles, or stagger. He may be overly sensitive to touch and sound, crash through fences, or wander blindly into obstacles. Shoulder and facial muscles may tremble. Then he may become severely depressed, with head hanging — perhaps resting his head on a fence or manger. In the final stages he goes down and can't get up. Some horses don't develop paralysis and manage to survive, but often with permanent brain damage. Some horses die within a few hours after symptoms begin, but most deaths occur in 2 to 4 days.

Prevention

Every horse should be protected by vaccination in the spring before mosquito season. In warm climates horses should be vaccinated again in late summer. Foals and yearlings may need more protection; young horses have been known to die from the disease despite vaccination. They should be vaccinated at 3, 4, and 6 months of age, then every 6 months thereafter until adulthood (i.e., 3 years of age). Because the source of the disease (birds, mosquitoes) cannot be controlled, the risk of sleeping sickness is always present, so vaccinate.

Strangles

Called "distemper" by old-timers, strangles is caused by *Streptococcus equi*, a bacterium. Distemper is a misleading term; distemper of cats and dogs is a different type of illness, caused by a virus. Strangles in horses causes inflammation of the upper respiratory tract and swollen lymph nodes that often abscess and rupture.

Young horses are more susceptible than adults, but any horse can contract strangles from a sick one or from contaminated feed, water, and surroundings (even a stall that housed a sick horse years earlier) if the horse has never been exposed and has no immunity. The disease begins with respiratory infection (cough, runny nose) and then produces swelling and abscesses in the salivary glands and lymph nodes of head and neck. Death may occur if an internal lymph gland abscesses and ruptures, especially if the infection has spread to other organs such as kidney, liver, or spleen.

To Use or Not to Use Penicillin for Strangles

Some veterinarians give penicillin in the early stages of strangles, but there is controversy over this practice; other veterinarians do not use penicillin in diseases that involve abscesses because it may temporarily delay their formation, only to have them develop later when the penicillin is discontinued.

Symptoms

After an incubation period of 3 to 8 days the horse suddenly shows symptoms, with high fever (103° to 105°F [39.4°–40.6°C]), loss of appetite, clear nasal discharge that soon becomes thick and profuse, and respiratory irritation. Infection in the throat may become so severe that the horse cannot swallow, and if he tries to eat or drink he may regurgitate through his nose. The horse may also cough.

The fever goes down after a few days, then rises again as abscesses develop in the lymph nodes. These swellings, which are hot and painful, may begin to drain fluid approximately 10 days after symptoms appear. If the nodes rupture they put forth a thick, creamy pus. If infection is severe the abscesses may appear anywhere on the body, particularly on face and legs. Lower legs may swell to several times normal size. If the horse develops purpura hemorrhagica (characterized by swellings under the skin due to damaged capillaries), he may die. Abscesses in the lungs cause pneumonia; an abscess in the brain causes meningitis — with paralysis and death. Abscesses in liver or spleen cause death if they rupture, which can happen weeks or months after the horse seems to have recovered from strangles.

How Strangles Is Spread

Bacteria are present in the nasal discharges of affected horses and can contaminate feed, water, bedding, and pasture. An infected horse can spread the disease for at least 4 weeks after contracting it. The bacteria may live for several weeks on the ground or bedding and may infect other horses months or even years after the sick one has been removed. After a horse recovers from strangles he'll have immunity for a while, but if the infection is persistent, horses in a group may suffer repeated attacks every 6 months.

Prevention and Sanitation

If a horse has been sick with strangles, the area in which he was kept should be thoroughly cleaned and disinfected. Old bedding should be burned. Any items used for that horse (brushes, tubs, blankets) should be carefully disinfected with chlorhexadine (Nolvasan) or a similar product from your veterinarian.

If other horses have been exposed, consult your veterinarian. Vaccination can be an effective means of prevention but should be given before a horse is exposed. A horse can have a serious reaction to the vaccine if he is already coming down with the disease. Vaccinating in the face of an outbreak is something you should discuss with your veterinarian. There is now an intranasal vaccine for strangles that is safer and less apt than the intramuscular vaccination to cause reactions.

Influenza

This is an infectious disease of the upper respiratory tract caused by a virus. The term *influenza* has sometimes been applied to several respiratory diseases, and there are other viral infections (including rhinopneumonitis and equine viral arteritis) that cause similar symptoms. Influenza causes respiratory disease, with fever, and can become serious if a horse is worked hard or stressed during the early stages. It is very contagious and hard to treat because the virus does not respond to antibiotics. In fact, antibiotics given to a horse with influenza are actually intended to prevent or control secondary bacterial invasions that follow the initial virus attack. When a horse is weakened by the virus, bacterial complications can cause pneumonia and, sometimes, death.

Symptoms

Influenza spreads rapidly among susceptible, unvaccinated horses — especially young animals that have not yet acquired immunity. Incubation period after contact is about 5 to 10 days. Symptoms appear suddenly. The horse is dull, depressed, standing with head down, and doesn't eat. He has a fever of 103° to 106°F (39.4°–41.1°C) that lasts for about 3 days. In many cases, eyes are sensitive to light and eyelids may be swollen. The eyes water and the nose runs.

The sick horse is often dull and depressed, with head hanging.

Prevention

Vaccination can prevent the disease, but this immunity is short-lived (3 to 4 months) compared to the natural immunity gained from having the disease itself; a recovered horse isn't susceptible to this virus again for at least a year. Horses at high risk for influenza, such as young individuals that have continued contact with other horses, should be vaccinated, starting at 8 to 9 months of age, every 2 to 3 months after an initial series of two shots given 4 to 8 weeks apart. After 3 years of age, vaccination every 6 months usually gives adequate protection.

Illness Following Flu Vaccine

Some horses develop a mild fever and muscle stiffness for a couple of days following vaccination for influenza, and they may go off feed temporarily. The inactivated virus in the vaccine cannot actually cause influenza, but in some horses an abnormal antibody response by the immune system makes them a little ill.

Rhinopneumonitis

"Rhino" is caused by equine herpesvirus (EHV), is highly contagious, and can affect horses in different ways — including respiratory disease, abortion, and nervous system disorders. There are two strains of the virus: EHV-1, which causes abortion, and EHV-4, which causes respiratory problems.

Symptoms

Respiratory symptoms include runny nose, cough, and fever. These appear most frequently in young horses (they have low immunity) and in those that have not been previously exposed. Symptoms occur 2 to 10 days after exposure.

When broodmares are kept adjacent to infected horses, such as a sick group of weanlings or yearlings, they are at risk of aborting even though they may never show respiratory signs of this disease. A mare may abort as soon as 2 weeks after infection or as late as several months after coming in contact with rhinopneumonitis. If she carries the infected fetus to term, it will be born sick and not likely survive.

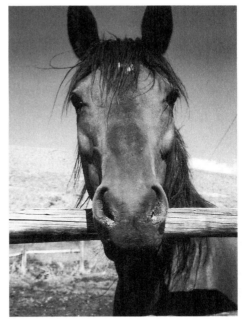

Signs of respiratory illness often include a runny nose.

A rare form of the disease, paralytic rhinopneumonitis, can cause collapse. Early symptoms are weakness in the hind legs, floppy tail, and dribbling of urine. Horses suffering from a mild case of this form often recover without permanent damage, but severe cases do not survive unless given intensive care.

Treatment

The main treatment for rhinopneumonitis is good care and lack of stress until the horse recovers. Most horses bounce back after 2 or 3 weeks. If a horse has high fever and is off feed, call your veterinarian. The horse may need antibiotics to prevent secondary bacterial infection that might cause pneumonia.

Prevention

Several types of rhino vaccine give good protection against the respiratory form but not long-lasting protection against abortion. Breeders trying to guard against rhinopneumonitis abortions in broodmares generally vaccinate several times during pregnancy. The average horse owner may not need to vaccinate against rhinopneumonitis, depending on circumstances. If your horse lives where other horses come and go, or if young horses are present, it's wise to vaccinate. If your horse is kept with older horses that rarely leave the farm, and if he is rarely exposed to unknown horses, vaccination may not be needed — especially if there's not much incidence of rhinopneumonitis in your area. Check with your veterinarian to see if your horse is at risk.

Pneumonia

Pneumonia is disease of the lungs. In the horse, it can be caused by a virus, bacteria, combination of the two, fungi, or migrating internal parasites. Bacteria in the lungs may be the primary invader, occur secondary to a virus, or develop as an after-effect of some other disease such as strangles. Any disease that lowers a horse's resistance and is accompanied by extra stress factors can open the way for pneumonia. The initial disease weakens the horse, then pneumonia sets in and becomes the actual killer.

Aspiration Pneumonia

Pneumonia can develop from moisture or foreign particles in the lungs, which cause irritation and infection. This may happen when a horse is given liquid medication or mineral oil and gets some in his windpipe — and on down into his lungs. It can also happen if a horse chokes on feed with blockage in the esophagus; fluid backs up and spills into windpipe and lungs. Anytime a horse can't lower his head to cough and get rid of a blockage (as when tied in a trailer), he is in danger of developing aspiration pneumonia if feed particles get into the windpipe.

Pneumonia Due to Septicemia

Most cases of pneumonia originate in the airways; pathogenic invaders come in through the windpipe. However, some lung infections are caused by blood-borne pathogens that enter the lungs via the bloodstream — such as in a newborn foal with septicemia from navel ill or some other serious infection.

Symptoms of Pneumonia

When a horse has pneumonia his respiration rate increases (the body tries to bring in more oxygen), he breathes rapidly and shallowly, and he usually has a cough. The lungs make abnormal wheezing or rattling noises that you can hear if you listen to his chest with a stethoscope. In most bacterial pneumonias there is toxemia — bacterial toxins in the bloodstream. In later stages of the disease the horse has difficulty breathing because much of his lung tissue has been damaged and is useless.

There may be a discharge from the nostrils. The breath may smell foul. The horse may have a fever and be quite dull, with elevated pulse. He may not want to eat and might be reluctant to lie down; it is easier for him to breathe while standing up.

Treatment

With prompt and adequate treatment in the early stages, bacterial pneumonia usually responds quickly and completely. Viral pneumonia may not respond at all or just temporarily (antibiotics control secondary bacterial invaders), then the horse relapses. A horse with pneumonia should be kept isolated from other horses and in a warm barn, out of the weather. He should have fresh, clean water and light, nourishing feed. Your veterinarian can help with treatment — especially if the horse is not eating much and needs to be fed by stomach tube or intravenously — and determine which antibiotics would be most beneficial. The veterinarian may even give the horse other drugs to aid breathing and reduce pain and inflammation so the horse can better fight the illness and maybe feel well enough to eat and drink on his own.

Prevention

The best prevention against pneumonia is to keep a horse in good health and to minimize stress. This means vaccinating for diseases such as rhinopneumonitis, influenza, and strangles if your horse might be at risk because these diseases can weaken a horse and make him susceptible to pneumonia. Minimizing stress includes protecting your horse from extremes in bad weather.

Equine Viral Arteritis (EVA)

EVA is an acutely contagious viral disease that affects horses of all ages. It is spread by direct contact and by breeding. Stallions can be carriers, shedding the virus and spreading the disease without ever showing symptoms themselves.

Symptoms

Symptoms are sometimes mistaken for those of a mild case of influenza; the horse may have elevated temperature and respiratory illness for a few days. Some horses experience swelling of the legs, muzzle, and eyelids. Stallions often have swelling in the sheath and scrotum. Other symptoms that sometimes occur are inflammation of membranes of the eye (conjunctivitis), skin rash, diarrhea, reddish gums, and nasal discharge. Fever may be mild or as high as 105°F (40.6°C), lasting for a day or two or as long as a week. Most cases are not life-threatening, but some horses become weak and unable to stand about a week after getting EVA; they may die unless given good care. The disease is most dangerous to foals and elderly horses.

The most serious complication of EVA infection is abortion in pregnant mares. At any stage of gestation, within 1 to 3 weeks of being exposed they may abort. A laboratory examination of the aborted fetus is necessary to determine the cause.

Stallions and EVA

Once recovered from EVA, horses are no longer contagious and are immune to the disease for several years. But nearly two-thirds of stallions continue to shed the virus in their semen, becoming carriers and spreaders of the disease, infecting susceptible mares.

Prevention

EVA is generally not life-threatening but is serious because of the risk of abortion. A modified live virus vaccine is effective in preventing outbreaks of EVA and can protect mares that will be bred to infected stallions. It is also wise to vaccinate disease-free stallions that breed other folks' mares because the disease is easily transmitted venereally.

Equine Infectious Anemia (EIA)

EIA ("swamp fever"), which is caused by a virus, occurs most often in horses pastured in wet places. Many cases occur in late summer and early fall; blood-sucking insects spread it from horse to horse. It can also be spread from carrier horses to susceptible ones by use of unsterile hypodermic needles and other such equipment.

Symptoms

EIA may appear in acute, subacute, or chronic form. The disease is usually characterized by high, undulating (up and down) fever, sweating, stiffness and weakness (especially in hindquarters), depression, anemia, jaundice, swelling of legs and lower body, and loss of weight even if the horse has a good appetite.

When a horse suffers an acute attack, symptoms develop suddenly. He may have fever (105° to 108°F [40.6°–42.2°C]), appear dejected, refuse feed, become anemic, have jaundiced mucous membranes, and sweat profusely in warm weather. There may be watery discharge from the nostrils. Attacks may last 3 to 5 days or longer, then he may seem to recover. Other affected animals become progressively weak, dying within 2 to 4 weeks after several fever cycles. Some horses recover with a lingering type of anemia that leaves them weak and useless for the rest of their lives.

Chronic cases are most common. Horses with chronic EIA have intermittent attacks of fever — usually a fluctuating low fever — and suffer from anemia and weight loss. They become weak during attacks, sometimes with staggering or paralysis. Many chronic cases show no symptoms over long periods, but their blood contains the virus and is a source of infection for other horses.

In most chronic cases, anemia develops, heart action becomes irregular, body and leg swellings appear and disappear, and muscular weakness varies from slight to so much that the horse cannot stand or walk. Sometimes the horse has diarrhea. Individuals that appear to recover from the first attack may be free of symptoms for days, weeks, or months, but sooner or later other attacks follow. The relapses often coincide with periods of stress such as hard work or another disease. In any of these attacks, the horse may die. Horses that survive become carriers; after a few attacks they may develop a sort of immunity and seem to be healthy, but they still carry the virus. They usually die of the disease eventually.

Transmission

The disease is spread from a carrier horse to susceptible ones through transmission of tiny particles of blood carried by biting insects or on instruments that puncture the skin, such as needles. Incubation period is 12 to 15 days, sometimes shorter but often longer — even as much as 90 days.

EIA can also be spread by simple contact through infected secretions and from contaminated feed and water. Mares can be infected when bred by a stallion that carries the virus. Foals occasionally become infected by drinking milk from infected dams.

EIA Resists Disinfecting

EIA virus can withstand ordinary disinfecting methods and even short periods of boiling. Contaminated instruments and needles must be cleaned and boiled for at least 15 minutes to kill the virus — a good reason to never reuse needles. The virus is present in all tissues, secretions, and excretions of the infected horse.

Prevention

Researchers have not yet been able to create a vaccine against EIA, so other attempts are made to control it. Blood testing of all horses that travel from one state to another, quarantining or destroying horses that react positively, and following strict sanitation measures (including the use of disposable syringes and needles) have helped reduce the incidence of this disease. Horses that are raced, shown, sold, or transported across state lines must be blood tested. (The Coggins test is named after Dr. Leroy Coggins, who developed it.)

Blood is drawn from the jugular vein for a Coggins test.

Leptospirosis

"Lepto" is a disease of animals and human beings caused by spiral-shaped bacteria. There are many kinds; some cause disease in dogs, others in cattle, pigs, horses, rodents, and humans. Leptospirosis can cause illness and abortions. It is often spread by a carrier — an animal that contracts the disease and recovers but still harbors bacteria and sheds them in urine, saliva, and other bodily secretions or excretions. The most common carriers are rats, mice, and other rodents, though wildlife, pigs, cattle, and dogs may spread leptospirosis for a while after recovering from the disease. The bacteria live for many days in water or damp soil but die in a few hours when dried out. Thus leptospirosis isn't highly contagious in dry climates or dry seasons.

Symptoms

In horses, the disease is not usually serious. A horse may have a fever, loss of appetite, dullness, or jaundice. Pregnant mares may abort. The bacteria may localize in the body and cause further problems in kidneys, joints, or eyes. In fact, leptospirosis is one of the main causes of moonblindness (periodic ophthalmia) in horses. They usually recover easily enough from leptospirosis but later suffer recurring attacks of eye irritation that may eventually cause blindness.

Prevention

There is no vaccine specifically for horses, but there is a cattle vaccine against five common types of leptospirosis. In areas where there is wet ground or ponds, horses can be at risk. Where leptospirosis causes abortion in mares or moonblindness in horses, some veterinarians recommend using the cattle vaccine to protect horses.

Potomac Horse Fever (PHF)

This is a serious disease of horses that first appeared in the Potomac River Valley in Maryland in 1979. Since then it has been found in nearly all states. It's caused by a rickettsial organism, *Ehrlichia risticii*, belonging to the same family of organisms that cause Rocky Mountain spotted fever, anaplasmosis, and typhus — all spread by insects, fleas, and ticks. PHF is transmitted by the larvae of flukes (parasitic flatworms) that live in freshwater snails. The disease occurs most commonly in summer and in horses that are pastured near streams or in irrigated fields.

Symptoms

The first sign is refusal of feed and water. There is usually fever up to 106°F (41.1°C). A thorough examination at that point would show low blood pressure and low blood count. Two to three days later the typical symptom is severe, explosive diarrhea. The horse is in extreme pain with diarrhea and/or colic and becomes seriously dehydrated, requiring intravenous fluids. Some horses founder, and some are euthanized to relieve their misery. Many horses test positive even though they have no history of illness, so the disease apparently is sometimes very mild.

Acute cases have a high mortality rate (unless treated immediately) or develop founder, which can cause more damage than fever and diarrhea. Most horses respond to fluid and tetracyclines given intravenously but may require constant care for a couple of weeks.

Prevention

Vaccination involves two doses the first year, several weeks apart, and a yearly booster. Sometimes PHF still occurs in vaccinated horses but is milder because they have some immunity. The best immunity is gained if vaccination is given before summer and in proper sequence (two shots the first year). Incidence of the disease ceases after hard frost. Whether you vaccinate will depend on where you live — find out if cases have occurred in your area, and determine the degree of risk. Ask your veterinarian to help with this. Risks for PHF can be reduced by keeping horses off irrigated pastures and away from streams and ponds; provide a different water source.

Suspect PHF in Summer

Many cases of PHF are never actually diagnosed because they are not blood tested, but any horse with diarrhea and fever during summer is generally assumed to have PHF. At that time of year most veterinarians presume the cause is PHF and promptly begin treatment — since early treatment is often crucial to survival of the horse.

Salmonellosis

Salmonella is a bacterium that causes disease in all species of animals, including human beings. Several types affect horses, including *Salmonella typhimurium*, which can also cause illness in cattle, sheep, pigs, ducks, dogs, and small animals. The feces of all these can serve as sources of infection for others. Farms with a lot of ducks or pigeons may have a problem with salmonella spreading from the birds to the horses.

Salmonella bacteria may live harmlessly in the horse's gut until stress enables them to cause disease, which can be mild or severe. Salmonellosis can be a primary disease — especially in foals — but in most horses it is often a secondary invader that attacks when the animal is stressed, suffering from some other disease or situation that lowers its resistance. The usual symptoms are inflammation of the intestine and diarrhea; mortality is high. The source of infection is a sick animal or carrier, wild or domestic, that contaminates feed, pasture, or drinking water with feces containing the bacteria.

Carriers

Salmonella bacteria often can't be eliminated from the horse's body even after he recovers, so he becomes a carrier and spreads it to other horses. This can become a chronic problem on a farm, with periodic outbreaks and high foal mortality. Young horses are most susceptible to fatal diarrhea, and adult horses are likely to become clinically normal carriers, breaking out with the disease only if stress or some other debilitating circumstance lowers their resistance and permits bacteria to multiply rapidly.

Symptoms

Infection results in rapid multiplication of bacteria in the intestine followed by invasion of the bloodstream, causing septicemia (bacterial toxins in the blood, poisoning the entire body), which can be rapidly fatal. The horse becomes restless and won't eat. He develops mild to moderate fever, and acute diarrhea may follow.

Newborn foals may develop severe septicemia, becoming dull and depressed with high fever and dying within 24 to 48 hours. Sometimes the infection spreads so quickly the foal dies from the overwhelming septicemia before experiencing any diarrhea. Adults more commonly get extremely watery diarrhea, sometimes bloody. There may be high fever at first and complete loss of appetite, but the animal may be very thirsty. Pulse and respiration rates are high. Pregnant mares may abort. The horse may have violent colic. Diarrhea causes severe dehydration and the horse loses condition rapidly. Some go into shock and die swiftly in spite of attempts to replace fluids. Others linger for weeks and then die or begin gradual recovery. Still others recover from the diarrhea within a few days, whereas other cases suffer chronic diarrhea for months.

Treatment

The most important treatment is fluids and electrolytes given orally and intravenously. The veterinarian may administer an antibiotic, although this is controversial; several forms of the bacteria have developed resistance to antibiotics. The goal of treatment is to keep the horse alive and give him good care until he can develop resistance and avoid relapse.

Prevention

If salmonellosis is suspected, the horse should be kept in strict isolation from other animals and all equipment used on him kept separate. Even boots and clothing worn while treating him, feeding him, and so on should not be worn where other horses might come in contact with contamination. You don't want to track manure from a sick horse into a pen where there are healthy horses. Salmonella outbreaks can be prevented by not overgrazing pastures (an overused, overgrazed pasture is more apt to be contaminated), keeping healthy horses out of an infested area, and quarantining new horses until you are sure they are healthy. Isolating sick and recovering animals is extremely important. To date, there is no effective vaccine for salmonella.

Botulism

Botulism is a highly fatal disease caused by toxins produced by a spore-forming bacterium, *Clostridium botulinum,* found in the environment and especially in soil. The bacteria live in the digestive tracts of plant-eating animals such as horses and cattle but do not usually cause disease. The bacteria multiply best in decaying plant or animal matter; under warm, moist conditions they create a lethal toxin. Botulism is responsible for "shaker foal syndrome," which causes affected foals to quiver, become weak, and collapse. Even with intensive care, most unvaccinated shaker foals eventually die of respiratory failure due to progressive paralysis.

In horses there are two strains of botulism (B and C) that cause "forage poisoning." Adult horses usually become ill after eating contaminated feed but can also ingest botulism spores from soil. The bacteria then produce toxins while multiplying in the gut. Foals usually develop botulism by eating soil containing the spores. Although the spores are commonly found in soil in the Northeast, horses can develop botulism anywhere in the United States if they eat toxins in feed or water that has been contaminated. Dead rodents in haystacks or small animals baled in hay are common sources. The toxin can also be found in spoiled grain.

Symptoms

Symptoms of botulism show up 3 to 7 days after the horse has eaten contaminated feed. Incubation period can be shorter if large amounts of toxin are eaten. Most horses die unless treated quickly; even then many do not survive. The toxin short-circuits the nerves controlling muscle function, causing paralysis and difficulty in swallowing, standing, and breathing. The horse has progressive muscle paralysis and is soon unable to chew or swallow, drooling saliva. Paralysis of the chest muscles results in suffocation. In foals with shaker syndrome usually the first sign is trouble swallowing; milk dribbles from the foal's mouth when he nurses. He becomes uncoordinated and lies down frequently. There is no fever.

Prevention

Vaccinated pregnant mares can transfer immunity to their otherwise defenseless foals through antibodies in colostrum. This gives foals temporary immunity for several weeks. Adult horses of any age can be given an initial series of three doses of toxoid at least a month apart, with an annual booster. Foals can begin their vaccination series as young as 2 to 4 weeks of age, with three inoculations at 2-week intervals. If an outbreak occurs, mares and foals can have an even tighter vaccination schedule, giving foals' initial series a week apart; a fourth dose might then be necessary to provide adequate immunity. A mare's vaccinations can be spaced more closely also, to give adequate concentrations of antibody in colostrum.

Take care to make sure all feed is free of contamination. A dead bird, snake, or rodent baled in hay can be deadly, and toxins can contaminate the hay around it. Any feed found with traces of dead animals in it should always be discarded. Be sure to get rid of the hay surrounding the animal carcass as well. The best rule for feeding is "when in doubt, throw it out."

Rabies

Many people think of rabies as a disease that affects dogs, skunks, foxes, or bats; but horses and cattle get it, too. This fatal viral infection of the central nervous system can occur in any warm-blooded animal. It's transmitted by bites of affected animals or by contact with saliva from an infected animal. If the virus enters the body through a wound or break in the skin, it attacks the nerves. Unlike other viruses that travel through the bloodstream, the rabies virus affects nerve tissue, moving slowly through nerve cells and eventually reaching the brain.

Incubation of Rabies

Incubation period depends on location of the bite and how long it takes the virus to reach the brain and spinal cord. A horse bitten on a hind foot may not show symptoms as soon as a horse bitten on the nose. Symptoms may not appear for weeks or even months after the bite. Once symptoms do show up, rabies is always fatal.

Symptoms

When the virus enters the spinal cord and brain, it causes encephalitis (inflammation of the brain) with destruction of nerve cells. One of the first and most common symptoms is difficulty in swallowing. The horse may try to eat or drink but cannot. Saliva may drool from the mouth. Inability to swallow may be misdiagnosed for choke, which results in handlers being exposed to rabies as the horse owner or veterinarian examines the mouth and throat.

Rabid animals show a variety of symptoms. In the *dumb or paralytic form* the animal is lethargic and depressed with head hanging, slack lip, dazed expression, and drooling. Temperature, pulse, and respiration are normal; there's no sign of shock or pain. The horse is uncoordinated; hind legs may knuckle under at the fetlock joints, and hindquarters may sag and sway if he tries to walk. The horse eventually becomes paralyzed, unable to get up.

In the *furious form* the animal becomes aggressive, tense, alert, and hypersensitive to sounds and movement. He may pace about or run wildly, biting at anything. He may snap at his halter rope or the person handling him and become aggressive and vicious. Acute change in personality may be the first noticeable symptom.

Death from Rabies

Death from rabies is usually due to respiratory paralysis; the muscles necessary for breathing no longer work.

No Set Patterns

Some rabid horses eat strange things such as straw or wood. Pulse, temperature, and respiration may be normal at first, then climb as the animal becomes frenzied. Some cases run a course of 6 to 7 days before dying, whereas others don't last more than 36 to 48 hours. Rabid horses don't always display aggressive behavior or drooling. They often have fever, depression, loss of muscle tone in the tail area, muscle tremors, and paralysis of hindquarters. A veterinarian is often called because the horse shows lameness or hind leg incoordination. Some are colicky. Rabies can be mistaken for many problems, and this makes it exceptionally dangerous; it may not be suspected soon enough to prevent human exposure. Precautions should be taken when handling a horse that might have rabies; wear protective clothing to avoid contact with his saliva.

Diagnosis

Positive diagnosis can only be made through laboratory tests on the brain after the animal has died, to confirm presence of the virus. If rabies is suspected, the veterinarian should submit the brain to a diagnostic lab for testing.

Treatment and Prevention

There is no treatment for rabies. Once symptoms begin it is always fatal, in man or animal. Because of the long incubation period (3 weeks to several months) human beings can be vaccinated soon after exposure. If vaccination is started early enough the body builds immunity in time to prevent fatal inflammation of the brain.

To protect against rabies, all cats and dogs should be vaccinated; they are often more at risk than horses for coming in contact with wild animals. Horses can also be vaccinated. This is generally adequate protection, but any horse exposed to rabies (e.g., bitten by a rabid animal) should receive a booster shot immediately and be quarantined for 90 days. An unvaccinated horse exposed to rabies is more at risk. Post-exposure vaccination is routine in human beings, but its efficiency in animals is under debate. Some public health officials recommend immediate euthanasia or putting the horse into strict quarantine after post-exposure vaccination — for 6 to 9 months with minimal human contact.

Has the Horse Been Exposed to Rabies?

The difficult part is being able to tell if a horse has been exposed to rabies. The bite of a rabid animal is often small and goes unnoticed, healing in a few days and leaving no clue to the problem until the horse shows signs of rabies. If you live where rabies occurs in wildlife, discuss vaccinating your horses with your veterinarian.

Piroplasmosis

Piroplasmosis is caused by protozoa and spread by blood-sucking insects or ticks. Incubation time is 7 to 20 days as protozoa multiply in the bloodstream. Death is usually due to lack of oxygen in tissues, a result of severe anemia. If the horse survives, it becomes a carrier for a year (or longer if it becomes reinfected).

Symptoms

Intermittent fever, loss of appetite, depression, weakness, and increased pulse and respiration rates are typical. As the disease progresses the horse shows evidence of jaundice, anemia, and dark-colored urine. Fever lasts a week; the total course of the disease is about 3 weeks. Abortion is common in pregnant mares. If the horse recovers, it takes a long time to overcome anemia and emaciation.

Treatment and Prevention

A few drugs such as imidocarb are helpful, but there are no vaccines. Prevention requires effective quarantine of horses and other measures to halt the introduction of the blood-sucking tick into new areas. Restricting imports or movement of animals from tick-infested regions has proved effective.

Anthrax

Anthrax is a serious disease that affects most warm-blooded animals. It comes on suddenly and runs its course quickly, ending in sudden death. The bacteria are present in the tissues of infected animals; if the carcass is scattered by birds or predators, soil and water may become contaminated. Spores are formed that can live a long time. The disease is often spread during floods, contaminating grass that grows in soil containing the bacteria. Grass grown in flooded lowlands may be infective, and animals fed hay grown on these lowlands may also get the disease. In hot climates anthrax is sometimes spread by blood-sucking insects, particularly horse flies. Close grazing of scratchy stubble (causing injuries to mouth tissues) and confined grazing in contaminated areas (e.g., around waterholes) can sometimes lead to spread of the disease.

Symptoms

Anthrax in horses is always acute. If a horse gets it from contaminated feed or water, he develops septicemia (bacterial toxins in the blood), enteritis (inflammation of the intestines), and colic. If infection is from an insect bite, the horse may get hot, painful swellings around the throat, neck, ribcage and abdomen, sheath, or mammary glands. There is high fever; the horse is depressed and may have difficulty breathing. Anthrax is usually fatal within 48 to 96 hours. Antibiotics and anti-anthrax serum are used in treatment; if given soon enough the horse may recover.

Prevention

In areas where anthrax is a problem, horses should be vaccinated in spring. Occasionally an animal develops severe or even fatal reactions to the vaccine. If an individual develops a reaction he should be treated with antibiotics and *antiserum,* a blood serum derived from an immune animal that contains antibodies against anthrax.

Vesicular Stomatitis (VS)

This contagious viral disease occurs in warm regions of South and Central America and Mexico but sometimes spreads north into the United States, requiring quarantines and restricted movement of horses and livestock. VS is not a fatal disease, but health officials are diligent to prevent its spread because symptoms are so similar to those of foot-and-mouth disease. Outbreaks must be accurately identified. Foot-and-mouth, a much more serious disease of cloven-hoofed animals, was eradicated from the United States in 1929.

Transmission

VS occurs during summer due to large insect populations and increased movement of animals to shows and competitions. The virus can be spread by insects and contact with saliva from an infected animal (or fluid from ruptured blisters). It is often spread by moving infected animals, failing to clean out vehicles used in transport, and being careless about human contact between infected and non-infected premises. Blister fluid or saliva on bits or in water buckets can infect susceptible animals. A farm that's had a sick animal cannot be considered free of VS until at least 30 days after the affected animal's lesions have healed. If a case occurs, the horse should be quarantined in a stall. Animals on pasture are more frequently infected. Isolating a sick one in a stall reduces the risk of spread.

Symptoms

The disease is rarely fatal itself but causes blisters in the mouth, tongue, lips, nostrils, teats, and feet. Blisters swell and break, and the skin sloughs away leaving painful ulcers; the animal stops eating and drinking (thereby losing weight) because of raw tissue and pain. Animals may die of starvation, dehydration, or secondary infection. Blisters on the coronary band cause lameness, founder, and occasionally loss of a hoof. The most common symptoms are drooling and high fever. Incubation time from exposure to first blisters is 2 to 8 days, sometimes longer. A close look at the mouth will reveal the blisters. In horses, blisters usually appear on the upper surface of the tongue. If there are no complications (i.e., no secondary infections) the horse usually recovers in 2 to 3 weeks.

Treatment and Prevention

There is no treatment for VS except to prevent secondary infection where the blisters have broken and raw tissue is exposed. Mild antiseptic mouthwashes and ointments help alleviate pain and speed recovery. Good sanitation and quarantine measures on affected farms can keep VS from spreading. Use disinfectants such as chlorine bleach or chlorhexadine (Nolvasan). A quarantine area should be as far away from other horses as possible. Avoid shared feeding equipment, or clean and disinfect it for at least 10 minutes between uses. If caring for sick horses, shower and change clothing and boots before handling healthy horses.

Health Tip for VS

Use protective measures when handling VS-infected animals to prevent contact with body fluids. The virus can be transmitted to human beings through skin or respiratory system, causing flu-like symptoms. In human beings VS causes acute illness: fever, muscle aches, headache, general discomfort, with blister lesions only rarely. When handling a sick horse, wear gloves to protect hands from saliva and blister fluid; never let these fluids come into contact with your eyes, mouth, or any broken skin. If you come in contact with fluids, consult your doctor.

To prevent VS, keep horses in stalls or dry corrals away from insects. Avoid stress or medications that suppress the immune system. Use insecticide and repellent daily on horses. Isolate new horses for at least 21 days to make sure they don't have VS. This is a reportable disease, and the premises must be quarantined if it occurs.

Lyme Disease

Lyme disease affects human beings and horses and is spread by tiny deer ticks. It was first diagnosed in Lyme, Connecticut, in 1976 and has since been reported in almost every state. It is caused by the *Borrelia burgdorferi* bacterium.

Symptoms

Symptoms include rash and swelling at the tick bite, fever, fatigue, and depression. Horses may also suffer laminitis, joint problems, and chronic lameness. Some are eventually euthanized because of chronic problems caused by the disease. Treatment with antibiotics is effective if begun early, but often the disease is mistaken for something else. Diagnosis is usually based on whether the horse is in a tick-infested region and a positive blood test.

Prevention

The best ways to prevent Lyme disease are to keep you and your horse free of deer ticks, keep pastures mowed, and remove brush and wood piles from pasture areas to decrease rodent-nesting sites. Precautions during tick season include careful grooming after a ride, and keeping your horse out of tick habitat (i.e., brushy areas) as much as possible. If you find a tick, carefully remove it with tweezers, not your fingers.

Ticks, Mice, and Lyme Disease

Not all ticks are infected. They must first bite an infected white-footed mouse. The ticks can be controlled with a selective pesticide that is available in biodegradable tubes containing cotton balls impregnated with insecticide. When these tubes are scattered in a tick-infested area, mice gather the cotton balls for their nests and the insectide spreads into their fur. When ticks make contact with the cotton or mice, they die — without causing any harm to the mice.

Equine Protozoal Myeloencephalitis (EPM)

This neurologic disease is caused by a protozoan, *Sarcocystis neurona*, that lives in opossums and possibly birds. Horses generally become infected by eating feed or water contaminated with feces from infected opossums. The one-celled organism then makes its way to the spinal cord, where it causes inflammation and nerve cell damage. This most often results in a vague lameness or weakness and incoordination.

Symptoms

The key to swift diagnosis is recognition of subtle changes in the horse's stance and way of traveling — little hints of weakness or incoordination such as toe dragging when tired, mild lameness in the hind legs, back soreness, choppiness of gait, or awkward stance. As the disease gets worse, symptoms become more obvious — muscle atrophy, seizures, facial paralysis (ears or lips drooping on one side, head tilting), or leaning to one side.

Treatment

Early diagnosis and treatment to rid the horse's body of the parasite can halt the infection in time to prevent permanent crippling, but if treatment is begun too late there can be irreversible damage to the nervous system. Current treatments are long and complicated, but researchers are working on new ones that should soon be available.

Prevention

Keep opossum and bird droppings from contaminating feed and pasture. Keep grain covered, keep birds out of your barn, buy heat-processed feed (high temperatures kill the organism that causes EPM), and reduce opossum population on your property. Live trapping is time-consuming but effective.

Summary

There are other infectious diseases that can affect horses, but most are not common and this book cannot cover them all. The best way to protect your horse from illness is to conscientiously give the best care you can, vaccinate against diseases in your area (if there's a vaccine available), and consult your veterinarian if the horse ever develops signs of illness or displays a change from normal temperament or behavior. Proper diagnosis and prompt treatment make a critical difference in his recovery.

NINE

NON-INFECTIOUS DISEASES OF HORSES

There are a number of problems in horses due to factors other than infectious organisms. Most of these diseases, such as "heaves," laminitis, navicular syndrome, and muscles tying up, can be prevented by proper care of the horse. Other problems, such as cancer or Cushing's disease, are not as preventable; they may occur despite the best care and responsible use of the horse.

Cancer

Cancer is not common in horses, but be aware of the possibility because many types are treatable in early stages. Benign growths enlarge slowly and rarely recur if removed. Malignant growths expand more rapidly and tend to spread to other parts of the body, infiltrating neighboring tissues or entering into the blood or lymph system.

A cancer can start from any type of cell (blood, bone, muscle), but more than half the cancers in horses originate in skin tissue. Most can be controlled if discovered in time and can usually be removed without major surgery. Some types of cancer are curable only if they are detected early. Some kill the horse within a few weeks; others take years to cause death.

Melanoma

This skin cancer can occur on any dark-skinned horse, most commonly on grays. The tumors originate in cells that produce skin pigment (melanin). Melanomas eventually affect 80 percent of all gray horses over 15 years of age. Most grey individuals develop the black lumps in their old age. Only a few melanomas progress to the point of killing the horse, but internal ones are fairly common in the intestinal tracts of older gray horses. You should suspect this if an older horse has chronic digestive or colic problems.

Location

In mares the lumps usually develop under the base of the tail or around the anus or vulva. The nodules generally start out benign but may spread inward along the mare's genital or rectal area, and the expanding growths in the pelvic lymph nodes can interfere with breeding and obstruct foaling. In male horses the melanomas may develop under the tail but are often found in the sheath, sometimes giving the whole sheath a swollen appearance.

Be Alert for Lumps

Most growths appear under the base of the tail; you are likely to find the bumps when grooming or taking the horse's temperature.

The black lump can be felt as a hard nodule in the skin. Not all bumps are melanomas; have a veterinarian take a look to be sure. Melanomas are usually smooth, raised areas without scabs or ulcers. Lumps may appear at any time after age 4 or 5 (but more frequently after age 15), and there may be few or many. Sometimes a single large tumor appears on a young horse; this type of melanoma is likely to become malignant more quickly than those on older horses.

External tumors usually don't grow malignantly, but internal melanomas involve the lymph system and are more dangerous, often spreading to other parts of the body. External lumps tend to grow slowly, whereas internal growths can spread rapidly. Sometimes a fast-growing melanoma creates a sore that won't heal, pushing the skin apart. It may cause bleeding from the affected area.

Treatment

Slow-growing external lumps are best left alone. A biopsy to confirm the diagnosis may activate the melanoma and cause it to grow or spread, sending cancer cells through the blood or lymph system into the lungs, where they become deadly very quickly. Removal of a melanoma is usually a last resort, done only if the lump is growing rapidly or causing a problem such as blocking the rectum or interfering with use of a saddle. The veterinarian freezes or burns the area after removing the growth, to kill any stray cells that might be left. There's been some success in treating melanomas with cimetidine (Tagamet), an ulcer medication for human beings, which helps the immune system fight the cancer. Another treatment involves creating a vaccine from a portion of the growth, which also stimulates the body to combat the cancer.

Check Melanoma Lumps Regularly

A slow-growing melanoma can become fast-growing, killing the horse in 6 months or less. A horse with melanoma lumps should be checked periodically to see whether the lumps have grown or spread. Make a habit of checking every time you groom or use the horse. External melanomas may never spread, but the potential for spread should never be ruled out.

Squamous Cell Carcinoma

Another skin cancer, this may appear on the eyelid and/or around the vulva or sheath if the skin is unpigmented. It is slow to metastasize, or spread. Intense sunlight stimulates its growth. Because skin pigment helps protect against ultraviolet rays, dark-skinned horses rarely get this cancer. It is more common in horses with pink eyelids (Paints, Appaloosas, Pintos, horses with white faces, those with white markings on the face with pink skin around an eye, or light-skinned horses). Tumors near the eyeball in light-skinned horses are common in sunny areas such as California and the Southwest.

Location

Because cancer of the eyelid can spread rapidly to nearby tissues, unless it is caught early you may need to have the eye removed to save the horse. Most squamous cell carcinomas occur on inner parts of the lower lid or third eyelid and appear as a single raised bump or raw surface — a runny sore. The eye is irritated and waters. Not every bump or reddened area is a cancer, but if the abnormality becomes larger instead of healing and going away, or becomes redder or more irritated, the horse needs immediate veterinary attention.

Health Tip

Protect horses with unpigmented eyelids from bright sunlight by keeping them stabled, providing shade during bright days, or painting a nontoxic dark substance around the eyes during bright, sunny weather.

Squamous cell carcinoma can also occur on other unpigmented areas, especially those that are thinly haired, such as under the tail, around the mouth, or on the sheath. Check these areas periodically for any abnormalities.

Treatment

The most common treatment is surgical removal. Other methods include freezing with liquid nitrogen, immunotherapy (injecting vaccine into the tumor tissue to stimulate production of anticancer cells), radiation, and hyperthermia (burning off the growths). When the cancer occurs on the eyelid, however, some of these methods pose a risk to the eyeball. Your veterinarian will choose a method that minimizes the risks.

Granulosa Cell Tumor

Granulosa cell tumor in the ovary of a mare causes problems with reproduction and disposition but is usually not malignant (although it can rupture, spreading cancer cells to other parts of the body). The granulosa cell is part of the ovary and produces the female hormone estrogen. If this cell becomes cancerous the tumor produces a mix of hormones, altering the mare's hormonal balance and causing personality change and bizarre behavioral symptoms. She may become irritable, aggressive, or vicious. She may even show stallion-like behavior around other mares, especially if they're in heat. Some affected mares tease and mount other mares. As the tumor grows, the mare does not come into heat and she cannot become pregnant.

Fatal Tumors

Granulosa cell tumors sometimes cause death. Normally each ovary is suspended from the roof of the abdominal cavity, but an ovary enveloped in a growing tumor may become so heavy that its attachment breaks and ruptures a blood vessel, causing fatal hemorrhaging.

Treatment

Usually the only effective treatment involves surgical removal of the tumor and ovary. Removal of one ovary does not interfere with the mare's ability to reproduce, if the remaining ovary is healthy. Most mares return to normal after removal of the tumorous ovary and start cycling again. If a mare shows a drastic change in behavior, have her ovaries checked. If there's a tumor it can be removed.

Other Cancers

Other cancers in horses include abdominal cancer and cancer of the blood. Some are rare and difficult to diagnose or treat. Signs of cancer include any unusual sore or abnormality — a growing lump or bump, a persistent sore, bleeding or discharge from any body opening, chronic cough (other than heaves), or persistent weight loss. Abdominal cancer in an old horse may cause dramatic weight loss.

Excessive weight loss can be an indication of abdominal cancer.

Lipomas

Lipomas are not really a cancer because they are almost always benign. Consisting of slow-growing balls of fat on thin stalks, these fatty tumors in the abdomen occur most commonly in older horses, especially fat individuals. Usually the tumor poses no problem unless the thin stalk becomes wrapped around an intestine, strangling the gut. In these cases the horse will suffer colic and a fatal blockage of the gut (the stricture causes death of a segment of intestine) unless the condition is promptly corrected by surgery. Many older horses have a number of these fatty tumors, but only rarely do they cause problems.

Chronic Obstructive Pulmonary Disease (Heaves)

The horse has a remarkable set of lungs that keep his blood supplied with oxygen. Anything that interferes with proper working of the lungs can limit his athletic career or compromise his health. A common problem that hinders a horse's ability to breathe is chronic obstructive pulmonary disease (COPD), called "heaves" by horsemen. It occurs when horses are confined indoors or in small pens and fed hay that is dusty or moldy; it can also be due to allergies.

COPD develops gradually and becomes worse, usually the result of prolonged feeding of dusty hay or an allergic reaction to hay dust, pollens, molds, or bacteria. A horse in a barn is more susceptible to heaves than one kept outside; there's more dust in the barn air from hay and bedding. Even an outdoor horse may get heaves if fed dusty hay because he constantly breathes dust particles as he eats.

Once a horse is sensitive to dust or molds, he relapses every time he breathes dusty air. Even if you keep him outside, he'll cough if you bring him in overnight or to saddle him. If his problem is related to dust, he must be kept in a dust-free environment.

Symptoms

This chronic respiratory disorder is characterized by loss of ability to perform (exercise intolerance), constant or intermittent cough, and watery discharge from the nostrils. As the problem progresses, the horse loses weight and has labored breathing. Because the condition is not due to infection, it doesn't respond to antibiotics. Similar to asthma in humans beings, breathing difficulty results from narrowed air passages and an increased effort for every breath.

How the Horse Breathes

When a healthy horse inhales, his chest expands and draws air in with a vacuum effect, enlarging flexible airways. Inhaling is an active process and exhaling is passive; collapsing of the chest wall pushes air back out. In horses with COPD, air passages are constricted and the horse must make an extra effort to force the air back out. This produces an audible wheeze.

There's a wheezing sound as the horse forces air out with two movements of the abdominal wall. Simple relaxation of the rib cage does not empty the lungs. The horse has to immediately follow it up with more effort, contracting his abdominal muscles and tensing his chest. In fact, many horses that suffer from COPD develop an enlarged ridge of muscle ("heave line") along the lower side of the abdomen from overworking these muscles while forcing air out of the lungs.

Listen to the windpipe to check for wheezing sounds indicative of restricted airways.

Treatment and Prevention

Horses with COPD can be treated with drugs that open the air passages as well as with antihistamines. Your veterinarian can prescribe medication. The best treatment is to keep the horse in a dust-free place and never feed dusty hay. If he coughs when fed hay, use pellets; non-dusty pellets are available for horses with heaves.

You can prevent heaves if you avoid dusty hay and bedding and keep horses outdoors as much as possible (unless the problem is due to pollens outside). Horses with allergies are especially susceptible to COPD. If a horse tends to cough when hay is dusty, shake it thoroughly to get rid of the dust (but don't shake it in the barn, as it will make the air dusty). Then sprinkle it with water to settle any dust that is left. For a horse with a serious problem, dunk each flake of hay in water (then drain it) before feeding so it is absolutely dust free. Bedding material should be as dust free as possible. Sometimes wood shavings are less dusty than straw.

Health Tip

Alfalfa is usually dustier than good grass hay, mainly because the leaves tend to shatter when dry. Horses that cough when fed hay usually do best with grass hay; alfalfa hay may make their allergies much worse.

Cushing's Disease

Cushing's disease, or Cushing's syndrome, sometimes affects older horses in their late 20s or into their 30s. Young horses are rarely affected. This problem is due to excessive output of adrenal hormones caused by overactivity of the adrenal glands (found in front of each kidney), often in response to a pituitary tumor. The adrenal glands secrete several hormones known as corticosteroids that affect the body's use of electrolytes and glucose. They also inhibit the immune system, a condition that slows the healing of wounds and results in lowered resistance to infection and disease.

Symptoms

The affected horse has an excess of hormones that raise the level of glucose in the bloodstream. This, along with the effect of the hormone on the kidneys, results in increased urine output, so most of these horses drink a lot and urinate frequently. They may also have a shaggy coat that doesn't shed, suffer from weight loss, experience chronic infection, or develop laminitis. They may become swaybacked and potbellied due to loss of muscle tone.

Treatment

There is no cure for Cushing's disease because the tumor in the pituitary cannot be halted. Yet, if the condition is diagnosed early the horse can be treated with drugs to alleviate some symptoms. The condition is eventually fatal.

Hepatitis

A healthy liver breaks down and filters out poisons, but occasionally in the process of protecting the body it becomes damaged or infected, resulting in *hepatitis,* or inflammation of the liver. Both infectious and non-infectious, the disease can be caused by a virus, chemical or bacterial toxins, or poisonous plants. Primary liver disease is generally the result of poisoning. Secondary disease of the liver occurs as part of a generalized disease process in the body.

Non-Infectious Hepatitis

Toxic (non-infectious) hepatitis, usually caused by a poison in the body, can be acute or chronic. Acute hepatitis comes on suddenly; chances of survival are poor because the liver damage is so severe.

Symptoms

Chronic hepatitis occurs from toxins or poisons in smaller doses; it comes on gradually and may not be noticed until symptoms worsen. This can happen if the horse is eating poisonous plants or contaminated feed over time or has a bacterial infection that creates toxins, which in turn damage the liver. Early signs are dullness and lack of appetite. Pulse, temperature, and respiration are usually normal, but the horse may have abdominal pain that may be mistaken for colic. Another symptom may be jaundice (yellow mucous membranes).

Toxic substances are processed by the liver and excreted, but in a damaged liver they build up in the bloodstream and affect the nervous system. Ultimately the affected horse may stagger or drag his feet. His mental condition deteriorates. He may stand with feet wide apart and head drooping, or he may have muscle tremors. He may become so oblivious to his surroundings that he walks into trees or fences.

In some cases the horse becomes violent and unmanageable, a danger to himself and anyone trying to handle him. He may have diarrhea, colic, or founder; he may also suffer from photosensitization (i.e., skin inflammation and sloughing in unpigmented areas) because his liver cannot filter out photosensitizing agents from plant material he has eaten.

Infectious Hepatitis

The disease (serum hepatitis, also called Theiler's disease) causes massive liver destruction. The virus is so hardy that chemical disinfectants cannot kill it. Because it can be spread from horse to horse by use of contaminated needles, you should always use sterile disposable needles and syringes (one for each horse). Discard each needle and syringe after one use.

Symptoms

Onset of symptoms is sudden, and the horse is almost always violent. He becomes mentally deranged and can't be caught; he runs wildly, crashing into fences, walls, or other obstacles, and soon dies. Hepatitis is easily misdiagnosed as some type of neurologic disease.

Serum hepatitis and rapid death of liver tissue are often the result of vaccination, occurring 30 to 90 days after the injection. A horse injected with vaccine made from equine tissue or derived from horses may occasionally be at risk from serum hepatitis. Cases have occurred up to 6 months after injections of encephalomyelitis serum, tetanus antitoxin, pregnant mare serum, and anthrax antiserum.

About 90 percent of all serum hepatitis cases are fatal, with the horse dying in 12 to 48 hours. Several antitoxins are derived from horse serum (a possible source of the virus); the most widely used is tetanus antitoxin, often given for immediate protection against tetanus following surgery, foaling, or a wound.

Treatment

Treatment with antibiotics, glucose, electrolyte solutions intravenously, and B-complex vitamins may be helpful in some cases of serum hepatitis.

Toxoid vs. Antitoxin

Antitoxin is not the same as toxoid. *Toxoid* is a modified substance produced in a laboratory, not made with horse serum. There's no danger of hepatitis with toxoid injection. *Antitoxin,* by contrast, is derived from blood serum and contains antibodies to counteract tetanus toxins. Though produced under rigid safety standards — testing donor horses regularly for hepatitis — antitoxin always carries the remote chance for error; it's possible for contaminated serum to reach veterinary shelves. There is no need to ever use tetanus antitoxin if you keep a horse's toxoid vaccinations up-to-date.

Prevention

To prevent hepatitis, always use sterile syringes and needles. Keep tetanus vaccinations current so you never need to use antitoxin. Be careful with chemicals that might contaminate feed or water. Keep horses well fed so they won't be tempted to eat poisonous plants. Check for poisonous weeds in pastures, feed, and hay. (See chapter 11 for information on poisonous plants.) Keep horses healthy with good care and a conscientious vaccination program against possible diseases. Most cases of hepatitis can be prevented with good management and forethought.

Laminitis (Founder)

Founder is a term for changes that take place in the feet as a result of *laminitis —* inflammation of the attachments of the hoof wall. Bones and tendons of the leg terminate inside the hoof in an area that serves as a cushion and attachment between the inner tissues and the outer horny shell. When a horse suffers from a disease or metabolic problem that upsets his body chemistry, many changes take place that sometimes affect his feet.

A condition affecting blood circulation may not cause permanent damage in other parts of the body, but in the feet — which are encased in solid walls — there can be terrible pain and damage to the sensitive laminae. Pain makes it hard for the horse to walk; in severe cases he reacts by lying down or rolling

coffin bone   sole

Normal foot with healthy horn and proper angle.

around. Lameness usually appears first in the front feet because they carry the most weight. Heat and pain occur in the hoof and around the coronary band. In acute cases of founder there may be pain in all four feet, causing the horse great distress. He shivers, sweats, breathes rapidly and shallowly, and has a rapid pulse. He stands with all four feet bunched up under his body, back arched upward and head hanging down. He doesn't want to move. If he does, his gait is stumbling and shuffling because of the pain. It is hard for him to get up and down, and he may lie flat on his side for long periods, sometimes hours at a time.

Several things can cause laminitis, including grain overload, lush pasture, concussion from fast speed on hard surfaces, increased weight on the feet, drastic changes such as drinking cold water when a horse is exhausted and hot and sweaty, infections, hypothyroidism, and complications from failure to shed the placenta after foaling.

Endotoxemia and Founder

Bacterial toxins in the bloodstream (which may occur with uterine infection from failure to shed the placenta after foaling, or when a horse overeats on grain) are common causes of laminitis. When a horse eats too much grain, excessive carbohydrates stimulate rapid multiplication of intestinal bacteria. The microbes that ferment the lactic acid proliferate, using up so much available oxygen that anaerobic bacteria (pathogens) begin to multiply more than normal. As they die their endotoxins (poisonous substances in the bacteria) are released and seep through the intestinal wall into the bloodstream.

The Hoofs Suffer First and Last

Once in the bloodstream, endotoxins damage the smallest blood vessels. Tiny arteries clamp shut to conserve blood pressure, but this deprives the tissues of nutrients and oxygen. The terminal ends of the hoof's arteries are especially vulnerable — the first victims of dying local tissue. As the horse's blood clotting system malfunctions, small clots form in many small vessels. As blood supply to the hoof is impaired, oxygen starvation causes excruciating pain followed by congestion and swelling. The laminae begin to die.

Even after the original cause of the condition has been corrected, the horse's feet may become deformed if the laminae have become separated from the hoof wall and allowed the sole to drop. The hoof wall spreads and develops rings and ridges, and the slope of the front of the hoof becomes concave. Sometimes the separation of hoof wall is complete and the horny shell comes off. In this case the horse may need special shoeing for the rest of his life.

hoof wall
with rings
and ridges

coffin bone

sole

A foundered foot with rings and ridges and a dropped coffin bone. When a horse founders the sensitive laminae, which bind the hoof wall to the coffin bone, lose their ability to hold the structures in place. With its anchor disintegrating the coffin bone sinks downward, rotating toward — or even through — the sole of the foot.

Treatment and Prevention

If you suspect founder, call your veterinarian immediately. Prompt treatment can help relieve the condition before the feet suffer permanent damage. The veterinarian will treat the primary condition (e.g., grain overload, uterine infection) and give medications to alleviate the circulatory problem and reduce pain and swelling in the feet.

Using and caring for a horse wisely is the best insurance against founder. Always feed a well-balanced ration and introduce new feeds gradually over several days; never overfeed or allow a horse to overeat on grain or lush pasture. Condition horses to their work gradually, never work them to the point of exhaustion, and always cool a horse out properly after a workout — gradually. Don't allow him to drink *cold* water until he is cooled out, and don't put him away hot. Keeping a horse fit and healthy, and giving prompt attention to any problems such as illness or retained placenta, will help prevent founder.

Navicular Disease

Navicular disease is a common cause of lameness in domestic horses but rarely affects the free-roaming equine. The practices of breeding certain horses with feet too small for body weight, using horses in inappropriate athletic activities, and keeping horses confined in unnatural conditions all lead to navicular disease. This is primarily a problem of front feet, because they are subjected to more weight and concussion than hind feet.

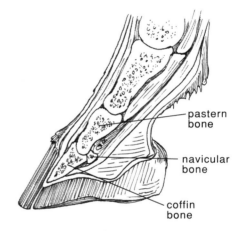

The navicular bone lies behind the coffin bone at the joint between the coffin bone and pastern bone.

The small navicular bone lies deep inside the foot, above and behind the coffin bone, and serves as a pulley for the deep flexor tendon that glides along its underside and as a bearing surface and shock absorber. It stabilizes the union of the coffin bone and short pastern bone, which are of dissimilar shape.

Anything that increases concussion to the foot, such as upright shoulders and pasterns, puts extra stress on the navicular bone. Jumping adds even more stress that squeezes the bone.

The first step to diseased bone is poor circulation caused by clotting of tiny capillaries in the bone itself. Reduced blood flow weakens the structure; the bone begins to deteriorate and die. As it loses health and strength, it can't bear weight properly and the horse suffers pain and lameness. Poor circulation may result from increased pressure (concussion, jumping, or working immature horses too hard) or from inactivity such as confinement in a stall. Standing in a stall reduces circulation in the feet and puts constant pressure on joint cartilages that are dependent on motion for blood circulation.

Young Horses and Navicular Disease

Confinement and lack of exercise in a young growing horse, especially if he stands around on hard surfaces, can lead to navicular disease. Keeping youngsters in stalls and overfeeding them (causing them to become overweight and inactive) are also detrimental.

Overuse of young horses subjects their immature feet to excessive concussion. Navicular disease usually starts between ages 3 and 6, although some horses don't show symptoms until later.

Long toes and unbalanced feet also put more stress on the navicular bone, as does excessive body weight. A horse may have good foot and leg conformation and still have stress on that bone if he is kept in a stall with sloping sides, putting feet off level; the forces on the navicular bone as he stands on an up-slope are the same as if he had long toes and low heels. Navicular disease is often seen in feet that are too low at the heel, putting an extra squeeze on the bone between the deep flexor tendon and the lower end of the short pastern bone when the horse is standing. Pointing the foot forward relieves the pressure temporarily; a horse with a sore navicular bone often puts that foot forward.

Symptoms

Mild tenderfootedness is noticeable when the horse comes out of his stall. During early stages the lameness diminishes with mild exercise, then gets worse if exercise is followed by confinement. Symptoms are usually not evident until the bone has already been seriously damaged. Lameness may appear in both front feet, but one may be worse than the other. Often the first sign of lameness is obvious when the horse is turning or walking on gravel or rocks. He takes shorter, lighter steps, trying to keep the weight off his front feet. He gallops with a short, choppy stride.

Normal foot (A) compared with a contracted hoof (B) due to navicular disease.

During early stages the lameness may come and go, but eventually it returns and gets worse. The horse stumbles a lot as he tries to put the toe down first and not take weight on the heel. Because of his shortened stride you might think the lameness is in the shoulder instead of the foot. The horse may point one front foot forward when standing still to relieve the pressure. He may also shift his weight from one leg to the other. If the foot is unshod it may be boxy, with high contracted heels, a worn stubby toe, and a small frog that is well up off the ground. If shod, the toe of the shoe will be excessively worn.

Treatment and Prevention

The best way to help a horse with navicular disease is to make management changes to halt the deterioration of bone. Nothing will return the damaged bone to its original shape and smooth surface, but changes in stabling and shoeing — and new medical treatments — can help him. The horse whose circulation is restored will be more comfortable and more sound. Treatment varies with the severity of the condition. Some medications can reduce blood clotting, and arthritis medication can help halt joint degeneration if the disease involves the joint.

Health Tip for Inactive Horses

Inactive horses have greatest damage to circulation in the feet. If a horse must be confined, give him a roomy box stall with a mound in the floor instead of a depression in the center. If he stands for hours on a floor that slopes downward to the center of the stall, his feet are in an unhealthy toes-up, heels-down position even if they are shod in perfect balance. The mound in the floor puts his feet at a better, less damaging angle.

Medication can prevent blood clotting or dilate blood vessels so more blood reaches the oxygen-starved bone, and surgery is an option to block nerves so the horse feels no pain. None of these treatments is without risk or possible complication. Your veterinarian should assess the horse's condition and advise you on the best course of action. It is always better to prevent or halt navicular disease before it progresses to the point at which surgery or drug treatments are needed.

Hoof care is important in preventing and controlling navicular disease. If shoes are left on too long, even a well-balanced foot will grow a long toe. For a horse with a serious problem, your farrier can use special shoes to give relief.

Rhabdomyelitis (Muscles Tying Up)

This problem usually affects a horse that has overexerted while being ridden but can occur in any out-of-condition horse that suddenly exerts strenuously, such as running and bucking in mud or snow. The condition has many names, including azoturia, Monday morning disease, tying up, being corded, or being set fast. It involves painful cramping of the large muscles in the rump and, sometimes, the thigh and shoulders. Symptoms range from minor discomfort to collapse and death. The muscle cramping generally occurs during or after physical exercise such as a fast gallop or a long ride, but it may also develop after stressful situations such as a fight with a farrier, frantic running during a thunderstorm, or a long trailer trip. Any muscle exertion beyond the horse's accustomed activity can trigger tying up.

There are two forms of tying up. Type A (Monday morning disease) occurs early in a workout or soon after exercise begins. Type B occurs after many hours of steady work and muscle fatigue. The causes and treatment are different for each type.

Type A (Monday Morning Disease)

This is most common in heavily muscled horses due to the build-up of lactic acid and waste products in muscle tissue working strenuously. Because the circulatory system cannot clear wastes away fast enough they damage the muscle cells, causing pain and cramps. When draft horses were commonly used, this was called Monday morning disease because it generally happened after a weekend — the draft horses were grained heavily and worked hard all week, allowed to rest on Sunday, and put back to work on Monday. Today's horses have similar problems. They are often fed a liberal grain ration but only exercised a few days each week. In most cases of Type A tying up, the horse has been idle for 2 or more days before his exertion and was fed a full grain ration on the days he was idle.

Symptoms

Symptoms develop 15 minutes to an hour after exercise begins. The attack is quite sudden. The horse comes to a stiff halt, begins to sweat, is reluctant to move, and may want to lie down. If you run your hand over the rump muscles, you can tell they're tight, stiff, and sore — similar to sudden thigh or calf muscle cramping in your own leg when you sprint or do something you're not used to.

It may look like the horse has a urinary problem because of his stretched-out stance. If he can walk at all, his hind legs are stiff and dragging. The longer the cramp lasts, the more intense the pain because the muscle is deprived of oxygen. The spasm squeezes the capillaries, hindering blood flow. There may be a peculiar odor to the horse's breath, urine, and sweat from wastes excreted through the lungs, kidneys, and sweat glands; his urine will be darker than normal.

In mild cases the symptoms disappear within a few hours if the horse is given immediate and complete rest and not moved. In moderate cases the horse is anxious, trembling, very stiff, and reluctant to move; these symptoms may last for 24 to 48 hours. If pain is extreme the horse will go down.

Is It Tying Up or Colic?

Tying up is often mistaken for colic because the horse may paw and sweat, but if you try to walk him, as you might for colic, you will make his condition worse.

Treatment

Prompt treatment usually relieves the muscle cramping. While waiting for the veterinarian, blanket the horse to keep him warm and relaxed, avoiding movement. Care for him where he is, or take him home in a trailer and keep him on his feet. The veterinarian will treat him with tranquilizers, pain relievers, and muscle relaxants to increase blood flow and relieve spasms in blood vessels. Once the spasms cease, the horse can relax enough to urinate and flush out toxins that have accumulated in his bladder. The veterinarian may also give vitamin E and selenium to help the muscles return to normal quickly. If muscle cramps are relieved in an hour or less, chances are good the horse will recover swiftly.

Prevention

To avoid tying up, don't feed grain on days the horse is not exercised. On days you ride, feed after the workout — after the horse is fully cooled out — to avoid glycogen overload in muscles. (The excess glycogen, an energy source, contributes to lactic acid formation when the horse is worked hard.)

Type B

This usually occurs at the end of a long ride or after a brief rest, usually in horses not conditioned for that much work and can also occur if cold water is put on the large muscles of their shoulders and hindquarters or if they get chilled by cold wind while hot and sweaty. Dehydration from long, hard work, especially in hot weather, causes inadequate circulation to the exhausted muscles.

Symptoms

Lack of oxygen and nutrients, and an abundance of trapped waste products due to poor circulation cause the same symptoms as Type A tying up. However, treatment is not the same.

Treatment

Treatment is aimed at restoring fluids and electrolytes, replenishing energy stores, and clearing out wastes by improving circulation. Slow walking increases blood flow to the muscles. The horse can have green grass and hay, but no grain. In mild cases the metabolic imbalance will be corrected if he can drink and eat. Alfalfa hay may help, as calcium depletion may be a factor. In more severe cases the horse needs electrolytes and intravenous fluids. Never use tranquilizers because they lower the horse's blood pressure and worsen the condition. The veterinarian should check blood samples to ascertain the degree of damage, as a horse's symptoms often do not indicate the extent of damage to muscle tissue.

Check for Brown Urine

For several hours after any tying up episode, check the color of the horse's urine. Brown urine indicates that muscle breakdown products are being excreted and is a sign of muscle damage. In severe cases, recovery may take weeks or months. The horse should not be used until fully recovered.

Prevention

Type B tying up can be prevented by not overworking a horse beyond his level of fitness and by avoiding sudden chilling of large muscles when the horse is extremely hot and sweaty.

Other Diseases

There are other diseases of horses that you might sometimes experience; to cover them all is beyond the scope of this book. Be alert to any unusual symptoms, and do not hesitate to contact your veterinarian if you think your horse has a problem. Some diseases occur only rarely, but the veterinarian may be able to help diagnose them and assist you with treatment.

TEN

PARASITES

All animals harbor parasites, which are organisms that live on or within another animal. The small creatures feed on the host, sucking blood or absorbing nutrients. The responsible horseman tries to control these pests to keep them at a low level so they cannot seriously damage the horse or interfere with his nutrition.

Internal Parasites

Most parasites are host specific; for example, horse parasites survive only in horses. Although no horse is free of internal parasites, over time horses have adapted to living with them; a young horse builds immunity to some worms after repeated infestations.

Most parasites need a host animal to complete their life cycle. An intestinal worm, for instance, lays eggs that pass out of the body with the horse's manure. The eggs then hatch into larvae that migrate to grass and are eaten by a horse with his feed. Once inside the horse, the larvae migrate through body tissues and feed on the host, becoming adults in the intestine, only to lay eggs and begin the cycle again.

Immunity to Worms

The immune system not only fights bacterial and viral infections but also responds against internal parasites because the invading worms release antigens in the bloodstream and body tissues. The antigens trigger the horse to produce immunity against the invader. Later, when more worms appear, the horse's immune system recognizes them and sets up a counterattack — reducing the females' egg-laying capacity and the males' fertility.

Young horses can build immunity to ascarids (roundworms) to keep larvae from maturing. Adult horses rarely carry large infestations of ascarids; these damaging worms are primarily a problem of young horses that haven't yet developed immunity. Bloodworms are a different story; horses do not build an effective immunity against them.

Parasites Can Be Deadly

Internal parasites are a common cause of death in horses that are not regularly dewormed and can contribute to health problems (e.g., many colic cases are related to bloodworm infestation). Loss of condition from a heavy parasite load and a lifetime of carrying worms is a major factor in other causes of death; the horse lacks reserves and disease resistance because of the constant drain on his body from parasites. Foals are especially susceptible to the effects of parasites.

How Horses Become Infested

Horses in confined areas are continually exposed to their own droppings, grazing the same places over and over. They become heavily infested unless manure is picked up daily or a regular deworming program keeps the egg-laying adults to a minimum.

Parasites thrive in a warm, moist environment. In springtime and early summer heavy infestations of bloodworms are likely to occur. The larvae hatch and swarm over pasture plants, waiting to be eaten. Sunlight, heat, and severe cold tend to kill larvae, but the eggs in manure are resistant to extremes in temperature. Any pastures that have been previously grazed by horses are a continual source of infestation, but proper pasture management and a good parasite control program can reduce the problem. For example, harrowing breaks up the manure piles and exposes the worm eggs, which are susceptible to sunlight. In addition, alternating cattle and horses on a pasture can break the parasite cycle. The horse worm larvae eaten by cattle will not complete their life cycle, and similarly the cattle parasites will not survive in the horse.

Horses grazing irrigated or wet pastures where manure is not picked up are highly likely to be reinfested with worms. The most worm-free environments are dry corrals or paddocks where manure is picked up regularly and stalls that are kept very clean.

Control of Worms

Control is aimed at preventing contamination of the horse's environment with worm eggs and larvae by means of pasture management, manure removal, proper feeding and watering, and deworming of horses at appropriate intervals to prevent reproduction of parasites. Another object of control, gained by using the newer dewormers such as ivermectin and moxidectin (Quest), is to prevent damage to the individual horse (e.g, by killing bot larvae before they get to the stomach or bloodworm larvae before they damage the arteries). Before the introduction of ivermectin, worms were vulnerable to drugs only during their final stage in the digestive tract. This prevented further contamination of paddock and pasture by reducing the egg numbers in manure, but damage to the horse had already been done by the larvae during their migrations through the body and arteries.

Ivermectin

Since its introduction in the early 1980s, this drug (available in several different brands) has become widely used because of its effectiveness against all internal parasites — except tapeworms — in larval as well as adult stages. It also kills skin parasites that suck blood such as ticks, lice, and mites. Unlike other drugs that poison parasites, ivermectin works by paralyzing the ones with gamma-aminobutyric acid (GABA) in their nervous systems. Ivermectin stimulates release of this acid and the parasite is unable to move or eat, so it dies. The drug has no effect on tapeworms because they have no GABA. Ivermectin is not harmful to horses or human beings.

There are many good dewormers that can be used in a control program. Knowing the parasites' life cycles and targeting them when most vulnerable will keep them under control. The most crucial time to diligently control roundworms and bloodworms is spring and early summer, when larvae are most active in warm, moist weather. Deworming horses just prior to the parasites' "egg-laying spree" and the spring rise of larvae onto forage plants (and then once or twice more at monthly intervals) will halt a worm population explosion. You can deworm every 2 to 3 months thereafter.

Bots

Most internal parasites of horses are worms, but the horse is also subject to infestations of fly larvae (grubs), especially the botfly, *Gasterophilus*. The larvae of this fly irritate the stomach and occasionally perforate the stomach wall. The adult fly is brown and hairy, the size of a small bee.

The botfly lays eggs on the horse's hairs, one to a hair, where they can be seen as yellow specks. One type of botfly lays its eggs on the front legs and flanks; the eggs hatch into tiny larvae that are transported to the mouth when the horse licks or nuzzles his legs. Other types of botfly lay eggs around the jaw or cheek, and these migrate directly into the mouth after hatching.

Once inside the mouth, the bot larvae burrow into the gums. After about a month they emerge and are swallowed down to the stomach, where they attach, feed, and grow for 8 to 12 months. They pass out in the horse's manure in spring and early summer, migrate into the ground, and pupate, emerging as adult flies after 3 to 5 weeks to begin the cycle again.

Bots in large numbers cause *unthriftiness* (thin body condition and poor hair coat), mild colic, and lack of appetite. Adult flies irritate horses; when attacked by egg-laying flies, they run frantically or strike out with a front foot. Horses can be dangerous when botflies are pestering them.

*Life cycle of the botfly: Flies lay eggs (**A**). Larvae migrate to mouth (**B**), pass to stomach (**C**), pass out with droppings (**D**), migrate into the ground to pupate (**E**), and become flies (**F**).*

Control

The crucial time for control is when botflies are most active — a 2- to 3-month late-summer season in northern climates or a year-round season in southern climates. Give a boticide 1 month after the first botflies appear and every 2 months thereafter, as long as bot season lasts. Do a follow-up treatment late in the season after hard frosts have killed all flies, to make sure that all the larvae within the horse are killed before winter.

Eliminating Bots

A good boticide program, combined with diligent egg removal from the hair coat, will eliminate a bot problem.

Roundworms (Ascarids)

The ascarid, *Parascaris equorum*, is the largest intestinal parasite of horses. The white worms resemble an earthworm in shape and can grow up to 12 inches (30.5 cm) in length. Adult worms live in the intestine, where a single female may lay more than a quarter of a million eggs per day. The eggs pass out with manure and after 2 or 3 weeks become infective if eaten by the horse, with the larvae migrating through the body tissues. They can be damaging to foals and young horses. Most pastures that have been previously used by horses, especially young ones, are heavily infested with roundworm eggs, which remain viable for several years under the right conditions — mild weather and lack of pasture management such as harrowing.

Ascarids are most common in foals. By the time a horse is 2 years old his immune system can hinder development of larvae. The ascarid robs the foal of nutrients and may cause mechanical blockage of the small intestine — and subsequent colic or death — if worms become too numerous. The most common and extensive damage, however, occurs during larval migration through the lungs. Ascarids are the main cause of lung damage and cough in foals under 6 months of age.

*Life cycle of the ascarid: Eggs are passed in manure (**A**); horse eats eggs with grass or hay (**B**); eggs hatch, penetrate intestine, and larvae travel through bloodstream and liver (**C**); larvae enter lungs (**D**); larvae are coughed up and reswallowed (**E**); larvae return to digestive tract to mature and lay eggs (**F**).*

After ascarid worm eggs are eaten, newly hatched larvae penetrate the intestinal wall and enter the bloodstream. Within 48 hours they reach the liver, where they remain for a week; then they travel via the blood to the lung, where they pass through small blood vessels, through tiny air sacs, and into the bronchial tubes. Respiratory problems develop 3 to 4 weeks after the foal eats ascarid eggs; symptoms include nasal discharge and, sometimes, a cough. The lung damage in a heavily infested foal can be permanent.

After spending time in the lungs and bronchial tubes, the larvae are coughed up into the mouth and are swallowed, returning to the digestive tract. There they continue to mature and become egg-laying adults. The interval between ingestion of an egg and development of an egg-laying adult is 2 to 2.5 months. Thus fecal examinations for parasite eggs are usually negative until the foal is almost 3 months old.

Control

Don't wait long before deworming the foal; ascarids have already damaged his lungs and other body tissues by the time they return to the intestine. Deworm the mare with ivermectin right after foaling, and make sure her udder, flanks, and belly are free of manure. Pyrantel tartrate (Strongid C) given daily in grain is effective against ascarids; the foal can be started on it as soon as he eats grain, and mares can be started on it a few weeks before foaling.

Worm Foals Routinely

Foals should be routinely wormed for ascarids even if they don't show symptoms. By the time symptoms appear, significant damage has occurred. A heavy infestation may stunt the foal's growth and make him rough-coated and potbellied; irritation from the worms may also give him diarrhea or colic. An extensive infestation can be fatal, as can deworming when the foal has a heavy load of worms — the bulk of dead and dying worms causes intestinal impaction and blockage. When deworming a foal with a heavy load of ascarids, it's safest to use a drug with a slower "kill rate," such as ivermectin, to prevent sudden blockage.

Bloodworms (Large Strongyles)

Large strongyles, called "bloodworms," "red worms," or "palisade worms," are active blood suckers that cause debility and anemia. Three species affect horses, varying in length from less than one-half inch to about 2 inches (5.1 cm). The most serious problems are caused by the smallest, *Strongylus vulgaris*. Some of these lodge in the walls of the blood vessels that supply the intestine and in the large arteries leading away from the heart.

Life cycle of the bloodworm: Eggs are passed in manure (**A**); larvae migrate onto forage plants (**B**); when eaten, larvae penetrate the intestine, enter the bloodstream, and live in major arteries (**C**); larvae travel to the liver (**D**); larvae return to intestine and lay eggs (**E**).

Eggs are passed in the feces of the host horse and hatch to produce larvae that are resistant to drying and low temperatures, living about 3 months or even a year or more. In warm, wet weather they migrate onto forage plants. When swallowed by a horse, they penetrate the intestine and begin extensive migrations through the body before coming back to the intestine to attach and lay eggs.

Damage to Blood Vessels

Too many bloodworms in an artery may block it or cause it to rupture. They cause inflammation, scarring, and roughening of the inside wall, as well as weak spots. This damage can lead to clots or aneurysms. If a clot breaks loose (becoming an embolus) and travels down the vessel, it eventually comes to a place too narrow to pass through and promptly blocks it, shutting off blood flow to the tissues supplied by that artery. Aneurysms (dilated sacs) bulge out from the vessel wall in a weak spot. If an aneurysm gets large or the blood vessel is weak, it may suddenly rupture, causing sudden death of the horse from internal hemorrhage.

Although any artery can be damaged by bloodworm larvae, *S. vulgaris* most often damages the mesenteric arteries that supply the intestine. When this happens a section of intestine suffers decreased blood supply, resulting in pain and colic. Severe blockage leads to death of a section of intestine. Another common site for blockage is where the aorta (i.e., the large artery leaving the heart) branches into arteries supplying the hind legs. If blockage occurs here the horse becomes lame in one or both legs and the affected limbs feel cold. If blockage is permanent the leg tissue dies from lack of blood, killing the horse.

Symptoms of Bloodworm Infestation

Mature bloodworms in the intestine (fastened to the wall) may cause diarrhea, anemia, weakness, and unthriftiness. Other symptoms of infestation are unexplained bouts of colic and lameness in the hind legs. Bloodworms deprive the horse of nutrients and blood, causing young individuals to grow slowly and working individuals to be inefficient.

Control

Bloodworms can be readily controlled by periodic deworming with ivermectin, typically every 2 to 3 months, the frequency depending on rate of reinfestation. This dewormer kills the immature forms wherever they are in the body, as well as the egg-laying adults in the intestine.

Small Strongyles

These worms, which cause debility and anemia, are blood suckers about ¼ to 1 inch (6 mm–2.5 cm) in length. They live in the intestine (mainly the cecum), ulcerating the mucous lining. Heavy infestations may leave little normal area in the intestinal lining. Ulcers made by these small worms are deeper than those made by large strongyles and occasionally cause fatal hemorrhage. All horses are infested to some degree with small strongyles.

The seasonal mass emergence of small strongyle larvae from the intestinal wall causes the most problems. Affected horses show a subtle decrease in performance, are slow to shed, and have rapid weight loss, diarrhea, mild colic, or severe inflammation of the intestine (i.e., enteritis), which can be fatal.

Control

The presence of immature worms embedded in the gut wall is hard to detect because they're not yet laying eggs (which show in fecal counts) and are beyond the reach of many dewormer drugs. Once the larvae emerge from the gut wall, however, a good dewormer takes care of them. Seasonal emergence from the intestinal wall takes place in late winter and early spring in northern climates, and in fall and winter in the South. For best control, time your deworming program to coincide with certain stages of the life cycle, hitting them when most vulnerable. Most veterinarians recommend targeting small strongyles diligently from April to August if you live in the northern part of the country where winters are cold.

Ivermectin is effective against small strongyles if given in three treatments at 8-week intervals. This kills large strongyles (bloodworms) at the same time, as well as bot larvae that might have been missed in the previous fall deworming. The recently introduced dewormer moxidectin is effective against encysted small strongyles in the intestinal wall.

Pyrantel pamoate, oxibendazole, and dichlorvos are effective against small strongyle larvae after they emerge from the gut lining but work best if given more frequently than ivermectin. Most veterinarians recommend five treatments with these drugs, given every 4 weeks. For horses in the southern part of the United States it may be necessary to maintain this deworming schedule every 4 to 8 weeks for a much longer period.

Many types of small strongyles have developed resistance to commonly used dewormers. Because worm populations are constantly adapting and changing, a parasite control program that worked a couple of decades ago may not be effective today. As a result the conscientious horseman must work with his or her well-informed veterinarian to change, adapt, and update the deworming strategy as needed.

Health Tip

Weanlings and yearlings need frequent and continual deworming because they are more vulnerable and adversely affected by parasites than are older horses; the youngsters have not yet developed much immune resistance to worms.

Pinworms

Pinworms are bothersome to horses but not as damaging as roundworms or strongyles. The parasite, *Oxyuris equi*, most commonly affects horses confined in small areas; it is not much problem in big pastures with lots of room. Pinworms inhabit the large intestine and cause intense irritation and subsequent disfigurement as the horse rubs his tail trying to relieve the itching. Many infested horses break off most of their tail hairs.

The irritation is caused by worms depositing eggs on the horse's anus and in the rectum. In some cases the parasites cause inflammation of the cecum and large intestine. Sometimes a secondary infection affects the raw sores created by constant rubbing. Irritation may cause weight loss because the horse spends more time itching than eating.

The mature pinworm is gray, white, or yellow. The female grows up to 6 inches long (15.2 cm) and has a threadlike tail. Mature females crawl out of the horse's large intestine and lay eggs by the anus in yellow clusters. The eggs hatch in about 3 days. The horse becomes reinfested by biting at areas on the body where eggs were laid or by eating feed contaminated with eggs and larvae from manure.

Control

A horse with pinworms can be treated with any good deworming drug. You should also thoroughly clean his stable area or pen because eggs may stick to stall walls, floors, and bedding.

Tapeworms

Horses in warm regions often become host to tapeworms, whose life cycle depends on orbatid mites, the intermediate host. Horses infested with tapeworms spread eggs in their manure. In areas where winters are mild and temperatures are rarely below freezing, the eggs hatch nearly all year round (because warmth and moisture promote hatching) and the host mites are active in pastures all year. The mites take in the tiny immature tapeworms, which undergo several changes inside the mites. Any horse that eats grass infested with the tapeworm-bearing mites becomes infested. The young worms continue development inside the horse.

Tapeworms cause problems if large numbers collect at the end of the small intestine, leading to colic due to partial blockage of the cecum. Tapeworms also rob the horse of nutrients, just like other internal parasites do.

Control

Tapeworms are resistant to many dewormers and are not affected by ivermectin, but pyrantel pamoate kills them if given at two to four times the dosage required for other worms. Another effective drug is a canine deworming product, praziquantel (Droncit). It can be given as instructed for dogs, with dosage adjusted to the horse's weight. It may take several treatments of praziquantel to completely rid the horse of these worms, even if he is kept off infested pastures.

Lungworms

Lungworms may be a problem if horses are pastured with donkeys or burros. Donkeys are natural hosts for this worm, *Dictyocaulus arnfieldi*, but can pass it to horses. Symptoms of lungworms in horses are coughing (especially during exercise), occasional difficulty in breathing, and lung irritation. Some heavily infested horses have no symptoms. Donkeys rarely show symptoms because they are rarely asked to exert enough to trigger coughing.

Control

If a horse develops a cough with no other symptoms *and* there is a donkey on your place, suspect lungworms. Deworm every equine, including the donkey, with ivermectin. If all equines are kept dewormed on a regular basis and ivermectin is used several times during the year, lungworms won't be a problem.

Warbles

Occasionally cattle grubs develop in horses instead of cattle, but they rarely do much damage other than causing irritation under a saddle. The late larval stages are sometimes found in the backs of horses, creating cysts or warbles. The grub comes out when mature and ready to become a fly. Fully developed larvae are about ⅓ inch (7 mm) thick and 1¼ inches (3.2 cm) long. They perforate the skin and fall to the ground to pupate before becoming flies. Adult flies are active during the heat of summer, laying eggs on the hairs of cattle (and sometimes horses). When the eggs hatch, the larvae burrow through the skin and migrate to the horse's back, coming to rest under the skin where they make a breathing hole to the outside and continue to grow.

Control

Warbles can be eliminated if nearby cattle are treated for grubs every fall, and if horses are routinely dewormed with ivermectin because this drug will kill the immature stages before they arrive at the horse's back.

The Battle Against Internal Parasites

A regular deworming schedule and diligent pasture management will keep a horse relatively free of internal parasites. Usually only 1 to 5 percent of the worm population is actually inside the horse; the majority of eggs and larvae are in the pasture. In small areas where horses are confined, cleaning up manure at least twice a week greatly reduces the chances of reinfestation.

Worm Resistance to Drugs

Some drugs that were traditionally effective against bloodworms and roundworms create worm resistance if used repeatedly, especially if given in doses that are not quite adequate to kill all adult worms. Because those that survive are the most resistant to the drug, their offspring are not particularly susceptible to it. The benzimidazole family of dewormers is a very safe and effective group of drugs, but it has the disadvantage of promoting worm resistance after being used repeatedly for several years. These drugs include cambendazole, oxibendazole, fenbendazole, oxfendazole, mebendazole, and thiabendazole. This family of drugs is not effective against bots.

On farms where benzimidazoles and febantel have been used, large and small strongyles may have developed resistance. Horses may carry a heavy load of parasites despite regular deworming. There can also be a problem if horses with resistant worms have been added to a herd. When a worm population becomes resistant to one benzimidazole drug, it also develops resistance to the other drugs in that class. If there's a chance that worm populations on your place are resistant to these dewormers, change to ivermectin, pyrantel pamoate, dichlorvos, or new dewormers containing moxidectin (which is related to ivermectin but gives longer-lasting protection).

Deworming Drugs

Some dewormers are used in purge doses on a periodic schedule to kill adult worms in the gut but do not kill the larval stages migrating through the tissues; after a time the worms become resistant to most of these drugs. Ivermectin seems to be an exception because it has not yet shown any resistance problems and affects not only mature worms in the gut but also the immature migratory larvae. It paralyzes rather than kills, but the worms eventually die because they cannot move or eat. Due to this mode of action, ivermectins are ineffective if you underdose the horse.

To prevent worm resistance, make sure drugs are given in adequate dosage. This is one reason it's better to give paste or tube application of purge doses rather than putting the dewormer in feed. Unless the horse eats the medication readily, he may get only a partial dose, which does more harm than good; surviving worms develop resistance to the drug. Most dewormers are safe for horses even in fairly large overdoses, so it's better to err on the side of overdose rather than to underdose when estimating dosage. The one exception to this rule is moxidectin; be careful *never* to overdose with this drug.

To combat drug resistance, rotate classes of drugs. If the benzimidazole drugs are used alternatively with an unrelated drug such as febantel, pyrantel pamoate, or dichlorvos, worm resistance won't become a problem. Dichlorvos is a good boticide.

Using Paste Dewormers

Many horses are fussy about eating purge doses of dewormers in feed. Pastes have made deworming much simpler but must be given correctly; otherwise the horse may spit out some or have a reaction to the dewormer. Always deposit the paste far back in the mouth, on the back of the tongue, so the horse can't spit it out.

A few horses are sensitive to the ingredients in dewormers. A local reaction in the mouth can occur in certain individuals if paste ends up under the tongue, between the tongue and cheek, or in some area other than the back of the tongue. Prolonged contact with mouth tissues sometimes causes problems, including excessive salivation (drooling) and swelling of the tongue and lips.

How to Give a Paste Dewormer

1. Stand beside the horse — not in front — and hold the halter so he can't toss his head or move around.

2. Put the syringe into the corner of the mouth, aimed far up toward the back of the tongue.

3. In one quick motion, insert the syringe far back into the mouth and depress the plunger, depositing the paste before the horse can toss his head. If he moves, don't depress the plunger; start over.

4. As soon as you've squirted in the paste, raise his head and hold it high until you're sure he has swallowed. You can tell when he swallows by the movement of his throat.

If a horse protests, tossing his head or trying to avoid the paste, train him to be more tolerant by giving him pleasant-tasting treats for a while until he looks forward to having a syringe put into his mouth. Use an empty, well-washed deworming syringe and give him applesauce or something he likes.

When it's time to deworm him, first give him the treat. Molasses and water works well for this, since it is liquid and won't stay in the mouth to interfere with the dewormer, which must be administered into an empty mouth so the horse can't spit it out. Then slip him the dewormer, and afterward reward him for his good behavior with another syringeful of the treat. With this type of program, very few problem horses continue to resist deworming.

After squirting in the dewormer, hold the horse's head up until he swallows so he cannot spit the dewormer out.

Daily Deworming

A deworming option is pyrantel tartrate given daily in the horse's grain to keep him continually dewormed. It is available in a pelleted alfalfa-molasses feed additive. The strategy of this deworming program includes first giving the horse a purge dose of ivermectin to clean out all major worms (larval stages and all), then keeping him on a daily dose of pyrantel tartrate to keep him free of the parasites. The drug works well for broodmares and young horses, but some athletic horses (primarily endurance horses) have trouble with it (e.g., occasional instances of colic or anemia). In these cases, something else should be used. Because pyrantel tartrate is not effective against bots, seasonal treatment with ivermectin or one of the other boticides is still necessary.

Plan a Good Deworming Program

Work with your veterinarian to plan a deworming program that will be effective for your horses in your climate and situation. Remember that young horses need more diligent protection than older ones. Warm climates present year-round problems with many parasites that are only a seasonal problem in colder climates. Your veterinarian can advise you on new drugs being developed. The battle against parasites is constant, but an understanding of their life cycles and the weapons provided by modern research can help you keep these pests from becoming a health hazard.

External Parasites

There are several parasites that feed on the horse externally, causing irritation, discomfort, loss of weight, and at times disease. Flies and mosquitoes are the most common during warmer months. Lice are a problem during colder months. Ticks are an occasional problem, in some locations, in spring and early summer. Also, some types of mites are more active in winter than in summer.

Flies

Flies cause annoyance and blood loss, and they spread disease. Their bites can result in skin allergies and hypersensitivity reactions. The annoyance may disrupt grazing as horses run frantically to get away from biting flies or seek shady areas in which to swish and stomp. An understanding of flies' life cycles and habits can help you control them. Flies go through four stages: egg, larva (maggot), pupa, and adult. The adult fly is the most troublesome stage, feeding relentlessly on the horse. The female fly must ingest blood before she can produce eggs, hence the biting.

Stable Fly

This fly is similar in size and appearance to the housefly. It has a long, slender proboscis that sticks up in front of its head. The stable fly rests on vertical surfaces such as fences, walls, trees, or structures near horses and goes to horses only long enough to get a blood meal. Usually feeding on the lower legs, flank, and belly — mainly in early morning or late evening — stable flies cause great annoyance. Their bites are quite

Stable fly

painful, making the horse not only stomp and kick to get rid of them, but also lick at the bite wounds. The bites often bleed freely. In large numbers these flies can cause considerable blood loss.

The stable fly lays eggs in rotting hay and straw and in horse manure. Several generations develop during summer. In areas with large numbers of flies there may be more than twenty-five flies per horse. This doesn't seem serious until you realize that the number observed on a horse at any given time may be only 2 to 3 percent of the total feeding on the horse that day. Large numbers cause anemia and loss of grazing time. Stable flies spread diseases, including anthrax and equine infectious anemia (EIA). They also act as an intermediate host for *Habronema microstoma*, a stomach worm that causes habronemiasis (summer sores). Stable flies are mechanical vectors for strangles and other diseases as well, physically carrying the germs from one horse to another.

Control. Remove rotting organic matter that the flies use as breeding sites. Clean out manure and soiled bedding daily from stalls, along with piles of grass clippings or wet hay around the barnyard. Keep flies from multiplying in compost piles by spraying with insecticide or larvicide or by covering with black plastic. Use repellents to keep the flies away from horses. Spraying barns, stables, and foliage where the flies rest also reduces their numbers.

Horse Flies and Deer Flies

These flies belong to the Tabanidae family, the most annoying group of biting flies that attack horses. The flies may be black, brown, yellow, or gray, varying in size from ⅜ inch (0.9 cm) to more than 1 inch (2.5 cm) in length.

Horse fly. When tabanids slash the skin, blood flows freely from the bite; the fly laps up the blood with a lobe at the tip of its proboscis. Only the female takes a blood meal. Males feed on vegetable sap or the juices of soft-bodied insects.

Tabanids are numerous in areas of permanently wet ground. Females lay eggs on plants growing near water or on objects projecting over water or marshes. The eggs hatch within 5 to 7 days. The larvae drop into the mud, where they burrow in and feed on organic matter or the juices of other insect larvae or earthworms. They remain in the larval stage during summer, fall, and winter and then migrate in spring to drier areas to pupate for 2 to 3 weeks, after which the adult flies emerge. Soon after emergence the female fly needs a blood meal. Some horses have local reactions to the bites, developing lumps in the skin. Large numbers of flies may leave the hair crusted with blood and the bites dripping blood. Horse flies drink more than their own weight in blood at each feeding.

Tabanids Carry Disease

Tabanids are often mechanical vectors, carrying blood-borne disease from one horse to another due to their habit of feeding on one horse and then immediately attacking another. Equine infectious anemia is often spread this way, and sometimes encephalomyelitis.

Control. Control of horse flies and deer flies is difficult unless swampy areas are drained or horses kept away from them when flies are most active. Some repellents are helpful. Sprays or wipe-ons may protect horses for 1 to 2 days. A shed (situated as far as possible from marshes) that horses can go into during the warmer part of the day when flies are active can help, as most tabanids prefer to be in the sunlight. These flies are most active on hot, sunny days and less active in cool or overcast weather.

Horn Flies

The horn fly is a cattle parasite that can also be a nuisance to horses living near cattle. These blood suckers practically live on the host animal, taking twenty to thirty blood meals per day. Constant irritation from the bites may cause skin problems such as abdominal midline dermatitis.

Horn fly

Horn fly bites often leave the skin raw and crusted. These flies generally do not transmit disease; their primary drawback is that they interfere with grazing. They sometimes crawl into the ears and cause severe annoyance. Heavy infestations can lead to weight loss or death in cattle, but extensive blood loss is not common in horses because they attract fewer horn flies.

Horn Flies on Cattle and Horses

Adult horn flies are half the size of a housefly and gather on withers, back, neck, shoulders, underline, and around the eyes of the host. When disturbed by a swish of the tail or toss of the head, they swarm up, only to immediately settle back down onto the animal. Cattle may have hundreds at once. Horses are much more sensitive; even as few as twenty horn flies may seriously annoy a horse.

Adult horn flies stay on the host most of the time, leaving only to lay eggs in fresh cattle manure. The eggs hatch within 24 hours and the larvae burrow into manure, completing development in 4 to 5 days. Then they crawl to a drier part of the manure or into the ground to pupate. Adults emerge and seek a host to drink blood, then lay eggs. In cold climates they spend winter as pupae. Though the mature flies can live equally well on horses or cattle, the eggs are generally laid in cattle manure.

Control. Apply insecticides directly to the host animal because the flies spend almost all their time on the host. Some horsemen fasten cattle insecticide ear tags to horses' manes or halters, but this is not a good idea because the insecticide dose in cattle tags is too strong for horses and is not recommended. There are some particularly effective repellents and insecticides that can be applied to the horse's legs, belly, and flanks to get rid of horn flies. Ask your veterinarian.

Black Flies, Buffalo Gnats, Sand Flies

These small (3/16 inch; 4.8 mm) gray to black flies are often present in large numbers after periods of flooding or in areas with many streams. Larval stages are passed in flowing streams. The flies congregate in swarms and attack all animals, biting the legs, belly, and head. Irritation is so great that many animals stampede or mill about nervously, injuring one another or trampling young ones. Hypersensitivity reactions to the bites can be a serious problem.

Black flies cause annoyance and itching. They feed inside the ears; on the chest, udder, and scrotum; inside the thighs, and under the belly. Their bites cause swelling, oozing blood, and bloody crusts. Swarms of these flies can travel 3 to 5 miles (4.8–8.1 km) to seek a host.

Control. Put horses in a barn or shed on mornings and evenings when flies are most active. Apply fly repellents and wipe-on products frequently, two to three times a day if necessary. Try ear nets. Apply petroleum jelly to protect the inside of the ears; the flies will not bite through it.

Midges

Tiny biting midges (also called "no-see-ums") sometimes attack in swarms. Thousands may feed on the horse at night. Their bites cause irritation and hypersensitivity reactions, and the horse may damage his skin by constant rubbing and scratching. The itching dermatitis (inflammation of the skin) causes the horse to rub his mane, tail, and withers; his face, chest, and belly may also be affected.

Control. Put horses indoors before dusk and keep them in until after dawn. If stabling is not possible, try frequent use of repellent sprays or wipe-ons in the late afternoon. Be sure that a horse showing an allergic reaction to bites, such as hives or swelling, is treated by a veterinarian.

Face Flies

Face flies are similar to houseflies in size and color. They feed on secretions from the eyes and nostrils — and on blood from the bites of other flies. They stay on the horse only a short time; the rest of the day they sit on vegetation, fence posts, and other nearby structures. Like horn flies, they lay eggs in cattle manure.

Control. Apply insecticide to the horse's head and neck. Also try fly masks or face screens.

Control of Flies

You can help your horses in their fight against flies through diligent management and careful use of insecticides. Provide shelter during peak fly season. Install screens on stalls, and coat the screens with insecticide. Run a fan in the stall area. Attach fringe to the halter or bridle browband to dislodge flies around the head when the horse shakes his head. Use mesh fly masks to keep flies from landing around the eyes. Obtain summer fly sheets (open cool-weave blankets) to keep flies from biting the horse's body.

Be aware that manure management and control of wet areas can limit fly breeding. Remove manure and wasted feed daily from stalls and pens; spread it thinly for quick drying, or compost it in a covered pile. Make sure there is good drainage in the barnyard and no moist areas — no leaky faucets or waterers.

Predator Wasps and Diatomaceous Earth

One way to break flies' life cycle is to introduce predator wasps. These tiny nocturnal wasps lay eggs in the pupae of house flies, stable flies, and some other flies that lay eggs in manure and rotting bedding. The wasps use the fly pupae as food, eating them before they can fully develop. The tiny wasps are harmless to people and animals. If you use this biological form of fly control, be careful when using insecticides because they also kill the beneficial wasps.

Another tactic is to use diatomaceous earth, an inert feed additive that passes through the horse and inhibits the flies' reproduction in manure. Follow directions on the product label.

Insecticides and repellents can be purchased in several forms. Be sure to use only preparations meant for horses because other types may be harmful. Always follow label directions. Be aware that if a horse sweats much after an insecticide or repellent has been applied, the effects won't last long and you'll need to apply more.

Topical spray repellents usually contain a small amount of insecticide and some sunscreen, coat conditioner, or other ingredients to help the repellent stay on the hair longer. Read labels and find out if the product is to be used full strength or diluted. Besides the standard sprays, wipes, and sticks (rub-on repellents), strips impregnated with insecticides can be attached to a horse's halter.

Vinegar as Fly Repellent

Apple cider vinegar can be added to a horse's feed or water as a natural fly repellent. The vinegar can be mixed with the daily grain, using ⅛ to ¼ cup (30–60 mL). The vinegar seems to make the horse smell different; some of the acid quality may be excreted through the skin pores. The flies still land on the horse but don't bite.

To start a horse on vinegar, put a few drops on the grain at first, then gradually increase the amount. If a horse doesn't like the flavor, add a little molasses to disguise the vinegar taste. Most horses don't mind it. Use only pure apple cider vinegar, and gradually increase the amount to the point at which it begins to work, usually about ½ to 1 cup (125–250 mL). The minimum workable amount must be determined for each horse.

Horse flies, mosquitoes, and even botflies dislike the smell or taste of vinegar. It can be used externally — rubbed over the horse's coat — as well as internally in the feed. The mild acid doesn't seem to bother most horses' fly bites or raw spots in the skin.

Methods of Fly Eradication

Some forms of insecticide have long-lasting, residual action for up to 6 weeks; these are designed for application on fly resting areas such as bushes or barn rafters. Fogs and mists can be used daily — expelled into the barn air or sprayed on the horse's body with a hand sprayer. Fly-killing strips work in enclosed areas such as tack rooms and feed rooms. Sticky baits and fly traps are effective in areas with large numbers of flies, but must be cleaned or emptied periodically. Electric fly attractors also kill flies. There are even chemical larvicides that you can feed to your horses; the chemical goes through the digestive system in a small pellet, with no effect on the horse, and out with the manure — where it kills any larvae that hatch in the manure from flies laying eggs there.

If a certain type of fly is a problem all summer long, the best control is to eradicate it early in the season before it appears in large numbers. This is especially true of stable flies. If you are using an insecticide, be sure to repeat treatment before the flies become numerous again so there won't be a chance for a large breeding population to develop. For information on fly control programs, contact your veterinarian or county extension agent. With proper care you can keep the fly problem to a minimum.

Know Each Type of Fly

In a fly control program, become familiar with the life cycle and habits of the type of fly you're trying to eradicate. That way you can attack it at a stage when it's most susceptible to the control measure you're using.

Mosquitoes

Although they are not flies these insects can be a serious nuisance, especially in swampy areas where water or frequent flooding creates breeding sites. Mosquitoes also spread disease, such as encephalomyelitis, and some types of tiny filarid worms.

In most species the female mosquito must have a blood meal before she can lay eggs. Males feed on plant juices. Eggs are laid on the water surface or on low ground that floods. Several batches hatch each summer. Strong winds can carry mosquitoes many miles. Snow mosquitoes appear in swarms in mountain and northern regions in spring; they overwinter as eggs and develop in pools of water when snow melts.

Control

Eliminate breeding sites, destroy larvae by treating breeding grounds with the proper insecticides, and chemically destroy adults. Some insecticides kill mosquitoes on contact, but there are no preparations that can be applied directly to horses for long-lasting relief. Daily application of certain products gives partial protection; for example, pyrethroid insecticides reduce mosquito numbers by about 75 percent. Observing horses just at dusk will give you a clue as to how many mosquitoes are bothering them.

Summer Sores

Summer sores (habronemiasis) are caused by the larvae of a stomach worm trying to develop in the horse's skin instead of the stomach after being brought to the skin by flies. Adult worms lay eggs in the horse's stomach that are passed in manure. The larval stage is eaten by larvae of the stable fly and horse fly; the tiny worms infect the fly. When infected flies feed on skin wounds, eye secretions, or other moist tissue such as the sheath, the worms are transmitted to the skin. This problem is most prevalent in warm climates. Skin lesions occur in summer after infected flies deposit the worms, but the worm larvae cannot mature and finish their life cycle in the skin. They can live up to 2 years in the skin, however, causing problems for the horse.

Summer sores appear on any part of the body but most often on the lower legs (where wounds are common), at the corner of the eye, and on the sheath. At first they look like infected wounds, but they don't heal. The area enlarges with granulation tissue, such as that created when a wound repairs itself, protruding from the skin surface — reddish brown (bloody sometimes) and often covered with coagulated wound drainage. After the wound is 2 or 3 weeks old it becomes circular, and may have areas of dead tissue within it. There is usually drainage and severe itching. Often the first clue is the wound growing larger instead of healing. If a wound is untreated in summer, it may develop into a summer sore and not heal until cold weather sets in. The skin lesions usually heal in the fall but often reappear the next spring because the larvae are still present. Some horses develop a hypersensitivity reaction or exaggerated immune response to the worms, resulting in swelling, hives, or even asthma-like respiratory symptoms. Treatment for a hypersensitivity reaction generally includes antihistamines and steroids.

Problems Resulting from Summer Sores

A horse with a summer sore on the sheath may spray urine instead of producing a normal stream because a lesion on the tip of the penis is obstructing normal flow. A summer sore on the legs may grow quite large, interfering with use of the horse until the wound is treated or until the lesion regresses.

Control

Take proper care of wounds so they stay clean, and keep flies away from them. Give ivermectin (two doses 3 to 6 weeks apart) to kill most of the worm larvae in the lesions. Practice fly control measures to reduce the population of horse flies and stable flies that spread the worms.

Lice

Lice are sometimes a problem on horses, especially in winter when hair is long. These tiny insects spend their entire life cycle on the host, although some species can survive for up to 2 weeks away from the horse. Eggs are laid on the hairs next to the skin. The life cycle varies from 2 to 4 weeks.

Grooming Tip

Remember that lice are transferred from one horse to another by direct contact; anything that comes in contact with an infested animal can carry lice to another horse. Blankets, grooming tools, harnesses, and saddle pads may remain infective for several days after being used on a lice-infested horse. Be sure to treat these with insecticide before using them again.

Lice are host specific; horse lice are not the same as cattle lice or human lice. You can't get lice from your horse. A horse can't get lice from cattle or chickens, only from another horse. Two kinds of lice infest horses; sucking lice and biting lice. The former feed on blood, and severe infestation can cause anemia from blood loss. Biting lice are smaller and more active, causing skin irritation and itching. The horse may rub and bite affected areas, and large patches of skin may become hairless or raw. Lice are found primarily under the mane or on buttocks, rump, and tail.

Although lice are very small you can see them on affected areas with close observation, especially if you raise the mane and rub your hand along the neck opposite the lay of the hair. Lice will be evident on the skin between the hairs. You can also see the eggs if you pull out a few tufts of hair (it comes out easily where the lice have been feeding) and hold the hair up to the light.

Most active in winter, lice multiply rapidly on thin animals. Some horses are more susceptible to heavy infestations than are others; individuals that are young, old, or have a suppressed immune system usually are most vulnerable. Lice rarely cause a problem during warm months because they are inhibited by sunshine and high temperatures.

Control

Treat affected horses with insecticides; several sprays and powders kill lice on horses. If weather is cold, use a dust rather than a spray so as not to chill the horse. When using any insecticide be sure to follow label directions, and treat the horse in an open area where there's no danger of contaminating feed, water, or feeding areas. Clipping the horse makes lice more accessible to treatment but is not recommended in cold weather unless you keep the horse indoors.

Administer a second treatment 2 weeks after the first one to kill lice that have just hatched or that weren't accessible to the first application (dusts and sprays don't kill the eggs attached to the hairs). Also, be sure to treat any brushes, blankets, saddle pads, or other items used on the infested horse.

Ticks

Ticks are sometimes found on horses, especially those pastured in brushy areas. Ticks carry several diseases and cause anemia in severe infestations. They are the main vectors (i.e., carriers and spreaders) of many protozoal diseases; the protozoa survive from one generation to another in the ticks by infecting their eggs. Ticks can also spread diseases caused by bacteria, viruses, and rickettsia.

Rocky Mountain tick

All ticks found on your horses should be removed and destroyed. Be careful handling them; the tiny deer ticks can carry Lyme disease, and the larger hard-bodied ticks can carry Rocky Mountain spotted fever. Both diseases are dangerous to humans beings.

Control

Try to keep horses out of tick-infested pastures in the spring. If horses do get heavy infestations, they can be sprayed with insecticide, preferably a product against which the ticks in your area are not resistant. Be sure the product you use is safe for horses. Amitraz, a chemical that is highly effective for cattle and pigs, is *not* safe for horses; it causes colic and sometimes death. Another way to help minimize a tick problem is to deworm with ivermectin every 6 weeks during spring and summer.

Life Cycle of Ticks

Some species spend their entire life on the host, whereas others pass different stages on successive hosts. Still others are parasitic only in certain stages. Eggs are laid in the soil, and young ticks crawl onto bushes and attach themselves to a passing host. Adult females suck blood or lymph from the host, then drop to the ground to lay eggs. Complete eradication of ticks is often difficult because many species live on several hosts, including wild animals, and adult ticks can live for weeks or months apart from a host.

Ear Ticks

Ear ticks are irritating to horses because they live deep inside the ear. Horses with ear ticks may become ear shy and head shy — hard to bridle because they resist any handling of the ears. They may shake their heads, rub their ears, or cock the head to one side.

Control. You can kill ear ticks by putting a half-and-half mix of mineral oil and insecticide, such as lindane, into the ear. Commercial mixtures are available from your veterinarian. The horse may have to be restrained and twitched, or you may have to use a Stableizer (see chapter 5), before he will stand still for you to put anything into his ears, as the ticks have made them sensitive. If ticks have been present in the ear for some time there may be a great deal of debris and secondary infection in the ear, requiring further medication.

Mites

Mites cause dermatitis (i.e., inflammation of skin). *Trombidiform mites* are harvest or grain mites that primarily attack harvested grain and infest animals only secondarily. However, they can transmit disease. Natural hosts for trombidiform mites are small rodents. Larvae drop off the rodents and develop into a nonparasitic stage that lives in grain and hay. They cause dermatitis in animals grazing in pastures or eating newly harvested grain. Horses are affected around the face and lips; the affected areas become itchy and scaly.

Mange mites infest the hair follicles of horses, causing chronic inflammation, loss of hair, and sometimes secondary infection (e.g., pustules or small abscesses). Some mange mites live deep in the skin, creating small nodules that feel like birdshot in the hide. These mites are most active in winter and early spring, usually affecting the area around the horse's face and eyes. This problem is sometimes confused with ringworm.

Sarcoptic Mange

One type of mite causes a severe, itching dermatitis (also called "barn itch" or "red mange") that makes affected animals too restless to graze and therefore causes them to lose weight. These mites are most active during cold, wet weather. Eggs are laid in tunnels in the skin made by adult mites; the hatching nymphs burrow into the skin and mature in approximately 17 days. Mite infestation can be spread from horse to horse or by materials such as bedding, saddle pads, grooming tools, and clothing. In moist, protected sites the adult mites can live as long as 3 weeks away from the horse.

When a horse has sarcoptic mange, affected areas of skin contain small red elevated areas. The area is intensely itchy and may be injured by constant scratching and biting. Hair is lost, and thick brown scabs form over raw surfaces with thickening and wrinkling of surrounding skin. The most common site on horses is the neck.

Psoroptic Mange

Another type of mite migrates to all parts of the skin, preferring areas covered with hair. The adults puncture the skin to feed on lymph and cause local irritation and itchiness. The skin weeps and forms crusts that spread, as the mites are most active around the crusts. In horses the large, thick crusts are found on parts of the body with long hair, such as the base of the mane and root of the tail, and on hairless areas such as the udder.

Chiroptic Mange

Also called "leg mange," this causes great annoyance to horses. The mite is most active in winter, spread by direct contact from horse to horse or by infested bedding or equipment. In horses the parasites occur almost entirely in long hair on the lower legs, rarely on other parts of the body. They cause severe irritation and itchiness; the first sign is usually stomping of the feet and rubbing of the back of hind pasterns on anything available. The horse may resent having his feet handled. In longstanding cases the skin may be swollen, scabby, cracked, and greasy. Scrapings from affected areas contain numerous mites.

Treatment and Control

Any itchy rashes on a horse should be examined by your veterinarian. He or she can usually differentiate between the various skin diseases by making a close examination and by taking skin scrapings to observe under a microscope. Several types of sprays and insecticides can be applied to the horse to eliminate mites, and the veterinarian can advise you on an effective treatment. Quarantine affected horses so the problem does not spread.

ELEVEN

TOXIC PLANTS AND POISONS

Most poisons that affect horses are found in toxic plants containing dangerous alkaloids. Other agents that are harmful to horses include chemicals in contaminated feed and lead-based paint on fences or barns where horses chew or lick. Some poisons are so deadly they kill a horse quickly even if he ingests very small amounts, whereas others are lethal only after many small doses accumulate to dangerous levels in the body.

Toxic Plants

Most poisonous plants vary in level of toxicity and conditions during which poisoning occurs, but some are extremely poisonous at all times. Others are harmless under normal conditions — in small amounts or at certain seasons — but become dangerous if eaten in large amounts or if chemically altered by freezing or enzyme activity (as when cyanogenic glycosides in chokecherry leaves turn into hydrogen cyanide). Certain plants absorb substances from the soil that accumulate to toxic levels (as selenium does in locoweed).

Horses usually avoid poisonous plants, but hunger, curiosity, or boredom may lead them to eat things they shouldn't. They may nibble shrubs through the paddock fence or plants along the trail when you ride if you aren't paying attention. They may also eat harmful plants inadvertently baled in hay or raked into lawn or hedge clippings. Even some plants and shrubs used in landscaping (e.g., yard or garden, barnyard or fairgrounds) or along a fence row, road, or trail may be poisonous. Most poisonous plants that might be eaten by horses are forbs — broad-leafed flowering plants. Only a few grasses are dangerous to horses.

How Poisonous Is It?

Poisoning depends on many factors, including
◆ Palatability (some plants are readily eaten; others aren't unless a pasture is overgrazed and grass is short or the horse is bored)
◆ Stage of development
◆ Growing conditions (drought may increase toxicity levels)
◆ Soil type
◆ Moisture content of the plant
◆ Portion eaten (some have highest concentration of the poison in the roots, seeds, or new sprouts).

If you suspect a horse may have eaten something toxic, showing behavioral changes or digestive upset, get veterinary help immediately. Some types of poisoning are irreversible and always fatal, but others can be resolved with prompt emergency treatment.

Plant poisoning should be suspected if horses experience sudden illness or death after being moved to a new pasture or fed different hay. Poisonous weeds in hay can be a serious threat because the hay-fed horse has no other food options, unlike a horse at pasture. Some horses sort out strange weeds, but greedy eaters may eat them despite their taste or smell. Even a not-so-hungry horse may eat harmful hay because the strange and unpleasant taste or smell of a live poisonous plant may be diminished in the drying process.

There is not space here to discuss all plants that are poisonous to horses. More than 500 plants in the eastern United States alone are toxic enough to kill a horse, and toxicologists are discovering more. This chapter touches on some of the common offenders.

Eradicate Toxic Plants

If you cannot identify certain plants or trees, take leaf samples to your county extension agent. Some toxic plants are easy to eradicate by taking out individual offenders growing along a fence or ditchbank, but others may require the use of herbicides. Your county extension agent can advise you on how best to get rid of them.

◆ **Garden flowers and plants** poisonous to horses include buttercup, narcissus (daffodil), lily of the valley, and delphinium. Wild delphinium is called *larkspur; tall larkspur* is highly poisonous to cattle and not as deadly to horses, yet if a horse exerts after eating delphinium he may die due to impaired muscle activity and breathing. Rhubarb leaves contain soluble oxylates that crystallize in the kidneys, causing kidney failure and death. Don't let a horse eat in your yard or garden; *never* let him eat any plant that is not a part of his regular diet.

Tall larkspur (delphinium): flowers and seedpods

◆ **Brackenfern** is part of the fern family. This large perennial herb grows in burned-over areas, woods, and shady places as well as on dry, sandy, or gravelly soil. Horses eat it when other feed is scarce or when it is mixed in hay. The fern contains an enzyme that inhibits absorption of thiamine (vitamin B_1) in the gut. Because the affected horse cannot absorb it, symptoms are those of vitamin B_1 deficiency: His nerve function deteriorates and he loses weight and becomes progressively uncoordinated and depressed. He stands with back arched, legs apart, and muscles twitching. Symptoms develop slowly over days or weeks after eating a large amount of fern, but if affected horses are not treated they eventually die. If the horse is taken off the offending feed and treated with intravenous injections of thiamine hydrochlorate when symptoms first appear, he will recover.

◆ **Castor beans** are grown as a crop (for castor oil) in California and the Southwest and as an ornamental plant in other regions. The seeds are the most toxic part, causing mild colic, diarrhea, sweating, and a thundering heartbeat. Only a few ounces of castor beans can kill a horse.

◆ **Cocklebur** is an annual plant that grows in lakebeds, lowlands along rivers, and disturbed soils of fields, pastures, and roadways. Poisoning occurs mainly when horses eat the leaves in early spring, as seedlings are more toxic than mature plants. Symptoms include rapid and weak pulse, labored breathing, and spasmodic contractions of the leg and neck muscles.

◆ **Crotalaria,** a member of the pea family, contains pyrrolizidine alkaloids that destroy the liver. In the Midwest the plant is called *rattleweed*; in Florida it may be called *showy crotalaria*. The effects of liver damage may not show up right away because the liver is a large organ that continues to function until about 60 percent has been destroyed. Then liver failure is quite sudden, with the horse dying in 3 to 5 days.

◆ **Death camus** grows on foothills and pastures in western and plains states. It takes less than 10 pounds (4.5 kg) of it to kill a horse. Symptoms are excessive salivation, weakness, difficulty breathing, and coma. Death occurs soon, within a few hours to 2 days after ingestion.

◆ **Equisetum** is called *horsetail, scouring rush,* or *jointfir*; several species grow in meadows and floodplains or wherever the water table is high. These plants may end up in meadow hay. Poisoning is cumulative and may occur after horses eat hay containing large amounts of horsetail or graze pastures short. Symptoms include excitement, muscle weakness, trembling, staggering, diarrhea, and loss of condition. As in brackenfern, the poisonous element in horsetail is the enzyme that destroys vitamin B_1. If diagnosis is made early, treatment with massive injections of thiamine can save the horse.

Stalk of death camus with flower (cluster of tiny blossoms) at top

◆ **Fiddleneck, fireweed, tar weed, buckthorn, and yellow burr weed** are common weeds in wheat and grain crops in the West and Midwest, causing liver failure if eaten. Fiddleneck may grow in pastures or hay fields. A single large dose, or small amounts eaten over days or weeks, causes poisoning due to pyrrolizidine alkaloids. These weeds aren't very palatable, but a horse may be poisoned by eating small amounts at pasture or in hay or grain over an extended period.

◆ **Lantana** is an ornamental flower commonly found in yards and along roadways, especially in the Southwest. Horses eat it if other forage is poor or unavailable; 20 to 30 pounds (9.1–13.6 kg) will kill a horse. Symptoms of poisoning are respiratory distress, diarrhea, and low blood sugar; the horse is dull and weak.

◆ **Locoweed,** called *crazyweed, milkvetch,* or *poisonvetch,* is native to arid and semi-arid rangelands; there are more than fifty kinds. It is a legume, part of the pea family with alternate leaves and pea-like blossoms. Its seeds resemble garden peas in a dry, leathery pod. Most locoweeds and vetches grow in selenium-rich soils and therefore have concentrated selenium in their tissues. Some types are unpalatable; horses eat them only when they are short of better feed, as in dry years or during a late spring when pasture grass is slow starting. Locoweed comes up early, ahead of the grass.

Locoweed blossoms

Other types are quite palatable and vary in toxicity, ranging from good high-protein forage plants to extremely poisonous ones. Some are addictive; once a horse starts eating these he seeks them out, even if better feed is available. In moderately toxic plants the poison is cumulative, building up to dangerous levels over a few days or weeks, but the most toxic plants cause death within just a few hours of eating. Nerve cells in the brain are destroyed, and even if a horse stops eating the plants he can never completely recover.

Signs of Locoweed Poisoning

- ◆ Nervousness
- ◆ Weakness
- ◆ Emaciation
- ◆ Depraved appetite, eating unusual things
- ◆ Impaired vision (because the optic nerve is affected, small objects appear large and spacial relationships are distorted — the horse may walk into fences or over cliffs)
- ◆ Incoordination
- ◆ Difficulty breathing
- ◆ Heart failure
- ◆ Respiratory paralysis
- ◆ Seizures

◆ **The nightshade family** includes black nightshade (also called *horse nettle*), jimsonweed or thornapple, ground cherry, and cultivated varieties — potato, tomato, and eggplant. Fruits of the cultivated plants are not poisonous, but potato skins are toxic to horses. The leaves and stems of potato plants and other members of the nightshade family contain the alkaloid solanine, which

Deadly nightshade: leaves, blossoms, and buds

causes nervous system impairment (weakness and incoordination), digestive problems, and abdominal pain. Green parts of the garden plants are toxic, and all parts of the wild nightshades are poisonous.

One to 10 pounds (0.5–4.5 kg) of black nightshade (horse nettle) will kill a horse. Deadly nightshade is a vinelike weed that looks somewhat like a tomato or potato plant with tiny purple blossoms and small red berries. Horses usually don't eat it but might nibble it in a pen where there is nothing else to eat.

◆ **Poison hemlock** is part of the carrot family and should not be confused with water hemlock, which is actually more poisonous but less often eaten. Poison hemlock grows 2 to 10 feet (0.6–3.1 m) tall with hollow, purple-spotted stems and a solid taproot. The leaves are finely divided, like carrot leaves. They smell like parsnip when bruised and have a nauseating taste. Poisoning is most common in spring when the tender, succulent, highly toxic new leaves appear; horses generally don't eat hemlock unless they are short on other feed.

Poison hemlock is found most often in pastures and along roadsides; less often near creeks and irrigation

Leaves of poison hemlock

ditches. Once eaten it causes incoordination, staggering, drooling, and slow heart rate. Death is usually due to respiratory failure, and treatment is generally of no help. Whether the animal lives or dies depends on how much he has eaten: It takes less than a pound (0.5 kg) to kill a horse.

◆ **Russian knapweed** is a member of the same family as yellow star thistle and causes similar symptoms if eaten over several weeks. Brain damage from this type of poisoning is irreversible, and affected horses starve from the inability to eat. They should be euthanized to spare them this slow death.

◆ **Senecio** and other species of senecio (groundsel, stinking willie, ragwort, and tansy ragwort) contain pyrrolizidine alkaloids that destroy the liver. A horse that eats these plants or contaminated hay over a period of several weeks will eventually show signs of liver failure. Because the plants grow in early spring, they are often found in first cutting hay. A horse is more apt to eat them in hay than at pasture unless good forage is lacking.

◆ **Saint-John's-wort,** called *Klamath weed* or *goat weed*, causes reaction to sunlight (photosensitization) and severe lesions in unpigmented areas of the body (e.g., areas with white markings). Saint-John's-wort is toxic in all stages of growth but is most likely to be eaten in spring when young and tender. Hay containing this plant can also cause problems. The toxic compound is absorbed into the bloodstream from the gut; when it reaches the skin it sensitizes unpigmented areas to sunlight. Light-skinned horses or those with white markings develop red, swollen, sore, and itching skin that peels or comes off in sheets. Affected animals must be moved out of the sun and treated. Photosensitization is discussed in chapter 12.

◆ **White snakeroot** grows in sandy areas of the Midwest. Because it often stays green in early fall after other plants become dry, horses may be tempted to eat it. Symptoms of poisoning are weakness, lethargy, stiffness, and incoordination. The plant contains the alcohol trematol; a lethal dose is 2 to 10 pounds (0.9–4.5 kg) of plant material. Enzymes in the plant damage the heart muscle, after which stress can cause heart failure. Trematol also damages the liver. It can be excreted in a mare's milk and poison a foal if a lactating mare eats the plants.

◆ **Yellow star thistle** is common along the Pacific Coast as well as in several western, plains, and eastern states — and it is spreading. Some horses develop a taste for it and eat enough to permanently damage the brain. Signs of poisoning include twitching lips, flicking tongue, involuntary chewing movements (ranchers call it "chewing disease"), and drowsiness. Animals have trouble eating and drinking but are able to swallow. The horse may hold his mouth open as if something is stuck in his throat. The muscles of jaw and lips become rigid, and the horse has trouble grasping or chewing food.

Grasses

◆ **Tall fescue** causes several problems in mares, including abortion, retained placenta, and failure to produce milk. The fescue grass itself is not toxic, but the fungus (i.e., *Acremonium coenophialum*) that infests many stands of it is.

Grasses and Photosensitization

Some grasses such as perennial rye grass and dried buckwheat can cause photosensitization, as can the aphid that lives on certain clovers; it is often eaten by the horse along with the plant.

◆ **Sorghum and Sudan grass,** which are common in pastures in the Southwest and Southeast, can be harmful. Their toxic effects are intensified during a drought or after a frost as the glycosides in the plants are changed to cyanide, a potent poison. The affected horse is starved for oxygen, breathes heavily, grows weak, falls into a coma, and dies within a few hours.

◆ **Torpedo grass** is a lush pasture grass in Florida and the South that is toxic to horses. Animals that eat this grass often develop severe anemia and die.

Trees and Shrubs

Many trees and shrubs cause problems for horses — even apple trees. If a horse eats too many apples he may develop colic. When horses have access to fallen apples in a pasture or paddock they may overeat, especially if the fruit is ripe. Some crab apples are especially troublesome. The horse may be fine until he exerts; then the combination of stress and apples in his digestive tract leads to an acute case of colic.

Health Tip

Many ornamental trees and shrubs are highly toxic. Make sure horses never have access to them.

◆ **Wild cherry** (chokecherry) grows in most areas of the United States. It has dark green leaves and clusters of white or cream-colored blossoms that produce dark red or black fruit. The leaves contain prunasin, which produces cyanide and hydrocyanic acid poisoning when damaged. The chokecherry is most deadly when the leaves are damaged. Because the poison interferes with oxygen transfer to the cells, the affected animal suffers from acute oxygen starvation; the horse breathes heavily, grows excited and agitated, then becomes weak and may suffer convulsions before going into a coma. Symptoms appear soon after he eats the leaves, and unless treatment is immediate he will die. Cherry, plum, peach, and apricot leaves are also toxic if wilted or otherwise damaged, so horses should be kept out of orchards. Cherry laurel (often planted in yards and gardens) is another potent source of hydrocyanic acid, as is crab apple.

The chokecherry grows wild in most parts of North America and is often found in pastures, lining fence rows, or growing along streams in arid regions. It may be merely a tall shrub in dry places, but it grows into a tree where there is plenty of water.

Chokecherry leaves and fruit

◆ **Rayless goldenrod,** a bushy, unbranched perennial shrub, is also called *rosea* or *jimmy weed*. Most common in the Southwest, it grows 2 to 4 feet (0.6–1.2 m) tall with yellow flowers. The toxic substance (trematol) causes muscle tremors, depression, stiff gait, and weakness. It can poison foals nursing mares that eat the shrub.

◆ **Black locust** is a legume with short thorns, creamy white blossoms, and thin brown seed pods. It grows east of the Mississippi and is used as an ornamental tree in other regions. Its leaves, twigs, and bark are poisonous to horses, causing colic, diarrhea, dilated pupils, weak and irregular heartbeat, depression, and death.

◆ **Red maple,** also called *swamp maple,* is native to the eastern half of the United States. It can be quite poisonous to horses, killing them within a day or two after they've eaten small amounts of wilting leaves in the fall or nibbled them off fallen or trimmed branches. The wilted leaves contain an unidentified substance that causes breakdown of red blood cells. Symptoms of red maple poisoning include rapid breathing, weakness, depression, brown urine, and very dark blood.

Red maple leaves

◆ **Oleander** is a common ornamental shrub in warmer regions that is deadly if eaten. One-fourth pound (113.4 g) or less will kill a horse; often just a few leaves can be deadly. Horses rarely eat leaves from the shrub itself, but if wilted or dried they are more palatable because they don't retain their bitter taste (but do retain their toxicity). Oleander contains two toxins, oleandrin and nerioside, that affect the heart much like digitalis, producing erratic heartbeat and diarrhea. Death may occur within a few minutes or hours. It's best to assume that any ornamental tree or shrub is unsafe for horses to nibble on. Magnolia, azalea, privet, boxwood, and rhododendron all cause illness in horses.

◆ **Black walnut** trees are often grown in the eastern part of the United States and in California. Wood shavings from these trees cause colic and laminitis in horses that are bedded on them. In addition, horses at pasture can be seriously affected if they have access to black walnut trees. Pollen in spring causes allergic respiratory reactions, and chewing of bark or branches leads to laminitis and colic at any time of year.

◆ **Yew** is a small tree or shrub — an ornamental evergreen with red berries (arils). Its leaves are stiff and needle-like, dark glossy green on top and yellow-green underneath. There are several species — English yew, Japanese yew, and American or Canadian yew. Yew contains taxine, a highly poisonous alkaloid even when dried. It has a depressing effect on the heart. About 4 ounces (113.4 g) (one mouthful) of yew needles can kill a horse. The animal dies of heart failure, often within minutes of nibbling on yew.

Yew leaves (needles) and arils (red berries)

Beware of Yews

Yew is poisonous to people and livestock. Symptoms include shortness of breath, muscle tremor, and sudden collapse. With horses, death may be so sudden that some of the plant is still in the mouth. Horse owners should avoid these shrubs in landscaping. Check your farm and neighboring fence lines to make sure there's not a potentially lethal hazard waiting around the corner for an unsuspecting horse.

◆ **Other trees** that are mildly toxic if chewed by horses are oak, Ohio buckeye, and the Kentucky coffee tree. The responsible horseman should look at *all* the plants in a pasture or paddock before putting any horses in. Check all trees and shrubs — even if they are on the outside of the fence line but within reach of the horses, or where branches or leaves can blow into the pens.

Prevent Accidental Plant Poisonings

◆ Never dump trimmings, pruned branches, or discarded garden plants near horse pens or pastures. Bored or curious horses may nibble on strange plants, even reaching under or through a fence to taste and eat them.

◆ Never let horses near ornamental shrubbery, plants, or flowers.

◆ Remove fallen tree limbs or leaves from pens and pastures immediately after a storm.

◆ Carefully inspect bedding straw when you put it in stalls, to make sure there are no weeds in it.

◆ When feeding hay, inspect it for strange plants and discard anything you aren't sure about.

Fungi

Fungi, a group of primitive plants that includes molds, are micro-organisms that live almost everywhere. Most prefer a moist environment and multiply rapidly in warm, humid areas or during wet seasons. Many are harmless, but some cause trouble when ingested.

Hay or grain can become infested by fungi while growing or may spoil and be contaminated during storage. Factors that affect severity of the problem include weather conditions during the growing season and exposure to moisture during or after the harvest. An animal may eat the same kind of hay or grain for years with no problem and then suddenly be poisoned as a result of unusual growing conditions that cause fungi to flourish in the feed.

◆ **Black patch disease** occurs on clovers and other legumes that are sometimes host to the toxic fungus *Rhizoctonia leguminicola,* which causes "slobbers" in horses. The fungus appears as spots on leaves and stems. Red clover, alsike, white clover, soybeans, and alfalfa are commonly affected. Because the fungus overwinters, it can persist in a field or pasture once established. When a horse eats affected plants his salivary glands are irritated, resulting in slobbering and drooling. The toxic alkaloids also cause fetal malformation or abortion in pregnant mares, along with diarrhea, frequent urination, weight loss, and even death, depending on how much is eaten.

◆ **Ergot** is a fungus that sometimes occurs on grasses, infecting the seed heads. Susceptible plants include wheat, rye, barley, bluegrass, red top, brome, and reed canarygrass. Ingesting small amounts of ergot daily can lead to circulation problems in horses and cattle. Blood vessels constrict and shut off blood supply to feet, ears, and tail. The horse may lose ear tips or ears and may develop gangrene in feet and legs, which may lead to death.

◆ **Moldy corn poisoning** is caused by *Fusarium moniliforme*, which produces deadly toxins. This fungus lives in soil. It is commonly found on moldy corn stalks or corn that has been rained on in the stalk or stored wet. Symptoms are often mistaken for those of encephalomyelitis. Poisoning can occur in horses that eat commercial feeds, including pellets. Horses ingesting a small amount of the toxin may suffer depression, weight loss, lack of appetite, and bleeding disorders. Larger doses are usually fatal. The horse circles, runs into fences, or stands with head pressed against a fence or wall. He may cross his legs, walk sideways, or circle aimlessly, losing control of his movements as his brain deteriorates. He may die within a few hours or linger for several days.

Beware of Corn

Corn is usually a good feed for horses, but only if it is of high quality and mold free. Corn that has not been adequately dried in the field, or in the drier after harvest, may spoil and mold when stored. Even fully dried grain may mold if it is stored where it can draw moisture, and condensation inside a metal bin can cause trouble unless all old grain is cleaned out before a new batch is put in. Corn is more subject to mold during growing and storage than are less starchy grains. Some of the visible molds are not as deadly as the aflatoxins, or poisons, produced by *Aspergillus* or other molds that may develop during a wet growing season. Often the deadly corn looks perfectly normal.

Even in shelled corn, bad kernels sometimes get mixed with good ones. Usually only a small percentage of individual kernels in a moldy batch contain high levels of toxin. This is why some horses develop fatal poisoning even though others on the same feed are unaffected. It only takes a few kernels of bad corn to produce problems if a horse eats them.

Fungal Toxins and Photosensitization

Fungal toxins, called "mycotoxins," can be deadly — damaging liver and kidneys, destroying red blood cells, or producing other serious effects. A damaged liver can lead to photosensitization.

♦ **Rusts and smut** are other fungi that may contaminate grain and pastures. Rusts are reddish brown patches that infect the leaves and stalks of grasses, and smuts are black powdery clumps that appear to replace the grass heads or grains. Eating smutty grain causes convulsions, paralysis, and death within a few hours. Rusts and smuts also cause colic — and respiratory problems if inhaled.

Beware of Wet Hay or Grass Clippings

Mowed hay that gets rained on before being baled may develop molds and other fungi. For this same reason piled lawn clippings are not safe for horses, because they have a high water content and spoil quickly. Within 3 or 4 hours the grass undergoes chemical changes in piles or clumps that can't dry quickly. It then heats and ferments, producing mold. The fermentation process may also produce nitrate, which can poison a horse, causing respiratory problems. A horse should not be allowed to eat grass clippings or graze a pasture that has been clipped by a rotary mower that tends to wad up the grass.

Blister Beetle Poisoning

Many species of blister beetle — the most common being the black and yellow striped — can cause poisoning. Blister beetles are small flying insects that are attracted to the blossoms of tomatoes, potatoes, soybeans, and alfalfa. The beetles travel in swarms and are baled up in alfalfa hay if it is cut while they are feeding on it. Their bodies contain a toxin, cantharidin, that causes severe blistering of the digestive tract. Ingestion of only 3 to 10 beetles will kill a horse. Cantharidin is just as toxic in a dead beetle as in a live one.

Blister beetle poisoning has been diagnosed in nearly every state, but hay grown west of the Mississippi is more frequently contaminated than hay grown in the East. Horse poisonings are most common in the Southwest. There are blister beetles wherever alfalfa is grown, but some species of the beetles are more toxic than others.

Symptoms

Blister beetle

Cantharidin causes extreme tissue damage in the gut. The affected horse goes off feed and has intense abdominal pain. Most horsemen treat for colic, not suspecting blister beetle poisoning unless several horses become sick from eating the same hay. The horse may survive if he eats only one or two beetles, but if several are ingested his chances of survival are diminished. The tissue damage causes shock and dehydration. Another sign of poisoning is frequent urination caused by inflammation of the urinary ducts. A burning sensation in the mouth from the blistering makes the horse want to cool his mouth with water, although he can't swallow.

Treatment

Treatment can sometimes save a horse if it is begun immediately and the animal hasn't ingested a large number of beetles, but often the true cause of his colic is not recognized in time. Treatment for blister beetle ingestion includes intravenous fluids to keep the horse from going into shock and drugs to minimize the pain.

Prevention

First-cutting alfalfa is safer than later cuttings, because adult blister beetles may not emerge until midsummer. Use only alfalfa cut before bloom stage; it is less likely to contain beetles that feed on pollen in the blooms. Beetle larvae eat grasshopper eggs, so there is usually a large beetle population following a dry year with lots of grasshoppers. Be cautious about buying alfalfa hay grown in areas that were dry during the previous year.

Chemical Poisons

Toxic materials commonly found around barns and stables include rodent poisons, insecticides, and lead-based paint on fences or buildings. Horses can also be poisoned by accidental overdose on medications, certain dewormers or vitamins, accidental consumption of seed grain or feeds not meant for horses, and ingestion of toxic substances such as antifreeze.

Antifreeze

Because antifreeze has a sweet taste, many animals try to sample it. When you drain it from a vehicle be careful not to let it spill on the ground, and be sure to rinse out the pan you drain it into. Even dried antifreeze crystals contain the poison, ethylene glycol, which is also used in brake fluid, windshield de-icers, and motor coolants.

Pesticides

When using pesticides never apply or spray them near feeds, and make sure the wind isn't blowing the spray into areas that shouldn't get contaminated (e.g., mangers, feed buckets, and water sources). Keep all pesticides far from horses and grain bins; a single bottle of pesticide can cause trouble if it gets spilled in the granary or where a horse might later clean up hay or grain off the barn floor.

Treated Grain

A common cause of poisoning is treated grain. Almost all seed grain and most other seeds are treated with fungicides to protect against fungi that damage or destroy seeds during storage or after sowing. The fungicide may contain mercury, which is very poisonous and causes colic, bloody diarrhea, acute circulatory failure, and death. If only a small amount is eaten the horse may survive, but if he eats a small amount regularly (as when feed is contaminated with seed grain) the effect is cumulative, resulting in eventual destruction of the nerve tissue and permanent brain damage that will ultimately kill the horse.

Safety Tip

It is extremely important to know the origin of all grain. Usually the feed from a commercial dealer is safe, but beware of using grain that comes in unmarked sacks or from an unknown source. Never keep any seed grain in the barn. That way it won't get fed to horses accidentally.

Overdose of Medication

Horses can be poisoned by accidental or inappropriate use of certain medications, dewormers, or vitamins, especially those prepared in a palatable base. If such a preparation is left where a horse can find it, he might eat the whole container and poison himself. Vitamins A and D are toxic in quantities well above the recommended dosage, as are some dewormers. Other medications, such as wound treatments and liniments are potentially dangerous if eaten or spilled in feed. Keep all medications labeled and out of reach of horses. Spilled medicine might entice a curious animal to smell and taste it.

Lead Poisoning

Lead is especially dangerous and can accumulate in the body over time. Even small doses, if repeated (as when a horse chews or licks a painted wall or fence), can prove fatal. The most common sources of lead-based poisoning are lead paint and old batteries. Don't let horses have access to old dumps that might contain batteries and other harmful materials. Fumes from wet paint or certain solvents are also toxic. Prevention of poisoning is by far the best course. Make sure your barn and stable area contain no questionable chemicals that might poison a horse.

Monensin (Rumensin) Poisoning

Many horses have died from eating cattle feeds containing monensin (Rumensin). This drug is often added to alfalfa pellets for ruminant livestock such as cattle and goats. It is fatal when eaten by horses. It causes damage to muscles, especially the heart muscle, often leading to heart failure. Symptoms of monensin poisoning in horses include rapid heart rate, weak and irregular pulse, blue mucous membranes, and abnormal lung sounds. Only one or two grams of the drug is sufficient to kill an adult horse. Some additives in poultry feed are also unhealthy for horses.

Prevention

To avoid poisoning of horses, never feed grain mixes intended for other livestock, even if you are temporarily out of horse feed. Also, keep horse feeds completely separate from other feeds, and be careful to feed in separate areas. Even the fine particles of dust left at the bottom of a feed trough or storage bin may contain enough monensin to kill a horse. Don't put horses in pastures where beef cattle might be fed mineral blocks or protein supplements containing monensin.

Twelve

Skin Problems

The horse's skin is his first line of defense. It is subject to problems caused by bacteria, fungi, biting flies, allergies, and the like. Chapter 10 discusses skin problems related to external parasites such as flies, lice, and mites. This chapter focuses on dermatitis related to allergies and hives, photosensitization, fungus infections, warts, and sarcoid tumors.

Dermatitis

Dermatitis means inflammation of the skin. It can be caused by many things including external irritants, burns, allergy, or infection. It can also result from systemic disease. The affected horse's skin may become itchy (and he rubs against fences or walls) or show redness, swelling, or lumps. There may be oozing or scaliness as well.

Skin problems in horses can be minor or serious. When in doubt as to the identity or cause of any abnormality, have your veterinarian examine the horse.

Allergies and Hives

An allergy is a reaction to a substance that would normally cause little or no response by the body. Some horses are allergic (hypersensitive) to insect bites, certain plants, drugs, vaccinations, or medications. If the horse's defense system works overtime, an inflammatory response may become worse with each encounter. Sometimes an allergic reaction creates raised areas in the skin. This condition is called "urticaria"; the bumps are called "hives" or "wheals." They are caused when small blood vessels near the skin surface dilate and seepage from the damaged capillaries takes place. The hives develop suddenly after the horse comes in contact with the allergen. If you are with the horse when it happens, you can see the bumps growing as you watch. They usually pop up on both sides of the body and sometimes on the head and neck.

Hives can be caused by insect bites or stings, contact with certain plants and/or pollen, and ingestion of certain weeds or plants. Moreover, some drugs or vaccines produce allergic reactions. For example, penicillin causes hives (or some other manifestation of allergy) in some horses.

Treatment of Hives

In mild cases the lumps subside without treatment within a few hours or by the next day. In severe cases they persist for several days and may be accompanied by diarrhea, fever, or breathing difficulties. The affected horse may need an injection of antihistamine or corticosteroids.

Hives may recur if the allergen is not removed, the diet is not changed, exposure to the causative insect bites or plants is not prevented, or the same offending drug or vaccine is used again. Many times the cause of hives is readily obvious, as when a horse reacts to a certain injection, drug, or feed change; walks through a patch of stinging nettle; or suffers insect bites. At other times it may be difficult to discover the cause. If hives recur or do not disappear within a few days, have your veterinarian do sensitivity tests (e.g., blood tests and localized skin tests) to determine the cause of your horse's allergy.

Horse with hives

Photosensitization

This is a skin disease in which unpigmented skin is damaged by an allergic reaction to something foreign in the horse's system, when the skin is exposed to sunlight. Photosensitization is most common during summer (long days and stronger sunlight) when horses eat pasture plants that may contain certain photodynamic agents (i.e., chemicals that cause a toxic reaction when exposed to sunlight). A horse with long winter hair is not as susceptible because the hair shields the skin from the sun.

Light Skin and Photosensitization

Areas most affected by photosensitization are pink or unpigmented skin, such as white face markings, and white legs. Pigmented skin is more resistant to the sun's rays.

If a plant containing photodynamic agents is eaten, these chemicals are absorbed into the bloodstream and travel to all parts of the body. If they accumulate in the skin, the sun's rays cause a reaction, with inflammation and cell death. This problem is triggered by a number of plants — especially fescue grasses and some species of clover — and usually occurs when they are lush, green, and growing rapidly. The disease can also occur when a horse's liver is impaired and unable to filter out the by-products of chlorophyll breakdown in the digestive tract. When these by-products accumulate in the skin, the reaction kills skin cells.

Symptoms

The ultraviolet rays penetrating the sensitized skin tissues trigger release of histamine, which causes dilation of capillaries and swelling of tissue. Irritation and itching may be intense due to painful swelling deep within the layers of skin. The condition may be mistaken for sunburn at first, as the area turns red and starts to peel. Then it becomes more swollen, breaking out in blisters and cracks, and the skin develops thick crusts and painful scabs. In severe cases the horse becomes quite ill, with patches of skin sloughing off. The horse may go into shock.

Treatment

Remove the affected horse from the pasture and put him in a barn or covered area out of direct sunlight. In serious cases your veterinarian may give laxatives such as mineral oil (a quart [1 L] or more by stomach tube) to help get rid of toxic materials still in the digestive tract. Antibiotics and antihistamines may be necessary as well. Corticosteroids reduce soreness and swelling but should only be used on the advice of your veterinarian. However, you may apply soothing ointments or creams to raw areas to help alleviate the pain and itching.

Fungal Infections

Many types of fungal infections lead to skin problems in horses, including the fungi that cause ringworm, girth itch, rainrot, and scratches. Some fungal infections, such as ringworm and rainrot, are most common in winter when daylight hours are less (sunlight tends to inhibit them), and several can be spread from horse to horse via tack or grooming tools.

Ringworm

Ringworm is caused by several types of fungi that send out spores to start new infections when rubbed into skin. Ringworm is spread from horse to horse by infected brushes, curry combs, cinches, saddle pads, blankets, and the like. Young horses, older horses, those deficient in vitamin A or zinc or selenium, and any

Ringworm

horse with an immature or compromised immune system may be quite susceptible. Some families of horses have less resistance than others. Occasionally, the horses that get ringworm most often on a farm are related.

After the fungus spores enter the skin, circular areas in which hair falls out or breaks off become noticeable about a week to a month later.

Treatment

A horse with ringworm should be treated to prevent spread. It is also important to isolate the horse and disinfect all grooming tools and other equipment used on the horse. Several topical medications kill ringworm fungi. You can obtain them from your veterinarian.

Prevention

Ringworm will crop up sporadically if the fungal spores are present in the environment, but incidence of cases can be minimized by keeping horses healthy (adequate vitamin A and selenium in the diet) and not sharing grooming tools and tack between horses.

Girth Itch

Girth itch is a fungal skin problem that occurs in the girth area, generally after the horse has been ridden with a saddle. When tender skin at the girth (especially behind the elbows and in the armpits) becomes irritated by the cinch or girth rubbing the skin, the fungus gets a start in the broken skin and then the skin begins to peel. Unless it is halted the infection leads to ever-widening raw spots, and the horse may become so sore that he can't wear a saddle until the infection is halted and the skin heals.

Don't Let Girth Itch Spread

Girth itch can spread from horse to horse if you use the same cinch, girth, or grooming equipment on different horses. The fungal spores are durable and can survive in the environment and on brushes and tack. If there is one infected horse on the farm, the problem can spread unless different equipment is used on each horse.

Treatment

Most good fungicides halt girth itch, but take care to use one that will not be irritating to tender girth skin. Tincture of iodine is too harsh, but "tamed" iodine (Betadine) is fine. Another product that works well for halting serious cases of girth itch is the garden fungicide captan (Orthocide) used on fruits, vegetables, and flowers. It is a wettable powder. Follow label directions, read label precautions, avoid getting it in eyes or inhaling it, and wear rubber gloves when applying it to horses. The solution can be sprayed or wiped on. It can also be used as a body wash if a horse has extensive fungal lesions or a skin problem such as rainrot.

Prevention

Don't use the same equipment on more than one horse (each should have his own brush). If you must use the same saddle, be sure to change cinches. If you use the same cinch or girth on more than one horse, wash it with a fungicide or bleach before using it on another horse. A neoprene girth is handy; you can wash it with warm water and chlorhexadine (Nolvasan) and have it dry and ready to reuse in just a few minutes because the rubbery material does not absorb moisture.

Use a washable neoprene girth if equipment must be worn by different horses.

Rainrot

Rainrot, also called "rain scald," often affects horses in wet weather. It's common in fall or in a wet climate. Patches of hair may come out in clumps; the horse may have red, raw spots on his back where the hair comes out. There may be random crusty patches in the skin that are sensitive when you touch or brush him.

Rainrot is caused by a fungus-like organism that lives in the soil. It is often present in dust particles. The combination of water and dirt, as when a dusty horse gets wet, makes an ideal environment on the skin for this invader. Dusty conditions before a rain can fill the hair coat with dirt, especially if the horse rolls a lot; then rain provides the moisture for the organism to multiply and infect the skin. Hair becomes matted in clumps that detach, creating bald spots. Scabs appear in the hairless areas, and the skin becomes raw and sore. The problem often lingers into winter, especially if the weather stays wet. A thick winter coat provides ideal conditions for the fungus to multiply.

A dirty horse is more at risk for rainrot when wet.

In early stages of this skin disease it is easier to feel the emerging bumps than to see them; they have not yet developed crusts that affect the hair. The bumps may be hot and tender, and the horse may be sensitive when you brush him. If your horse reacts adversely to being groomed, look more closely at his skin.

Rainrot is not contagious from horse to horse, but it may appear in several horses at once if they are kept in the same dusty pen. It can also be spread by dust and dirt on grooming tools and saddle pads, so be sure to keep equipment clean.

Treatment

Treatment consists of cleaning up the skin and hair, using an iodine shampoo (or a human dandruff shampoo or antiseptic pet shampoo), and then applying a mix of equal parts tamed iodine and mineral oil to the crusty areas. Daily bathing for several days may be necessary to get rid of fungus-laden dirt in the hair coat. Clean the skin thoroughly. Use a soft brush or your fingers to massage the skin as you wash it, gently working the scabs loose so you can pick them off. After the horse is clean and dry you can use the iodine–mineral oil mix or an ichthammol salve (ointment made from coal tar) on affected areas. The horse will get over the infection rapidly if you keep him out of the rain during treatment or use a blanket to keep water off his back while the skin is healing.

Prevention

Rainrot is rarely a problem in dry weather because the organism needs moisture to multiply and cause sores. Thus the best prevention is to keep the horse clean and dry. In climates where rainrot is a recurring problem you should follow a program of regular grooming, with periodic vacuuming to get as much dust and dirt as possible out of the hair, and a bath twice a month with a medicated pet shampoo.

"Scratches"

"Scratches" is a skin problem on the lower legs that affects horses most often when they walk through wet areas; the fungus (and sometimes bacterium) is present in boggy mud or in the dust of a paddock. If the horse has a nick or scrape on the skin, the fungus can enter. The infected leg area becomes crusty, scabby, and thickened. In severe cases the skin may ooze or the lower leg may swell, and the horse may become lame. This condition affects horses with unpigmented (white) skin more readily than dark-skinned ones.

Treatment

Traditional treatments for scratches include astringents such as methylene blue, iodine and glycerine mix, ointments made with zinc oxide, nitrofurazone, and steroids. A better treatment involves thoroughly scrubbing affected areas and applying a mix of one part nitrofurazone ointment (antibiotic), one part DMSO (dimethyl sulfoxide), and one part dewormer paste containing a benzimidazole (thiabendazole, fenbendazole, cambendazole, oxibendazole, oxfendizole, or mebendazole — these are good fungicides as well as dewormers).

DMSO reduces swelling and inflammation and also helps the fungicide penetrate the area deeply, taking the medication into the underlying tissues. The nitrofurazone combats bacterial

Treating scratches: Wash the affected area before applying medication.

infection and buffers the DMSO so it won't burn or irritate the skin. The wormer paste kills the fungus. Applied daily, this mixture clears up most cases of scratches quickly. Bandaging is not necessary. In fact, it's actually detrimental; you don't want to hold the moisture in.

Prevention

Scratches can often be prevented by keeping the horse in a drier environment so he doesn't have to walk through mud. Serious cases can be prevented by treating small lesions when they first appear, never allowing them to become extensive and deep.

Warts

Warts are a common problem in horses, especially young ones (yearlings and 2-year-olds), and are caused by a virus that lives at the surface of the growth. They spread from animal to animal by direct contact. On horses they are often spread by grooming equipment or tack used on more than one horse, or by a person who touches an infected horse and then a susceptible one. Small cuts and scrapes in the skin can lead to wart infection if the virus is present. Once established on the animal, warts can last for several months up to year. After exposure a susceptible animal may break out with warts a few weeks later. Infected horses often pass the virus to other horses before their own warts are evident because of the long incubation period.

Warts are unsightly but not a serious problem for the animal unless they grow large enough to interfere with breathing. In horses they are usually found in clusters around the muzzle, but occasionally they appear on other parts of the body or inside the ear. They are distinguished from other skin growths by their dry, rough appearance. They may be pale or gray in color.

Warts and Immunity

Once warts start to appear, they usually develop rapidly over several days and continue to grow and spread for several weeks. Warts affect young animals because they have not yet developed immunity to the virus, and they occur in older animals that have not been previously exposed and have no immunity.

Treatment

Warts usually run their course and then disappear after the animal builds immunity. When left alone they often go away after a while. If they must be removed for health reasons, have your veterinarian remove them surgically. You can hasten their disappearance by crushing them with forceps or twisting off the larger ones; in fact, the veterinarian may choose this method. It encourages the immune system to fight them because some of the wart tissue makes contact with the bloodstream, stimulating the animal to make antibodies.

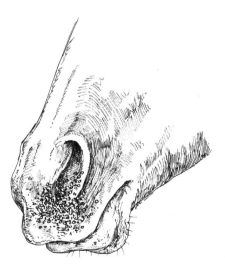

Horse's muzzle covered with small warts

Sarcoid Tumors

Sarcoids are also caused by a virus. Some horses develop them following an injury, others for no apparent reason. The tumor first looks like proud flesh (excessive granulation tissue that protrudes from a healing wound; see chapter 16) but continues to grow, producing a problem that may ultimately threaten the horse's usefulness.

Sarcoid tumors do not travel through the bloodstream or lymph system to other parts of the body, but they do continue to spread and grow locally. They are moderately malignant, in that they usually grow back if removed — often in larger form. If not removed early a tumor may become so large that it disables the horse. Because removal often stimulates regrowth, some veterinarians leave small, non-growing sarcoids alone.

At first a sarcoid is hard to differentiate from proud flesh. The virus attacks young fibrous cells in the proud flesh and causes them to multiply rapidly. This excessive growth cannot be controlled by medications used to control proud flesh. Wounds in areas where there is motion (e.g., joints or moving parts on the legs) seem most susceptible to sarcoid tumors. The growth is usually mushroom shaped; the base of the attachment is generally smaller than the outer surface. Wounds that are not properly treated to prevent contamination, irritation, and delayed healing are quite susceptible to sarcoid.

Treatment

Sarcoids can sometimes be successfully removed with surgery, followed first by cauterization of the area to control bleeding and kill stray tumor cells, and then by radiation treatment. The area is kept clean and bandaged while it heals. Other removal techniques include freezing and burning. A vaccine made from the tumors has also been effective in stimulating the immune system to build its own defense. The treated tumors usually shrink and eventually dry up and fall off. The vaccine doesn't always work, but it is one of the most effective treatments available and rarely leaves blemishes.

THIRTEEN

DENTAL CARE AND MOUTH PROBLEMS

Most problems in a horse's mouth are related to his teeth and can be resolved by periodic dental care such as filing off sharp edges. Sometimes a horse gets a sore mouth due to problems in the soft tissues such as gums or palate (roof of the mouth). If you have your veterinarian check a horse's mouth and teeth once or twice a year, and address any problems promptly, small problems will not become serious and he will be healthier and more agreeable when you handle him.

Sore Mouth

There are a number of problems that can affect the soft tissues of the mouth and cause a horse discomfort. As a result, he may have trouble eating or resent the use of a bit.

Gums

Irritation to gums can be caused by rough feed, tartar build-up, gaps between the teeth where food may catch and exert pressure, swellings from injuries and bruises, weak gum tissue in older horses, irritation from equipment such as an ill-fitting bit, or lip chains and twitches. Gum injury sometimes leads to gum disease. The bruising may result in gingivitis (an irritation of the gums), which if not resolved can lead to tooth or bone problems.

Tartar Build-Up

Often tartar build-up starts as gum deterioration, putting pressure on surrounding gums. The most common place for build-up is on the canine teeth — the small, sharp teeth that appear a short distance behind the front teeth and are present in most male horses and a few females. Horses at pasture rarely develop tartar. Finely ground feeds such as pellets and mashes tend to cause more build-up on the teeth than do hay and roughages; the latter actually clean the teeth and gums.

Lampas

Lampas is a swelling of the fleshy lining of the horse's hard palate — the roof of his mouth just behind his top front teeth. You may not notice the swelling until you examine the horse's mouth. The condition is often first discovered by a veterinarian who looks at the teeth to determine why the horse is not eating well or spilling feed out of his mouth.

Lampas commonly occurs in young horses when they are introduced to hard feeds such as grain or pellets. Because they have not yet learned how to handle it in

Lampas; swelling of the roof of the mouth behind the front teeth.

their mouths, they bump and bruise the tender roof of the mouth right behind the front teeth, which causes swelling and pain. As a result they may go off feed and lose weight.

Soft Feed Soothes Lampas

If a young horse develops lampas as a result of eating hard or abrasive feed, try switching to a very soft pellet or softening his grain by soaking. Good pasture grass is the most ideal and natural feed for a young horse; turnout on pasture — with less dependence on grain or hay — may quickly solve the problem.

Tongue Injuries

Occasionally a horse bites his tongue (as when hitting his head while falling) or has it injured by a bit. If a horse steps on his reins when wearing a bridle and jerks his head up, he can seriously injure the tongue. The horse may bleed profusely for a while from a cut tongue, but it usually heals. Sometimes the injury makes it difficult for him to eat or interferes with future use of a bit. Have your veterinarian check the horse.

Embedded Seeds

Sharp seeds such as those of cheatgrass and foxtail can become embedded in a horse's mouth and create a sore or abscess. If the horse has a sore mouth he moves the food around more than usual while chewing or even spits it out. A sore mouth may make him uncomfortable when wearing a bridle. If the horse has any obvious swellings at the mouth or cheeks, or has trouble eating, call your veterinarian. The horse may have an embedded seed or a tooth problem.

How Teeth "Grow"

Sometimes horses have tooth problems because of the way their teeth develop. Whereas the teeth of human beings and other carnivores grow rapidly and then stop once the permanent teeth are in, the horse's teeth keep erupting slowly through the gums throughout his life to compensate for the constant wear of chewing feed. The teeth do not actually become longer but are gradually pushed through the jaw bone out of the sockets. At first the crown of the tooth wears down, then the neck, and then the root. The very old horse may end up with short stubs that eventually fall out.

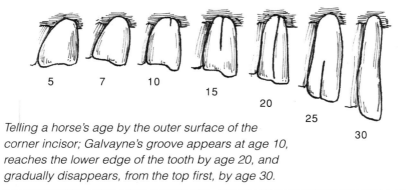

Telling a horse's age by the outer surface of the corner incisor; Galvayne's groove appears at age 10, reaches the lower edge of the tooth by age 20, and gradually disappears, from the top first, by age 30.

The Teeth

Horses have two kinds of teeth — incisors in front (used for biting off grass), and premolars and molars in back (used for grinding feed). Between the incisors and the premolars is the interdental space, a gap of several inches. A foal's teeth emerge soon after birth. First come the central incisors and initial set of premolars. By the end of his second week the next two sets of premolars appear. The second set of incisors comes in between 4 and 6 weeks of age, then the third set between 6 and 9 months of age. The first set of wolf teeth may come in at about 5 or 6 months of age — they are technically first premolars, just ahead of the large second premolars — and are often shed around the same time as the temporary first premolars. Some horses have wolf teeth and others don't. Time of eruption is variable.

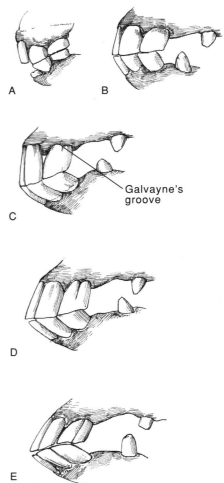

The first permanent molars appear at the back of the mouth behind the baby teeth, between the ages of 9 and 12 months, and the second set comes in (replacing baby teeth) at around 2 years of age. The baby teeth begin to shed at about 2.5 years of age, replaced by permanent teeth in the same order they appeared. All the baby teeth are replaced by the time the horse is about 4.5 years old. At that time he has 36 teeth (12 incisors and 24 molars). In addition, he may have as many as 4 wolf teeth and 4 canines.

Teeth of horses at different ages; notice how the angle of the teeth changes over time. A. Teeth at 2.5 years. B. Teeth at 5 years (full mouth). C. Teeth at 10 years; notice Galvayne's groove beginning at top corner incisor. D. Teeth at 15 years; notice Galvayne's groove extending halfway down tooth and incisors angling forward. E. Teeth at 20 years; notice Galvayne's groove extending down length of corner incisor and incisors protruding farther forward.

Canine Teeth

Not all horses have canine teeth. Male horses have them and so do a few mares, but theirs are much smaller. Canine teeth, if they come in, emerge behind the incisors at about 4 years of age. A horse cutting canine teeth may be irritable and go off feed or salivate a lot. The lower canines are usually larger than the upper ones and may be more than an inch long and fairly sharp. When putting your fingers in the interdental space to paste worm a horse or to encourage him to open his mouth for the bit, be careful to avoid the canine teeth.

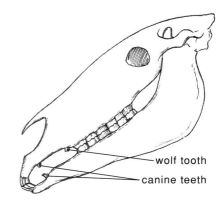

Canine teeth appear a short distance behind the incisors, top and bottom. Wolf teeth are located just in front of the back teeth, often just in the upper jaw.

Canine teeth appear early and are shed with the baby teeth. Permanent canines appear when the horse is between 4 and 5 years old. Canines rarely cause problems and are best left alone unless too sharp, in which case they can be smoothed with a dental rasp by your veterinarian.

Wolf Teeth

There is often confusion between wolf teeth and canine teeth. Wolf teeth sit farther back, next to the molars, and are harder to see. You must open the mouth and look toward the molars to find them. Most wolf teeth appear in the upper jaw and come in when the horse is 5 or 6 months old. Wolf teeth may be large and well formed, sometimes as large as a molar, but they are usually small and shallow rooted. They can cause bitting problems and many horsemen have them removed. If they cause pain and head-tossing, or a bad attitude about the bit, it's best to have them taken out.

Tooth Problems

Good teeth are crucial to the horse's health and comfort. If he has a bad tooth it may cause pain that interferes with chewing. Indeed, the condition of the teeth affects the horse's entire performance. A horse with a bad tooth may resent the bridle, travel with his head crooked, hold his head stiff, or become clumsy because of the awkward way he travels. Horses rarely suffer from tooth decay or cavities but are plagued by a number of other tooth problems.

Teeth Tell the Horse's Age

To keep the grinding surface rough, teeth are made up of three materials that vary in hardness and wear away at different rates. It is this rate of wear that enables horsemen and veterinarians to tell a horse's age by looking at the teeth; different areas of the surface wear away at different stages in the horse's life. The age can also be determined, in a young horse, by noting which teeth are present.

How teeth wear at different ages. **A.** *At 8 days old, central incisors are in.* **B.** *At 2 to 3 months, central and lateral incisors are in.* **C.** *At 8 to 10 months, corner incisors appear.* **D.** *At 2 years, centrals and laterals are wearing.* **E.** *At 2½ to 3 years, permanent centrals are coming in and show no wear.* **F.** *At 3½ to 4 years, centrals are fully in, showing wear, and the permanent laterals are coming in. (Note that the corner teeth are still baby teeth.)* **G.** *At 4½ to 5 years, all permanent incisors are in; also, male horses will have their canine teeth by this time.*

Retained Caps

Human baby teeth come loose as the permanent teeth come in. In horses, as a permanent tooth erupts, the baby tooth deteriorates and grows hollow, becoming a *cap*. Sometimes these do not detach from the gums as they should when permanent teeth are pushing them out. This condition, which is known as a *retained cap*, can cause inflammation of gums, painful mouth, and a sinus problem if the retained cap is in the upper jaw. If not removed, it may cause the new tooth to grow in at improper angle or become impacted.

Sometimes caps are retained when there's not enough room in the jaw for cheek teeth to expand. Signs that a horse has retained caps are bad breath, drooling, tilting the head back while eating, and dribbling feed from the mouth. Loose caps may move and irritate the horse and provide entry for infection. Bacteria from the infection may find their way into the bloodstream and cause serious trouble, so don't neglect any tooth problem.

Impacted Teeth

A young horse (about age 3) may develop bony protrusions on the lower jaw as the permanent cheek teeth come in. First and fourth molars may pinch the second and third ones as they emerge, temporarily inhibiting their upward growth, especially if the horse has a narrow jaw. This impaction creates bony lumps on the bottom of the lower jaw.

Almost always impacted teeth gradually force their way up, and the problem eventually corrects itself. The second molar comes in at age 3; the third emerges at age 4. After the teeth come into their proper places,

"Tooth bump"

"Tooth bumps" on the lower jaw of a young horse indicate that the permanent molars are coming in.

the bony protrusions on the jaw smooth out. The lumps are gone by the time the horse is 5, 6, or 7 years old. Sometimes the impaction doesn't correct itself and the teeth must be surgically removed. If a young horse has "tooth bumps" that rapidly become larger or are quite tender when touched, have your veterinarian examine the jaw and mouth. X rays may be necessary to see if the teeth should be taken out.

Uneven Wear

Often sharp points develop on the molars due to uneven wear. Remember that the horse's incisors and molars continue to emerge upward, with the rate of growth balanced by an equal rate of wear. If lower and upper teeth are not in perfect opposition, uneven wear creates sharp points — usually on the inside of the lower cheek teeth and on the outside of the upper cheek teeth. This is because the upper molars are slightly farther apart than the lower ones, because the upper jaw is wider than the lower jaw. When the molars do not meet evenly, they don't wear properly.

Signs of Tooth Trouble

◆ Difficulty chewing
◆ Quidding (dropping wads of chewed food)
◆ Drooling
◆ Weight loss
◆ Pain in the mouth or sides of the face

Sharp Ridges

Sharp ridges on teeth interfere with the horse's ability to grind his food, and they may lacerate the tongue and sides of the mouth. The sharp edges can cause great discomfort, making cuts along the tongue or inner surface of the mouth. A horse with a jagged tooth may not chew his food completely; it passes through his digestive tract partly undigested, leaving him prone to colic and impaction. In serious cases the horse may starve because chewing is so painful that he refuses to eat adequately.

Malocclusion: Parrot Mouth and Sow Mouth

Jaw and mouth conformation contribute to the likelihood of a horse having tooth problems. A common abnormality is parrot mouth, or overbite — the top teeth (molars as well as incisors) are not aligned with the bottom ones and protrude over the lower teeth. As a result the last lower molar also protrudes up behind the last upper molar. The overlapping tooth grows long because it has nothing to grind against, creating a "hook."

The opposite condition, sow mouth, occurs when bottom teeth project farther forward than top ones, but this is not as common as parrot mouth. In a horse with sow mouth, the last upper molar protrudes down behind the lower one, eventually creating a hook. These sharp hooks sometimes become so long that they puncture the opposite gum. Your veterinarian or equine dentist can file them down or cut them off with nippers or shears.

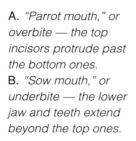

A. *"Parrot mouth," or overbite — the top incisors protrude past the bottom ones.*
B. *"Sow mouth," or underbite — the lower jaw and teeth extend beyond the top ones.*

A   B

Shear Mouth

Another conformation abnormality in some horses is an exceptionally narrow lower jaw. In these horses the molars do not meet squarely and the grinding action eventually wears the teeth at an angle, making the grinding surfaces steeper until the teeth slide past one another — a condition known as *shear mouth*. If this is discovered early, before the teeth start sliding past one another, regular dental care to make sure the grinding surfaces remain normal will keep the problem in check.

Lost Tooth

A horse may occasionally lose a tooth. When this happens the opposing one has nothing to grind against and keeps growing. It may get so long that it grows into the opposite gum unless it is cut off or periodically rasped down. A horse may also lose the central core of a molar, which creates a channel down into the root, opening the way for infection. When this happens the tooth becomes abscessed, infection may break into the sinus cavities, creating a foul pus discharged through the horse's nostrils. A veterinarian must remove the abscessed molar or drill the central core. Sometimes a horse's teeth become damaged or broken, in which case the broken pieces must be removed.

Health Tip

Whether or not a horse will have tooth problems depends on the individual and also the type of feed he has been eating over the years. In general, horses have fewer tooth problems if they've been on pasture most of their lives rather than fed hay.

Dental Check-Ups

As a horse gets older, dental problems are more likely to occur — so you should have his teeth checked. Horses between the ages of 2 and 4 may also have problems. During this period it's wise to have the mouth examined two or three times a year. If necessary, premolar caps or wolf teeth can be removed and sharp edges filed. Most horses, however, get by with a dental examination just once or twice a year.

Floating

If there are any sharp edges, they can be rasped smooth. This filing is called *floating*; the rasp is called a *float*. The chewing surface itself is not rasped, just the rough uneven spots and sharp hooks that develop along the edges of the molars. The veterinarian or equine dentist puts the float into the mouth with the rasping surface next to the tooth that needs filing. He or she then moves it firmly back and forth, slowly at first, then more rapidly as irregularities are rasped down. If the file is kept in close contact with the tooth edges, it will not bump or injure the tongue or cheek. Most horses tolerate the procedure since it is not painful if done properly. There are nerves in the horse's teeth, but they are deep below the grinding surface, so filing off the sharp edges causes no pain.

The veterinarian or equine dentist examines the mouth by grasping the horse's tongue (through the space between incisors and molars) and holding it outside the mouth, which forces the horse to keep his mouth open. The veterinarian then checks the teeth while feeling for sharp tooth edges. Some veterinarians use a gag or mouth speculum to hold the mouth open, but nervous horses resent this and tolerate the procedure better if the veterinarian merely holds on to the tongue. In some cases a nervous horse may need to be tranquilized or a Stableizer used (see chapter 5).

Old Horses Need Dental Care

As a horse becomes elderly, his teeth may start to fall out or wear down to the gums. It may become difficult for him to eat coarse roughage; green grass and soft feeds are easier to chew than hay. It is a good idea to feed older horses by themselves so they don't have to compete for feed with younger horses that eat more rapidly. If your older horse eats too slowly or begins to lose weight, it is time to have his teeth checked again.

FOURTEEN

EYES AND EARS

The horse depends on keen eyesight and hearing to become aware of predators and danger. His eyes are located at the widest part of his head to give a good view of everything around him and most things behind him. In this position they are vulnerable to injury and sometimes get bumped if he tosses his head when startled in close quarters. They may also be flicked by branches. A blow to the eye rarely ruptures it, however, because of its design; there's a fatty cavity behind it, into which the eye can retreat without bursting. Yet eyes and ears are still vulnerable to problems, and the horse owner should be aware of these.

The Normal Eye

To detect a problem, you must first know what a normal eye looks like. Then you can notice abnormalities. The normal pupil is a horizontal oval, large in dim light, and constricting to about half its original size in bright light. The pupil of a blind eye may fail to constrict at all. The upper edge of the pupil has projections from the iris (the colored portion of the eye) that make a tree-like pattern in the pupil — like an awning over it. This keeps out bright light. The cornea has a smooth, clear surface.

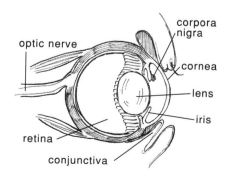

The equine eye

Eye Problems

A scrape or blow may result in pain and watering or a blue spot on the eye. Cataracts make the lens cloudy. Tiny slivers or grass seeds can become embedded under the lid and scrape the eye, causing an ulcer. Infections sometimes damage inner parts of the eye and lead to impaired vision or blindness. You should never ignore an eye irritation in a horse. Minor problems often become serious if they are neglected.

Watch for Signs of Trouble

Some of the simplest eye problems often produce the most obvious symptoms, whereas more serious problems may be harder to notice. Early recognition and proper treatment usually make the difference in outcome. Check your horse's eyes closely whenever something seems amiss, and get professional help for diagnosis and treatment if you are unsure of the problem.

Every visible eye disorder shows at least one of several possible signs:

1. Squinting (keeping the eye closed)

2. Tears or pus

3. Constriction of the pupil

4. Discoloration behind the cornea (the transparent membrane covering the visible part of the eye)

5. Damage to the cornea

Signs 1 and 2 are most common, but often the least serious. If either or both of these signs are combined with any of the others, however, consult your veterinarian promptly.

Squinting and Tears

If these are the only signs of trouble, the horse probably has conjunctivitis, that is, local irritation of the normally pinkish lining of the eyelid (the conjunctiva). This most common and least serious eye problem is usually caused by dust, smog, or some other irritant. If irritation leads to infection, the eyelids may become swollen and there may be yellow mucus or pus at the corners of the eye. Conjunctivitis can be treated with eye ointment or eyedrops.

Sometimes a horse has a watery eye from a blocked tear duct — clogged by mucus or swollen shut by inflammation. If the eye is constantly running or mattering, your veterinarian may have to flush out the blocked tear duct from the nose end, using antibiotic solution and sometimes steroids.

Corneal Injury

A common eye problem is corneal injury from a tree branch, a slash across the eye by another horse's tail, a blow from a whip, or anything that rubs or scrapes the outer covering of the eye. This causes severe pain. If the eye becomes infected the horse will need an antibiotic ointment. If it does not become infected the scratched eyeball may heal within 24 to 36 hours. Large injuries and those that are slow to heal should have daily treatment for a while to help prevent or eliminate infections that might damage inner tissues.

Lacerated Eyelids

Nails, barbed wire, or sharp protrusions in a stall can catch and tear a horse's eyelid. Any eyelid laceration should have immediate attention by your veterinarian. Most of these rips can be repaired. The veterinarian will tranquilize the horse, wash the torn skin, and suture it back together. Most eyelid injuries heal well unless a chunk of eyelid has actually been torn away.

Cataract

Loss of transparency behind the pupil is a cataract, which can be due to low-grade infection or old age. Sometimes a horse is born with cataracts. There is no pain, but haziness dims the animal's clarity of vision. By the time the older horse's lens becomes cloudy the problem is already well advanced. Cataracts generally result in blindness, but sometimes they can be removed — especially in young horses whose lenses are still soft.

Embedded Seeds and Slivers

Foreign bodies that get caught in the conjunctiva (i.e., the delicate membrane that lines the eyelids and covers the exposed surface of the white of the eye) are usually seed awns or other bits of plant material from bedding, feed, or tall pasture. Large seeds and pieces of hay or straw in the eye usually travel to the corner, where tears wash them out. Sometimes they get trapped under the eyelid and cause direct irritation and scraping of the cornea, which leads to erosion or ulceration of that transparent covering. During an eye examination the veterinarian can usually see large foreign particles and flush them free or remove them.

Microscopic Slivers

Tiny particles less than 3 millimeters long are very hard to locate and remove. They easily become embedded in folds of the eyelid lining. When caught there, they create a sore by scraping against the eyeball each time the horse blinks. Symptoms include sensitivity to light, holding the eye closed, discharge and weeping, and constriction of the pupil — all of which may lead the veterinarian to think the horse has a corneal injury, until conventional treatment (topical antibiotics) fails to work. (See under "Burdock" for information on veterinary eye exams.)

Beware of Burdock

Fall and winter sometimes bring unexplained eye problems in horses and cattle — irritation, inflammation, and corneal ulcers. These are generally caused by microscopic barbed "slivers" embedded in the eyelid — bristles from the seed heads of the burdock plant.

Burdock

Like cocklebur, the burdock produces stickery burrs that adhere to hair or clothing. When ripe, they release hundreds of microscopic barbed slivers that are difficult to remove from clothing, skin, or an eye. The slivers are most likely to get into the eye in fall or winter after the seeds ripen. The inflammation persists and does not respond to treatment. The cornea may become inflamed and ulcerated; the eye may turn cloudy and have a white spot or bulge on it.

If a horse has an inflamed eye that is not responding to treatment, you

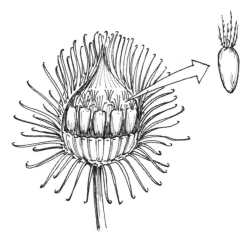

Cross-section of a burr (burdock plant), showing enlargement of tiny seed head and its microscopic barbed slivers

should suspect a burr fragment, especially if there are burdock plants on your property or you have noticed burrs in bedding or hay. Most injured or inflamed eyes respond to topical antibiotics and show improvement within 3 or 4 days of medication, but if an eye does not improve you should take the horse to a clinic that has specialized equipment for eye examinations. Technicians will use a special dye to stain the eye, making it easier to locate the sliver with a powerful magnifying lens. Then it can be removed with small tweezers.

Backyard horses, horses at pasture where burdock grows in wet or shady areas, and draft horses with lots of hair at the fetlocks are at risk for burdock slivers. If there are any burrs caught in your horse's mane, tail, or fetlocks, suspect a burr sliver as the cause of his eye problem. Because burrs hang on the dead plant all winter and into spring, horses may come in contact with them even before the new plants have put forth seeds. A horse can also get a burr sliver at any time of year if his bedding contains shredded burrs that got baled with the straw.

Moonblindness

This is the most common cause of blindness in horses. Long ago people thought the recurring attacks were related to phases of the moon, hence the term *moonblindness*. The main causative factors were thought to be heredity, dark damp stables, unwholesome food, and marshy pastures; the actual cause, however, is an infection. Moonblindness (periodic ophthalmia, or iridocyclitis) is inflammation of the inside of the eye due to an abnormal immune response. It is generally triggered by local or systemic infection. One of the common causes is leptospirosis, a bacterial disease spread by rodents and other animals that contaminate feed and water with their urine.

Often the initial infection or eye injury goes unnoticed until the reaction occurs in the eye. Attacks of inflammation often recur, and unless it is treated properly in early stages this eye problem may lead to blindness in the affected eye.

Symptoms

Early symptoms are watering of the eye and constricting of the pupil; the horse holds his eye shut because it is sensitive to light. The membranes are red and swollen, and the cornea becomes opaque and cloudy. Pus may form in the lower part of the eyeball. Attacks last for a few days or weeks, then the eye seems to get better. The problem recurs, however, a few weeks or up to a year later, and each attack further damages the eye. Close observation of the eye between attacks shows it is not normal; there is a bluish ring around the cornea, and the pupil may stay somewhat constricted except in advanced cases — when it is widely dilated and cloudy.

Causes

One of the most common causes of moonblindness is leptospirosis (see chapter 8). Other causes are streptococcus infections such as strangles, and a filarid worm (a parasitic larva) whose immature stage is spent in the bloodstream of the host horse; the worms are spread to others by biting flies. Occasionally, moonblindness is a side effect of some other problem; the eye condition is not contagious. It may not occur for a year or more after the horse has leptospirosis or other infections; a second or third exposure to the bacteria may bring it on, with a more serious reaction in the eye.

Treatment

A horse with an attack of moonblindness should have veterinary attention. Administering medication and keeping the horse in a dark barn during treatment are important. A preparation of sulfacetamide and prednisolone acetate (human eye medication) is effective. Put two or three drops in the eye twice daily for 10 days to 2 weeks. A drop of atropine in the eye at the beginning of treatment, to keep the pupil dilated, is also helpful. Treatment is most successful if begun early, before the eye has been severely damaged by several attacks. Untreated horses eventually become blind.

Prevention

Moonblindness is somewhat preventable by protecting horses against leptospirosis, streptococcal diseases such as strangles, and parasitic larvae. Because leptospirosis is often a mild disease in horses, many people do not realize their horse has had it unless it produces serious side effects such as abortion or moonblindness.

Night Blindness

Moonblindness should not be confused with night blindness, which is an inherited abnormality in some horses, predominantly Appaloosas. It is inherited as a recessive trait, which means that both parents must be carriers of the abnormal gene for it to show up in the offspring. Most horses affected with night blindness begin to show nighttime visual impairment as foals. In some individuals vision is affected in daylight as well. The problem does not worsen (as with moonblindness) and is unlikely to result in total blindness. Affected eyes look normal; electronic tests are necessary to make a diagnosis. There is no known treatment for night blindness. You can prevent it by not breeding any mare or stallion that has been known to produce a foal with this disorder.

Applying Eye Medications

When giving eyedrops or ointment, have someone hold the horse's halter to keep his head still. Then gently insert the ointment or the eyedrops into the front corner of the eye nearest his nose. The horse will likely close his eye, but you can usually hold the eyelid apart enough at the corner to insert the medication. If the horse is uncooperative, use a Stableizer to keep him calm and relaxed while you apply the medication (see chapter 5).

Ear Problems

The horse's ears are very sensitive and delicate. The ear canal at the base drops straight down about 2 inches (5.1 cm), then turns sharply before reaching the eardrum. This sharp turn keeps insects, dust, and dirt from getting down against the eardrum. The fine inner hairs also provide a protection against foreign material. If a horse becomes sensitive about having his ears handled because they hurt, he may have infection or irritation within the ear. Flies or crawling insects may have gotten into the ear canal, or dirt and trimmed hairs may have fallen down inside. You should never trim the fine inner hairs.

Infections and Trauma

Horses rarely suffer hearing impairment but occasionally lose their hearing due to damage to the eardrum. Penetration by a sharp object, a hard blow to the head, or infection in the middle ear can cause a punctured eardrum.

Inflammation of the gutteral pouch sometimes causes swelling beneath the ear. This pouch is a blind sac that opens into the eustachian tube between ear and throat. The gutteral pouch may fill with pus if it becomes infected, causing chronic nasal discharge.

Occasionally inflammation of the salivary gland below the ear causes pain and swelling. This makes the horse carry his head in an abnormal position and shy away from having his ear handled.

Middle Ear Infection

Infection is the most common ailment of the middle ear. It may be caused by bacteria or viruses that enter the ear through the bloodstream or from the throat through the eustachian tube. Pus may develop in the ear, and the horse may have a fever, carry his head at an odd angle, or experience difficulty chewing. Middle ear infection makes him ear shy. At first the infection may not produce any symptoms except ear rubbing and head tossing. As it gets worse the infection builds up pressure inside the middle ear, affecting the horse's balance and behavior. An untreated infection can eventually lead to skull fracture, convulsions, or death.

Effects of Partial Paralysis

Infection in the middle ear irritates surrounding tissues and can cause abnormal bone growth or damage facial nerves, paralyzing the muscles on that side of the horse's face. The ear and eyelid droop, the muscles of the face and lip sag, and the horse may have trouble swallowing.

The eardrum is tough and unlikely to break, so infection cannot drain to the outside; instead it may spread to bones surrounding the ear at the base of the skull. Inflammation causes new bone growth, fusing the joint between the base of the skull and the bones that extend to the tongue. Eventually the movement of the tongue breaks the new bony fusion, causing a skull fracture and subsequent contamination of tissues around the brain. This kills the horse.

Difficulty in Diagnosis

It is hard to accurately diagnose an ear infection. The head tossing, ear rubbing, and other expressions of pain are often mistaken for other problems or bad manners. Once bony enlargements develop, they show on X rays; otherwise the head tossing, pain at the base of the ear, and — in some cases — eating problems are the only outward signs. Middle ear infections, if discovered early enough, should be treated with antibiotics. Surgery will relieve stress and pressure on the ear bones and prevent eventual skull fracture.

External Ear Problems

Most ear problems are minor ailments involving the external ear; they cause discomfort rather than hearing loss or threat to life. External ear problems include frostbite, lacerations, bites from another horse, skin infections, warts, tumors, and insect irritations. Skin infections and insect bites may cause the horse to shake his head in annoyance.

Flies

Flies and other external parasites sometimes get into the ears and cause discomfort. Horn flies, which are very small, crawl in past the protective ear hairs to suck blood. Black flies leave bloody scabs and crusts inside the ear where they have bitten. Some small flies leave crusts that itch and become sore. You can relieve this soreness by applying medications that contain corticosteroids to the insides of the ears.

To protect the delicate insides of the ears from fly bites, gently smear a small amount of petroleum jelly on the inner surfaces of each ear as far down as your finger will reach, but without contacting the ear canal structure at the bottom. The flies can't bite through it and will leave the ears alone. The jelly is also soothing if the horse already has itchy fly bites. Ear nets drenched in fly repellent also keep flies from getting into the horse's ears.

Ear Ticks

Several ear ticks, including the spinose ear tick, occasionally get into the ears. The larvae of the spinose ear tick burrow into the ear canal and grow there for several months, causing discomfort. A horse with ear ticks shakes his head a lot, rubs his ears, and has a droop-eared, unhappy look. Some types of mites also cause problems in ears.

To get rid of ear ticks or mites, use an insecticidal ointment from your veterinarian. It goes directly into the ear canal with a special syringe or an oil can that has been thoroughly cleaned with soap and water. The long neck on the oil can gets the medication deep into the ear. If the horse is sensitive about having his ears handled use a twitch or Stableizer so he cannot move his head during treatment (see chapter 5). The ointment kills ticks and mites and keeps the horse from becoming reinfested for about 2 to 3 weeks. Worming a horse with ivermectin also eliminates blood-sucking parasites and controls ear ticks, lice, and some types of mites.

A horse with ear ticks or a foreign object in his ear will have a droop-eared, uncomfortable appearance.

Ear Mites

Excessive infestations of ear mites eventually destroy a horse's hearing. The inflammation and build-up of debris in the ear canal interfere with the animal's perception of high-pitched sounds. Mites and ticks also make the ear secrete too much earwax. Your veterinarian may need to administer a special solvent to dissolve the excessive wax so the horse can shake it out of his ear.

Warts

Warts are a common problem in ears, especially in young horses. Horses usually don't mind ear warts, so they are best left alone unless they become large and begin to bother the horse. Most warts eventually go away on their own. However, they cause discomfort when irritated. If your horse suddenly becomes difficult to bridle for no apparent reason, look in his ears for warts or other irritations. Your veterinarian can remove the warts.

Foreign Objects

Water, small twigs, leaves, or even pebbles can fall into an ear under certain circumstances and cause instant and constant pain. It can be hard for a horse to dislodge a foreign object from his ear because he cannot turn the ear completely upside down. All he can do is shake his head from side to side, which may drive the object in even farther. The hairs inside the ear usually keep foreign objects from penetrating too deeply or quickly, so the horse can shake them out. If a horse gets water in his ears, he is usually able to shake most of it out and the rest evaporates.

Pay Attention to Ears

If a horse starts doing a lot of ear rubbing and head shaking or becomes reluctant to have his ears handled, try to discover the cause of discomfort. Have your veterinarian examine any swelling, lumps, or discharges. Don't ignore a horse's attempts to tell you he is uncomfortable. A thorough examination of his ears, and proper diagnosis and treatment of the problem, may be crucial to his future well-being — or even his life.

FIFTEEN

DIGESTIVE TRACT PROBLEMS AND COLIC

Colic is one of the most common causes of death in horses. The word *colic* simply means abdominal pain; it refers to the symptoms rather than to a specific disease or cause of pain. Many things can cause abdominal pain in horses, so there are many kinds of colic. The term is generally used to describe a problem originating within the digestive tract, but the same painful symptoms may also occur with uterine contractions in a foaling mare or with malfunction in the kidneys, bladder, or other internal organs. This chapter examines sources of pain within the digestive tract.

Horses are particularly susceptible to colic. In fact, they are more likely to colic than many other animals, partly because they cannot vomit to unload the stomach of harmful contents or burp as easily as a cow to relieve pressure from gas. The horse's digestion is not as efficient as that of a cow, so he is more likely to have problems with poor-quality feeds or feed changes. Colic sometimes occurs because of the way the digestive tract is constructed — some portions can become displaced and adversely affect or twist other parts.

Colic Can Be Fatal

Digestive tract colic in horses is sometimes fatal because of blockage, twisted intestine, or shock due to digestive tract shutdown and complications. Not all cases of colic are fatal, but a horse with signs of abdominal pain, even though mild, needs constant watching. Call your veterinarian to determine the cause of pain.

Signs of Colic

The horse may be merely uncomfortable or in severe pain, depending on the seriousness of the problem. Mild colic may just make him restless; he may switch his tail, look at his belly, paw, stamp his feet, or lie down. Severe colic will make him noticeably uncomfortable — getting up and down, staggering, or throwing himself on the ground to roll violently or thrash around.

A horse with gut pain may stagger, with legs buckling.

Causes of Colic

Pain accompanies many types of digestive problems — inflammation in the intestines, pressure from gas, spasms, or obstructions. The pain is caused by distention of the stomach or intestines. This may be persistent when the tract is filled with too much feed, gas, or fluid, or *intermittent* when there is localized, periodic pressure from spasms and increased activity in certain parts of the digestive tract.

Spasmodic Colic

The intermittent cramps of spasmodic colic are sometimes due to overexcitement and nervousness but more commonly result from eating spoiled or moldy feed, making a sudden change in diet (e.g., going from hay to green grass, from grass hay to alfalfa, or adding a lot of grain or other feed the horse is not used to), taking a big drink of cold water when the horse is overheated from exertion, or doing anything else that suddenly upsets the digestive tract. The affected horse suffers from short attacks of abdominal pain. He may roll, paw, and kick at his belly for a few minutes, then seem normal for a while until the next wave of pain hits. Sometimes noisy intestinal sounds are audible when the gut is hyperactive.

Symptoms of spasmodic colic often disappear without treatment; the horse may be fine in a few hours. If attacks of pain do not subside, he should have medication to relax the digestive tract and ease the pain. Your veterinarian can administer or prescribe pain medication. Flunixin meglumine (Banamine) is often given for colic. If this does not ease the pain, the veterinarian may administer stronger painkillers, along with a tranquilizer. If the horse rolls too violently during an attack of pain he might twist an intestine, which is much more serious than colic — and fatal unless surgically corrected.

Twisted Intestine

Whenever an intestine becomes twisted or kinked, pressure on blood vessels at the site of the twist block the blood supply to that part of the gut. Also, food cannot pass through the twisted area. If the problem is not corrected quickly, that portion of gut will die from inadequate blood supply. The resulting toxins, shock, gangrene of the bowel, or rupture of the intestine will kill the horse.

Causes

A primary cause is violent rolling due to the pain of colic from some other problem in the gut such as impaction, excessive gas caused by indigestion, or parasite damage that hinders proper working of the gut. Regardless of the original cause of colic, twisted intestine becomes the fatal sequel if the horse rolls too violently in his distress and flips part of the gut over on itself or displaces it. Even after-pains from foaling (uterine cramps) may make a mare so uncomfortable that she

A horse with abdominal pain will often roll (left). *A horse with colic may keep getting up and down* (right).

rolls violently. Any abrupt action by the horse that makes his intestines shift can cause a twist if conditions are right. If part of the gut is quite full and the horse rolls frantically or plays hard (e.g., bucking and running, making fast stops and spin-arounds), the heavy gut may not turn as fast as the horse and thus become twisted.

Gas can also cause a twist even if the horse is standing quietly. A normal by-product of digestion, gas usually passes on through. If the horse has indigestion, however, excessive gas may distend part of the gut. The distention may itself create a twist or displacement, or lead to pain and rolling, which can result in a twist.

Bloodworms play a major role in causing colic and twisted intestine in horses that are not regularly dewormed. The adult worms attach to and feed on the intestinal lining, causing ulcers and irritation; these in turn cause abnormal contractions and indigestion, which occasionally leads to colic. Even more damaging are the actions of worm larvae within the arteries that supply the intestines. These larvae often create blood clots, which can block an artery and cause death of the portion of gut that is supplied by the artery. Bloodworm damage also decreases the flow of blood to the intestines, resulting in abnormal intestinal movements.

When Rolling Is Dangerous

A colicky horse with gut partly distended from gas or food overload is more likely to suffer a twist from rolling than is a healthy horse that rolls just to scratch his back or get rid of flies. The healthy horse doesn't have the overfill and distended gut. In contrast, the overfull gut has more weight and swing to it; rolling may cause it to turn over on itself.

Endotoxic Shock

If blockage from the twist is complete (i.e., food can no longer pass through) and blood supply to that part of the gut is crimped off, the horse suffers severe pain and goes into shock very rapidly. The damaged gut wall (where the intestine is twisted) and the bacteria within the tract will create a fatal endotoxemia unless the condition is promptly treated. If blood supply to a section of intestine is impaired, cells in the gut lining die and slough off, causing the protective barrier to be lost. Toxins then gain access to the bloodstream and circulate through the body, so the horse goes into shock. This can be difficult to reverse.

Diagnosis

Spasmodic colic from indigestion causes waves of pain. The horse shows symptoms of colic and increased pulse rate when the pain hits and is relatively comfortable between spasms. Twisted intestine creates constant pain and consistently elevated pulse. Because blood supply to the gut is shut off, the horse soon shows signs of shock; his mucous membranes (e.g., gums) become pale

Signs of Endotoxic Shock

◆ Increased heart rate
◆ Weak pulse pressure
◆ Dehydration
◆ Cold feet and ears due to circulatory failure
◆ Paleness or reddening of the mucous membranes

and dull or dark muddy red instead of normal healthy pink. He becomes dehydrated as he loses circulating blood volume. Immediate veterinary attention is essential to combat shock and keep his condition stable. Actual diagnosis of twisted or displaced intestine can't be made until the horse has been surgically opened, but sometimes this condition can be felt during rectal palpation. If a twist is suspected, surgical correction is the only hope for the horse and should not be delayed.

Treatment

Most intestinal twists cannot correct themselves, so surgery is the only hope for the horse. Whether he survives the surgery depends on how much intestine is involved, how much it is twisted and damaged, and how long it has been that way. The earlier the surgery is performed, the better chance the horse has; once he goes into endotoxic shock his condition swiftly deteriorates. Also, the longer the wait before surgery, the more chance of the blocked gut rupturing. The horse's condition is also critical. If he is exhausted from pain he may not have a chance, especially if he must be transported a long way to a medical facility.

Every colic should be treated promptly to prevent rolling and the risk of twisted intestine. If a twisted intestine is suspected, the treatment given by a veterinarian may prolong the horse's life and keep his condition stable until surgery can be done.

Prevention

Horses are less apt to suffer twisted intestine if the underlying causative factors can be prevented. This includes minimizing bloodworm damage with a good deworming schedule, and instituting good feeding practices to prevent colic. Feed only good quality feeds; avoid spoiled, moldy, or fermentable feeds that create a lot of gas. Follow a regular feeding schedule to keep the horse from overloading his stomach and digestive tract.

Avoid overloading on grain, especially just before exercise. During exercise the blood supply is shifted to the working muscles and away from the gut, which may result in indigestion as toxins build up due to grain sitting too long in one place in the tract. Feeding a lot of grain right after a hard ride can also cause indigestion and colic because the gut is nearly shut down at that point. It's safer to feed grain at least 4 to 8 hours before a ride, depending on how strenuously the horse will be used, or 2 or 3 hours after the horse has been cooled out and is back to normal.

Enteritis (Inflammation of the Intestines)

Gut pain can be due to inflammation of the intestines. Because of the inflammation the horse does not absorb fluids traveling through the gut and suffers from pain and diarrhea. This happens when there are too many parasites in the digestive tract or when sand, poison, or infection irritate it. Indeed, bacteria such as *Escherichia coli* or *salmonella* produce toxins that are injurious to the intestines. If this is the case, the horse will need immediate veterinary attention and intravenous fluids.

Sometimes the intestinal irritation is caused by a protozoan or a chemical agent such as lead, arsenic, too much phosphorus, mercury, copper, salt, oxalates, nitrates, or poisonous plants. Enteritis also can develop if the horse eats too much soil or sand.

In acute enteritis with abdominal pain the horse kicks at his belly, rolls, and shows other signs of colic. Enteritis is always a serious matter. If a horse develops diarrhea, have him checked by your veterinarian immediately.

Sand Colic

Sand colic is caused by an accumulation of sand or small gravel the horse has swallowed, usually over a long period. This can happen when feed gets blown full of sand during a windstorm or when the horse eats hay on sandy ground (he picks up sand while eating the last wisps or leaves of hay). It can also happen on sandy pastures where plants pull up easily and sand clings to the roots. The sand irritates the gut and accumulates within the digestive tract. Being heavier than the horse's feed, the sand lodges in the cecum or lower loops of intestine instead of passing on through, and may cause a fatal colic.

Diagnosis involves checking the horse's manure for bits of sand — diluting the manure with lots of water and seeing if any sand settles out. Accumulations of sand in the digestive tract can be removed by treating the horse with mineral oil (the veterinarian may give 1 or 2 quarts [1–2 L] by stomach tube) and feeding him a product containing psyllium powder. The psyllium swells when wet and traps the sand, carrying it out through the gut. Bran has been traditionally used for this purpose for many years, but the psyllium works better.

Flatulent Colic

This type of colic is caused by excessive gas in the digestive tract. The build-up of gas may be due to intestinal obstruction or highly fermentable feeds such as lush green grass or too much clover and alfalfa. Excessive gas in the small intestine is generally the result of some kind of obstruction, whereas gas in the large intestine is usually due to fermentable feeds or obstruction.

Intestinal Obstruction

Occasionally the intestine becomes completely blocked, causing dehydration, colic, and death unless the block can be removed or dislodged. Many things cause intestinal obstruction. Sometimes a horse eats indigestible material such as hay twines, rubber fencing material, fibers from old rubber tires or rubber tubs, and hay nets. If the indigestible object or fragment does not pass

A horse with an impaction (intestinal obstruction or constipation) will not eat or drink much and becomes dehydrated. This horse is gaunt and "drawn up" in the flanks, both visible signs of impaction.

through, it may block the intestine. If it doesn't cause a blockage immediately it may sit there for a while, and become the start of an enterolith (intestinal stone), a concretion that may grow until it does block the tract.

Enteroliths

These intestinal stones develop when foreign material builds up layers of minerals around itself, like a pearl in an oyster, creating a mass of minerals, salts, and enzymes. Some horses carry enteroliths in their guts for years without any problems. The stones, some larger than a grapefruit, are only discovered after the horse dies from another cause. Other horses suffer blockage of the gut, which must be relieved by surgical removal of the obstruction.

Impaction

Obstruction can also be caused by extremely dry feed or improperly chewed food becoming impacted in the intestine. Some feed, such as dry beet pulp, makes the manure too firm and leads to impaction. A horse that bolts his feed without chewing may become impacted, as can a horse with bad teeth that cannot chew properly. Eating dry feed without adequate drinking water is a common cause of impaction, especially in winter when horses don't drink enough due to cold weather or frozen water. In these cases the horses become constipated; what little manure they do pass is small and hard, covered with mucus.

If food in the cecum (the large fermentation pouch located between small and large intestine) is too dry for proper digestion, it just sits there fermenting and swelling and does not move on through. The fermentation creates gas and pressure, producing mild persistent pain. Excessive gas causes acute pain and colic.

When impaction occurs in both the cecum and large colon, it is more difficult to alleviate. Several days of treatment and medication are necessary to lubricate the impacted material and get it moving. Sometimes a serious impaction can only be corrected by surgery.

Constipation

A simple fecal impaction may be due to a combination of dry feed, lack of exercise, insufficient water, or eating of bedding. This type of impaction may build up over several days. The horse becomes increasingly sluggish and eventually stops eating (because he is "full") or just picks at his feed. He may lie down a lot. He stops passing manure or passes only small amounts. As the overfull intestine puts pressure on his bladder he may frequently try to urinate, but he passes only a small amount of urine.

Often the blockage can be moved with lubricants — mineral oil administered by stomach tube, and water to soften the impacted mass. The earlier the problem is discovered, the better the chances for simple correction without having to resort to surgery. Fluids given intravenously may be necessary to rehydrate the horse.

What to Do Until the Veterinarian Arrives

If a horse shows signs of colic, call your veterinarian and describe the symptoms, pulse, and any gut sounds. The veterinarian may advise you on emergency treatment measures and tell you to keep the horse on its feet while you wait for help. Tend to the horse in an open area with soft ground so that if he does crash down to roll he cannot injure himself. If the horse is not trying to go down, you don't need to force him to walk; let him rest whenever you can. If he wants to roll or collapse to the ground, keep him moving so he'll be less likely to actually go down. Sometimes the exercise of gentle walking helps the horse pass a bowel movement, which relieves some of the pressure and pain.

SIXTEEN

FIRST AID AND MEDICAL TREATMENTS

Injuries and illness in horses require prompt and proper care to ensure recovery. Always try to assess the severity of a problem to determine if you can treat it yourself. If it's a serious or potentially serious problem, you'll need veterinary help. What you do for the horse as you wait for the veterinarian may make all the difference, especially in wounds that are bleeding profusely, fractures, or some types of colic.

Be prepared to give the veterinarian accurate information over the phone about the horse's vital signs and symptoms. This will help him or her determine whether the horse needs immediate emergency treatment or professional diagnosis, or whether the situation is something you can handle and monitor with veterinary advice.

Emergency Situations

Some situations such as bleeding, fractured leg, or colic carry a large responsibility for the horse owner; your first-aid actions may well mean life or death for the horse. In most instances you have time to call the veterinarian and then give first aid until he or she arrives. However, with serious bleeding or possible leg fracture you should give emergency treatment first and then call the veterinarian (if you are alone) because time may be crucial. If possible, send someone else to call while you work on the horse.

Tetanus Shots Are Important

Minor injuries can be taken care of by the horse owner, so you must know how to treat them. In addition, after any wound that breaks the skin, the horse should have a tetanus shot if his vaccinations are not up-to-date. To be effective, it should be given within 24 hours of sustaining the wound.

Dealing with Wounds

There are basically five types of wounds: bruises, abrasions, incised wounds, lacerations, and punctures. Fortunately most wounds are not life-threatening, but most will need some attention.

Bruises

A bruise is a surface injury that doesn't break the skin, but the severity of the blow may damage the muscle underneath. A horse may get a bad bruise from a fall, a kick, or running into something solid. Bleeding under the skin, as well as lymph seeping from injured tissues, causes swelling. Small bruises are best left alone; they resorb in a few days and the swelling disappears. Large bruises may need to be drained because there is too much fluid to be resorbed. After shaving and disinfecting the area, your veterinarian will insert a large-diameter sterile needle into the lower part of the swelling so fluid can drain out.

A bruised leg benefits from soaking in cold water or ice. Reducing the swelling immediately after the injury can keep a knot from forming on the bruised bone. A bony lump can be an unsightly blemish at best or, more seriously, may interfere with movement of tendons.

Draining a fluid-filled bruise

Abrasions

An abrasion is scraped skin that may bleed or ooze and be slow to heal. Usually the best treatment is to clean the scraped area and apply antibiotic ointment to combat infection, keeping the area soft instead of crusty as it heals.

Incised Wounds

An incised wound is a clean cut. The usual problem, if the cut is deep, is excessive bleeding. Horses have a lot of blood, and losing even a pint or two won't be fatal. Small vessels readily clot and stop bleeding. If the bleeding does not cease after a few minutes, or if the wound is spurting profusely, you should intervene. When a 1,000-pound (453.6 kg) horse loses more than 2 gallons (7.6 L) of blood, the situation could become fatal.

Artery or Vein?

A cut artery squirts bright red blood with each beat of the heart. A cut vein oozes dark blood continually (the blood hasn't been through the lungs yet for oxygen). There can be significant blood loss from a cut major vein, but the bleeding is slower than from a major artery; you have more time to control blood loss from a vein.

Pressure Bandage to Control Bleeding

Don't panic. Apply the cleanest pressure bandage you can create quickly; put it directly over the spurting artery or flowing vein. Towels, shirts, quilted bandages, and the like will work. Your barn first-aid kit should always contain clean cloths, gauze sponges, or old sheets torn into strips and rolled as bandages.

You can make a pressure bandage by placing a roll of bandage or a hard object such as a piece of wood or even a rock (wrapped and padded with clean cloth) against the cut, then wrapping firmly with bandage or strips of old sheet. Continue wrapping until all bleeding stops. It may take many layers wrapped fairly tight. Applying a firm bandage will allow you plenty of time to contact the veterinarian. Leave the bandage in place until he or she arrives. Don't use a tourniquet; it can cause permanent damage because it completely blocks blood flow to the area below it. Always call your veterinarian if you can't get the bleeding to stop within several minutes.

Lacerations

A laceration is a cut with torn, irregular, or jagged edges. Healing is slow because of extensive tissue damage, and there is usually danger of infection. If there's no emergency bleeding problem, clean the wound and take a closer look to determine if stitches are needed. Don't medicate the wound until you have cleaned it. Plain warm water is best for cleaning a wound; many antiseptics are irritating

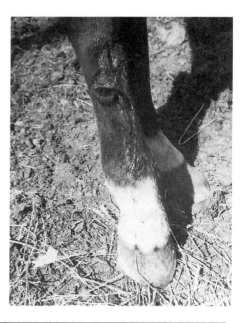

Laceration on hind leg, with skin torn away

Cleaning and Assessing an Old Wound

You might not discover an injury while it is still fresh and clean; it may be hours old, with caked blood and dirt in it. If it's swollen and painful, the horse may resent the handling it takes to clean it. You may have to resort to simply holding a running hose on the wound, if the horse will stand still. Use a Stableizer, if necessary (see chapter 5). The cold water will dull the pain as well as soften and loosen the dirt and dried blood. Then you can scrub the area gently with a cloth and examine the wound as you remove the debris. Hair and dirt can disguise a deep wound, making it look superficial.

Don't worry about new bleeding as you scrub. Unless the bleeding becomes extensive, go ahead and finish cleaning the wound before applying a bandage. Once cleaned, you can decide whether the wound is something you can treat by yourself or whether you should call the veterinarian. Any cut over a joint or into a tendon should have veterinary attention, as should any deep cut below the pastern.

If bacteria get into a joint wound, infection can cause permanent damage to the joint. Tendon injuries are slow to heal, as are cuts in the heel. Damage to the coronary band may cause permanent changes in hoof growth. Injuries to other body parts may need stitches. If you have any questions about a wound or how to treat it, call your veterinarian.

Same injury as opposite, a few weeks later; we treated it with alum (left) *to avoid proud flesh. Four months after the injury, it is nearly healed* (right).

to tissues. When in doubt, don't use them, especially if the horse might need stitches. A disinfectant that burns the tissues will make it impossible for a veterinarian to stitch the wound; it won't be able to heal properly.

Running water from a garden hose is often effective for cleaning a wound, even if the water is cold; cold water can help numb the pain. If necessary you can use a washcloth, gauze sponges, or even paper towels to gently scrub the wound and remove hair, dirt, dried blood, and dead tissue. Continue scrubbing until you are down to clean, pink flesh with all dirt and debris removed.

Proud Flesh

In deep cuts below the knee or hock, proud flesh may develop unless steps are taken to prevent it. Proud flesh is an unsightly rubbery growth of excessive granulation tissue — the tissue that grows to fill in and close up a wound. As the wound fills in, this tissue keeps growing, extending above the surface and preventing the skin from growing over the wound — essentially making too much scar tissue. It's usually best to use a healing medication (from your veterinarian) the first few days until the wound gap begins to fill in, and then change to a caustic medication to retard and prevent growth of proud flesh until the wound has healed and the skin has grown in from the edges.

Alum, an inexpensive white powder used in making pickles, is available at grocery stores and works well to prevent proud flesh. Its astringent action retards growth of proud flesh and is much less painful than traditional caustic remedies. Just sprinkle the powder over the wound, enough to lightly cover the surface. You can use your fingers — a pinch of alum is usually enough.

Stitches

Some wounds and cuts heal more readily, and with less scarring, if the edges can be sutured together. Others are best left to heal by letting new skin grow in from the edges. For successful stitching the wound has to be relatively fresh, with good blood supply to the tissues that are to be put back together. If the wound may require stitches, *do not* medicate it in any way until the veterinarian has seen it. Stitching should be done within 2 hours of injury for a head wound, within 6 hours for a leg wound, and within 8 hours for a wound on the upper body.

The wound must be clean for stitching to work. If dirt and foreign material or dead flesh are still in the wound, the chances of successful suturing are greatly reduced due to the likelihood of infection. The skin pieces must also be still intact, not torn so badly that the edges can't be stitched back together. Stitching works best if the wound only involves skin or the top layer of muscle. A deep muscle cut usually can't be sutured; it must heal on its own from the inside out. Likewise if a joint or tendon sheath is also damaged, suturing may not be possible. Deep lacerations usually need to drain, and there may be some sloughing away of damaged tissue, so it is better to leave them open.

Even if a wound is stitched, the injured area may need additional bandaging to support the stitches and keep the area from opening up again until the tissues have begun to grow back together.

Punctures

A puncture wound doesn't bleed much and may heal first on the outside, leaving a pocket of infection that will abscess or send infection through the horse's body if it gets into the bloodstream. To treat a puncture, first open the hole a little more and clean it with hydrogen peroxide or a disinfectant recommended by your veterinarian (such as dilute povidone-iodine [Betadine]). Then put sterile cotton or gauze soaked in antiseptic into the hole. Change

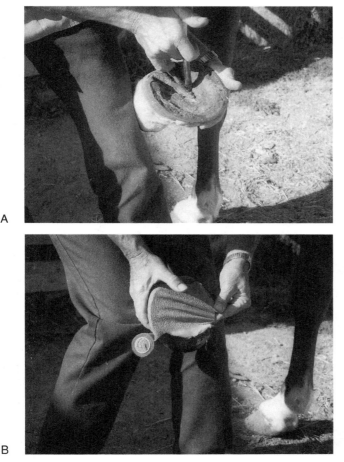

A. *Treat a puncture wound in the foot with iodine.*
B. *Wrap and bandage the treated foot to keep the puncture wound clean between treatments.*

this dressing daily until the wound has healed.

Never use tincture of iodine or other strong medication on open wounds or punctures in the muscle. These strong disinfectants burn the tissues, and the flesh may slough away. Tincture of iodine should be used only for treating thrush, ringworm, and punctures in the foot and for disinfecting a newborn foal's navel stump.

Kick Wounds

Because horses that live together often squabble, kick wounds are common. A flying hoof has enough force to shatter bone, but most of the time the damage is not that severe. It's important to evaluate the injury in terms of the horse's disability rather than the size of the wound. For example, a blow from the edge of a hoof may seem only to graze the skin but in fact may crack or chip the bone underneath. A kick wound may be an open laceration or cut, swelling, or bruise. It can also cause lameness. Sometimes there is irreversible damage under the skin to nerves, blood vessels, muscle, tendons, or ligaments. Call your veterinarian if there is excessive swelling, lameness, a large area of torn flesh, or exposed muscle tissue, bone, or tendons.

You can prevent kick wounds by separating horses at mealtime, introducing newcomers to the herd gradually, taking hind shoes off horses that are going to live together at pasture, and separating individuals that just don't get along.

Fractures

Most fractures occur below the knee or hock when a horse takes a misstep or falls, gets cast on his back against the wall in his stall, or is kicked by another horse. Whenever a horse won't put weight on a leg, suspect a possible fracture. Play it safe and assume a fracture until proven otherwise, or you may jeopardize the horse's future. Do not move the horse. Have the most experienced horseman available take charge of the horse and control him until the veterinarian arrives; a horse in pain may unintentionally hurt a young or inexperienced handler. Don't give any medication until a veterinarian has examined the horse; painkillers may enable the horse to put weight on the leg, causing more damage and potential, more serious injury. If the horse will stand quietly, do nothing until the veterinarian arrives.

Supporting a Fractured Leg

If the leg must be supported, do it properly and carefully. Improper splinting or bandaging can make the problem worse, and trying to bandage a frantic horse is dangerous. A fracture below the knee or hock requires a heavy support wrap, such as ten or more quilted leg wraps under a firm bandage, or terry-cloth towels or pillows wrapped tightly with bandages to hold them in place. No matter how high the break, start low on the leg and work up it with the wrapping. The finished bandage should be at least 10 to 15 inches (25.4–38.1 cm) in diameter. You can add additional stabilization with 3-inch-wide (7.6 cm) boards of appropriate length (making sure they are well padded) placed on the outside of the bandage, not next to the horse. Once bandaging is complete, don't move the horse unless absolutely necessary; wait for the veterinarian to examine him. The horse may or may not have a chance for recovery, but often the difference between success or failure depends on the first aid given.

Help Your Injured Horse to Relax

While waiting for the veterinarian, keep calm. Your horse will do best if he is as relaxed as possible, and he'll stay more secure and relaxed if you are. He can sense your nervous worry and will become more upset and difficult to manage if you are uneasy, which may make his problem worse. Part of your first-aid job is to comfort and console him, being mindful of his mental state as well as his physical condition. If he senses confidence, he'll stay more relaxed and easier to handle as you treat him or wait for the veterinarian. Talk soothingly, and stroke him if that is something he likes.

Head Injuries

The horse is prone to head injuries because of his emotional nature. An upset horse will fling up his head, increasing the risk of striking a stable door, trailer top, or the like. Simple injuries include skin lacerations that must be sutured. A blow to the face (such as a kick) can fracture the facial bone over the sinuses, even though the skin may not be broken. The horse will need reconstructive surgery to realign the bones and antibiotics to prevent sinus infection. If facial nerves are injured his face may become paralyzed on that side, causing the lip and nostril to droop — and also the eyelid and ear if the damage is high.

A common head injury occurs when the horse rears over backward, as when breaking his halter or rope while tied. If he lands on his poll (the top of his head) a fracture may occur at the base of his brain, causing permanent damage or death. After hitting his head the horse may be unable to stand or may get up and stagger. Other signs of serious head injury are bleeding from the nostrils and involuntary rapid eye movements.

A horse suffering from severe head injury must have immediate veterinary attention. If the skull has been injured or bruised, prompt treatment with dimethyl sulfoxide (DMSO) given intravenously may keep the brain tissue from swelling and prevent permanent impairment.

Choke

Choke occurs when a mass of food or a foreign object becomes stuck in the esophagus. A horse can choke on an apple or piece of carrot if he swallows it without chewing it first. He may also choke on hay or pellets if he doesn't chew them enough; a wad of food may be too dry and bulky to slide down easily.

The choking horse drools saliva, and feed may come out of his nose. When he tries to drink, water flows out of both nostrils. When he coughs, he sprays food from nose and mouth. If the blockage is not relieved he becomes dull and depressed, standing with head down. Sometimes a choking horse shows signs of discomfort similar to colic.

When a horse suffers complete blockage — being unable to swallow or cough it up — he is in danger of getting material into his windpipe and lungs because saliva and feed fill the esophagus and overflow into the windpipe. This can lead to aspiration pneumonia, which is almost always fatal; because foreign particles cannot be removed from the lungs, they serve as a constant source of irritation and infection.

First Aid for Choke

Some horses panic, slinging the head and throwing themselves around in attempts to avoid the discomfort. Call your veterinarian as soon as you discover the problem. Try to keep the horse calm until the veterinarian gets there. Don't let the horse drink. Let him put his head down so material can drain out of his nose and mouth. The horse is in no immediate danger because he can still breathe. Just keep him quiet (so he won't injure himself), either by slowly walking him or holding his halter and talking calmly, and gently massaging the throat area to help move the blockage on down.

Shock

Sometimes when a serious injury occurs the body reacts by going into shock. Measures should be taken to prevent or treat shock as soon as it is suspected. Horses often die from shock rather than from the trauma that caused it. Shock is a condition in which circulation in the smaller blood vessels and capillaries shuts down. The horse in shock has pale mucous membranes, subnormal temperature, falling blood pressure, feeble but rapid pulse, and shallow respiration that may be fast at first and then slow. He may be restless and anxious or very dull. He sweats but has cold, clammy skin.

Hemorrhagic Shock

If shock is due to excessive blood loss, the horse should be kept warm with blankets and leg wraps while you try to stop the bleeding and get veterinary help. His circulating blood volume must be restored with intravenous fluids or blood transfusions, and medication should be given to control the shock.

Secondary Shock

If shock is due to colic, injury, severe burns or some other serious injury to body tissues, or toxic infection, there is no actual blood loss; but circulatory failure in the small vessels is triggered by pain and tissue damage. In these cases, fluid from the bloodstream seeps into other tissues and the mucous membranes become pale because no blood reaches them. There is only a short time in which this condition can be reversed — by giving the horse intravenous fluids, adrenalin, and steroids to constrict the small blood vessels (stopping any further loss of fluids through the capillaries), thereby preventing irreversible damage to tissues from lack of oxygen.

Call Your Veterinarian to Treat Shock

It is very important to carefully monitor all vital signs when a horse is injured or colicky. If a horse ever appears to be going into shock, call your veterinarian immediately. If you have steroids such as dexamethasone on hand, ask if you should give the horse an injection. The sooner the shock can be treated, the better the chance of saving the horse. While you wait for help, also try to keep the horse warm and quiet. Blanket him, if possible, to retain whatever body heat he has. Do not move him. Good first aid may make the difference between life and death during that critical short time when the effects of shock are still reversible.

Burns

A burned horse needs immediate first aid, especially if the burns are deep. A severely burned horse may go into shock, collapse, and die. Shock may occur soon after the burn accident or several hours afterwards, brought on by circulatory failure due to loss of fluid from the many small blood vessels injured in the burn.

First-aid treatment for a serious burn includes applying ice packs or cold packs, even cold water from a garden hose, as soon as possible to reduce the depth of the burn and minimize tissue damage. A serious burn from high heat keeps penetrating the tissues for a while, and the sooner you halt this the better. The horse needs painkillers and antibiotics to prevent infection in the damaged tissues. The veterinarian will treat the burned areas and show you how to care for the horse in the following days. Raw areas must be cleaned and treated daily to prevent infection.

Toxemia and infection are likely to follow an extensive burn because the damaged skin cells die and produce toxic substances. Germs may also invade the raw and damaged tissues and start multiplying, producing toxins in the process. This is why a burn should be treated carefully and kept very clean. An extensive burn that covers more than 5 percent of the body surface requires immediate attention — treatment with painkillers and steroids to prevent shock as well as treatment for the burn itself.

A   B

A. Treating a horse that was severely burned in a barn fire, losing the skin off much of his body. B. The same horse a year later after skin grafts and healing, with some hair growing back.

Rope Burns and Friction Burns

If a lead rope or longe line gets caught under the horse's leg or tangled around his body it can cause a rope burn, whether the "rope" is soft cotton, nylon, or a leather lead shank. The same kind of "burn" can also occur when a horse skids along the ground after falling or slides on the backs of his legs.

Rope burn is often harder to heal than a cut or wound, simply because it is a burn. The friction from the rope grating along the flesh (or the flesh sliding along the ground) produces heat as well as scraping and laceration. The scraping damages more skin tissue than does a simple cut, and the heat from the friction tends to kill the skin cells. This kind of damage takes a long time to heal.

Embedded Particles

Another problem in rope burns is irritation from particles of rope; the tiny fibers rubbing into the skin are like little needles breaking off and embedding in the damaged flesh. The swollen, rope-burned area has the added insult of these foreign particles, which can delay healing. For these reasons a serious rope burn may require more medical attention than a bad cut in order to heal properly without extensive scar tissue and persistent inflammation. Consult your veterinarian for advice regarding treatment.

Snakebite

Horses are occasionally bitten by snakes. The seriousness of a bite depends on the type of snake, amount of venom injected, and location of the bite. For example, a rattlesnake bite is generally not fatal if the horse is bitten on the leg, because a large animal such as a horse is not as adversely affected as a small animal. If the bite is on the nose or face, however, swelling caused by toxins may interfere with the horse's ability to breathe and he may die of suffocation. If you suspect a horse may be snakebitten, consult your veterinarian immediately. The horse may need antivenom (depending on the type of snake) or treatment to reduce swelling and tissue damage and to control infection.

Porcupine Quills

Quills in the nose are a common problem when curious horses get too close to a porcupine. The best way to remove quills is to have one person hold the horse by the halter and distract him while another person pulls out the quills one at a time with needle-nosed pliers. A quick, straight jerk works best. If you pull sideways they'll break off and be harder to get out. Once you get them all out, feel the skin to see if there are any broken-off quills. Also feel inside the lips in case there are some in the mouth. Do not clip off the ends of the quills; this pushes them deeper into the tissue and makes them more difficult to remove. If the horse is too upset to stand still while the quills are being pulled, use a Stableizer (see chapter 5) or have your veterinarian give him a tranquilizer.

A horse with porcupine quills in his nose

Urinary Stones

Occasionally a horse suffers from urinary stones, which are formed by crystal-lized mineral salts in the urine. A stone may block the outlet from the bladder or lodge in the urethra, causing discomfort, colic, and painful urination. This is more common in male horses because their urethras are long, narrow, and curved — more easily obstructed. The affected horse may take a stretching stance to urinate but hold the stance a long time, passing only small amounts of urine. If a stone is caught in the urethra the veterinarian may be able to insert a catheter to remove the obstruction. If the stone is in the bladder, surgery may be necessary.

Sheath Swellings

Sometimes the sheath of a gelding swells or becomes sore, usually due to a build-up of debris (a mixture of dirt and urine) inside the sheath. This may lead to infection and urinary problems. Glands in the sheath lining also produce a dark secretion called *smegma*. Sometimes these secretions build up into a soft, waxy deposit or dry hard flakes. If the lining of the sheath becomes irritated, pain and swelling may make it hard for the gelding to let down his penis to urinate.

Signs of Sheath Problems

A horse with a sheath problem may try to urinate without letting down his penis. His sheath may become swollen, or there may be flakes or deposits of smegma clinging to the sides of the penis or hind legs. If the horse has constant discomfort, he may be cranky just because he is hurting. Pain may make him colicky after a ride because he needs to urinate but has trouble doing it. If your gelding becomes grouchy, it may mean he's hurting. The older the gelding, the more likely he has a painful build-up that needs to be washed out.

Any difficulty in urinating calls for careful examination. Swelling of the sheath can be caused by problems other than dirt build-up (e.g., recent castra-tion, local injury, local erupting of a skin problem, bladder stones), so it is important to determine the cause.

Cleaning the Sheath

Some geldings manage fine without ever having the sheath cleaned; others need periodic care. A build-up that is not washed out may form a clay-like ball of debris at the end of the penis in a small pocket just inside the urethra. This "bean" can cause infection or interfere with urination if it becomes larger than the end of your thumb. You can feel into this pocket with the end of your finger to tell whether there's a bean there.

A horse with a bean may urinate in any of several ways. He may spray urine in an obstructed stream, or just dribble, or start and stop suddenly due to the discomfort; he may try several times to urinate before he finishes the job. The bean may become as large as a walnut and cause infection. You can work out a small bean with your finger, but a large one may be so painful that the horse resists. In fact, a large bean may have to be broken into several pieces for removal. If this is the case you might need help from your veterinarian, especially if the horse must be tranquilized. The tranquilizer may be needed to make the horse relax enough to let down his penis.

Washing the Sheath

The few horses that develop beans regularly need routine cleaning of the sheath to prevent build-up. Some need cleaning every few weeks, others once or twice a year. It's a good idea to get the horse used to being handled in this area whenever you groom him, so he won't be ticklish and likely to kick. When you are ready to wash him, squirt warm water into the sheath with a soft-tipped rubber syringe to soften and loosen debris. With your hand lubricated with mineral oil, put your fingertips together and gently enter the sheath, using your hand to scrub away deposits and debris. You can check for beans at the same time.

An easy way to clean the sheath if your horse is used to being hosed with water is to irrigate the sheath with a garden hose, running with low pressure, if the water is not too cold. You can get him used to the hose by putting a slow trickle on his legs on a warm day, working up the legs gradually as he gets used to it, and eventually directing the water into his sheath. A simple irrigation with the hose can wash away a lot of debris; for some horses this is adequate cleaning if it is done periodically. If your horse won't tolerate water from a hose, use a large syringe (without needle) to gently squirt warm water into the sheath.

An even easier way to wash him is when he relaxes his penis after a ride. He may leave it down long enough for you to clean it gently and quickly with a soft, wet cloth when you unsaddle him. If he lets down or urinates, take gentle hold of his penis as he finishes and quickly clean it. A few quick cleanings like this after a ride may be enough to prevent serious build-up. You can check for beans at the same time because the end of the penis is visible. A little care in this area now and then can prevent problems later.

First-Aid Kit

Keep all first-aid supplies together in a handy place, except for injectable antibiotics and epinephrine that must be kept cool. Reserve a place in your refrigerator for those. Also keep cold packs handy in your freezer for use in an emergency as cold compresses. In a pinch you can use a package of frozen vegetables. Other items should be at room temperature (not in an unheated barn or tack room in sub-zero weather). Also, be sure to keep on hand clean buckets (for carrying wash water) and towels. If you don't have your veterinarian's phone number memorized, keep it posted in the barn or by your phone.

A first-aid kit should contain:

- Wound dressings such as nitrofurazone ointment (Fura, Fura-Zone), povidone-iodine, Scarlet Oil or wound spray (a mixture of various oils and alcohol)
- Chlorhexidine (Nolvasan) ointment or zinc oxide (Desitin) for fungal skin problems
- Povidone-iodine solution in a spray bottle
- Tincture of iodine for thrush, quicked hoof, or hoof puncture
- Dexamethasone
- Phenylbutazone (an anti-inflammatory to relieve pain and swelling) in tablets or paste form
- Chlorhexidine solution, hydrogen peroxide, and furacin solution (this must be mixed for you by a veterinarian); these three ingredients may be mixed to make a flushing solution for a puncture wound or for cleaning a deep, dirty wound
- Disposable syringes and needles (various sizes for injectables; large ones for flushing and other purposes)
- Gauze pads
- Strips of clean old sheets rolled into bandages
- Sterile nonstick wound dressings
- Cotton batting or cotton leg quilt
- Conforming stretchy crepe bandaging that sticks to itself (e.g., Vetrap)
- Stretchy adhesive tape
- Duct tape
- Bandage scissors
- Animal thermometer
- Stethoscope
- Flashlight and fresh batteries
- Pocketknife
- Twitch or Stableizer
- Petroleum jelly
- Bottle of mineral oil
- Protective boot for a hoof
- Bottle of alcohol
- A watch with a second hand if you don't wear one
- Disposable razor
- Well-washed squeeze bottles (e.g., ketchup, dish detergent) for applying warm soapy water or rinsing a wound
- Well-washed deworming syringe for administering dissolved tablets.

Estimating the Horse's Weight for Proper Dosage

Most medications and dewormers are given according to body weight, so you'll need to know approximately how much each horse weighs. Although most medications have some leeway for safety, overdosing can be wasteful or even dangerous (depending on the drug) and underdosing may not provide adequate treatment. If you know the horse's weight beforehand you can figure out the proper dosage. Weight tapes are helpful but only give a rough estimate.

The most accurate way to determine a horse's weight is by using livestock or truck scales. At a truck weighing station you should first weigh the towing vehicle and horse trailer empty, then weigh it again with your horse in it (be sure the horse is the only additional cargo). Subtract the difference to determine the horse's weight.

If you can't weigh the horse on a scale, the next best alternative is to use a formula based on two body measurements: heart girth and body length. Measure heart girth just behind the elbow, taking the reading right after the horse exhales. Measure body length from point of shoulder to point of buttocks in a straight line. Don't use a cloth measuring tape because it may stretch. A metal carpenter's tape measure is accurate but noisy and might spook the horse. A plastic coated tape is best, but if you don't have one you can use a piece of cord (or string with no stretch) and mark the spot on the cord with a pen. Then measure the cord with a carpenter's tape or yardstick.

Multiply heart girth (in inches [cm]) by itself (heart girth x heart girth), then multiply that figure by body length (in inches [cm]). Divide the total by 330 (12,018) to determine the approximate weight of the horse. For example, if a horse measures 75 inches (191 cm) around heart girth and 64 inches (163 cm) in body length, you would multiply 75 x 75 x 64 (191 x 191 x 163) to come up with 360,000 (5,946,403). Dividing that by 330 (12,018), you arrive at an approximate weight of 1090.9 pounds (494.8 kg).

Alternative Treatments

Many horse owners are interested in alternative therapies for horses — treatments other than medications, drugs, and surgery. These treatments, which are outside the realm of traditional veterinary practice, include herbal therapy and homeopathy, chiropractic and physical therapy, massage therapy, and acupuncture. In 1992 the American Association of Equine Practitioners (AAEP) acknowledged these unconventional therapies as valid alternatives when carried out or overseen by a licensed veterinarian, to make sure they are used ethically and not misused.

Herbal Therapy and Homeopathy

Both herbal therapy and homeopathy make use of natural substances to treat specific ailments. Herbal treatments are used like traditional medicines, aimed at destroying disease agents or easing a specific ailment. A number of products are available, such as Cough Free (Sure Nutrition, Bensalem, PA) to alleviate coughing, and yucca pellets (Farnum, Phoenix, Arizona) for relief of pain and inflammation in joints. Homeopathy (which literally means "like cures like") involves treating illnesses with substances that in larger doses would produce the same symptoms of the illness. The highly diluted substances are given to promote the body's self-healing efforts by stimulating the immune system.

Chiropractic

In chiropractic therapies, spinal adjustments are performed to relieve pressure on nerves. The theory behind this type of treatment is that the body's ability to maintain good health relies on the nervous system, and that misalignments of vertebrae cause disease by putting pressure on the spinal cord. Chiropractic treatment is aimed at returning proper balance to the total body. The same is true of physical therapy, in which the horse's limbs and body are manipulated to rebalance the whole animal and do away with stiffness, soreness, and lameness.

Back Pain

The horse's spine includes roughly 170 joints, each of which has only slight capability of movement — creating a rigid backbone. Back pain occurs if a vertebra is out of line, suffers inflammation, or is pulled by muscle tension and spasms.

Equine chiropractors can often help horses with chronic back and neck soreness. The chiropractic evaluation and diagnosis include a history of the horse's lameness or stiffness, gait abnormalities, saddling objections, or poor attitude about work; they also take into account unnatural head and neck position, back outline (e.g., humps, sways, or curves in the spine), and deviations from normal gait when moving. As trouble spots are identified, the chiropractor adjusts them with rapid thrusts of the hands along the backbone.

Massage Therapy

More than 60 percent of the horse's body weight is muscle tissue. Muscles can become tight, less flexible, and vulnerable to damage when fatigued or stressed beyond their capability. This can lead to muscle spasms, stiffness, and pain. After locating the area of soreness, a massage therapist can use deep palpations and body manipulation to break up muscle spasms and restore proper circulation of blood and oxygen to the affected areas. This reduces pain and restores normal motion and mobility. Once the tension in his muscles is reduced, the horse's comfort and general attitude greatly improve. Massage can help prevent as well as treat muscle injuries, especially in athletic horses being worked hard.

Acupuncture

Acupuncture is a healing science dealing with the individual animal as living energy rather than as a catalogue of various signs of illness. This ancient Oriental healing art is now used to treat many equine health problems, ranging from arthritic joints to heaves and colic. It has proven beneficial in relieving pain and chronic conditions (e.g., arthritis, respiratory ailments, allergies, and intestinal disorders) that have not responded well to standard medications, and in correcting reproductive problems. Effective veterinary acupuncture is based on both natural and scientific aspects of healing. No one is sure why and how it works, but acupuncture has been effectively used in human medicine by the Chinese for more than 3,500 years.

In Chinese cosmology everything maintains a balance. The two sides of this balance are yin and yang (e.g., dark and light, front and back, cool and warm). When an animal is healthy its yin and yang aspects are balanced. Part of this balance involves the circulation of vital energy, which follows certain pathways through the body. The Chinese believe that all illnesses and physical problems are due to a disruption of the yin/yang balance and this flow of vital energy, with a resulting pattern of disharmony. Stimulating or unblocking the disharmony allows the normal energy flow to be reestablished. The energy flow is connected to all vital organs, muscles, bones, joints, and so on. Thus acupuncture treatments in specific locations along the energy pathways can influence all areas of the body.

Acupuncture Points

The Chinese developed a systematic method of stimulating relationships between external body points and internal organs as they relate to normal body functions. Acupuncture treatment consists of stimulating certain points by means of needle insertion, massage, pressure, or heat. Modern veterinary acupuncture also includes other methods of stimulation such as weak electrical currents, injection, cold laser, and ultrasound. The stimulation of certain acupuncture points can produce changes in blood vessels, nerve impulses, and hormone levels. Its action can be more specific than that of most drugs — and without side effects.

Most veterinary acupuncturists combine the practice with traditional medicine for more complete benefit to the animal being treated. The two methods work well together, being very different but complementary concepts of medicine. For instance, a horse with respiratory infection can be treated not only with antibiotics but also with acupuncture — to stimulate the immune system, ease the pain and discomfort so the horse will feel better and eat better, and improve the function of the lungs.

Traditional Medicine and Alternative Therapies Are Complementary

Alternative therapies are not magic bullets, nor are they better than conventional medicine in controlling disease. One type of therapy is not a cure-all. There is a place for antibiotics, surgical repair, and alternative treatments, using the complementary skills of a number of practitioners. Today the horse owner can take advantage of specialists who are well trained in both alternative therapies and conventional medicine. For readers who want to learn more about alternative therapies, a good overview is *Healing Your Horse: Alternative Therapies* by Meredith L. Snader et al., published by IDG Books Worldwide, 1993.

Euthanasia

Anyone who owns horses must eventually deal with the hard reality of losing one. After all, horses do become old and unable to function, or seriously ill with no chance of recovery (as in an acute and terminal case of colic), or fatally injured in an accident. You may someday face the difficult and unpleasant task of putting your horse down and ending his pain.

Making the Decision

With an old horse, there is time to think about it and plan for it in advance, realizing that at some point you must determine when life becomes more pain than pleasure — and mercifully end it for your equine companion. You'll also have to decide where and how that life should end, who will be responsible for it (usually the veterinarian), and what to do with the remains afterward.

The hardest situation is when you must deal suddenly with an unexpected euthanasia decision, as when a young horse is critically and irrepairably injured, suffers incurable disease, or is moving into the realm of hopeless and painful colic; at this point it becomes a matter of mercifully ending his life rather than prolonging the suffering. Another difficult situation is when the horse's condition is not life threatening but his future athletic ability or general health is seriously compromised and he faces a long retirement. If the prospects for his future are continous discomfort or pain, can you justify keeping him alive just because of your own personal feelings and reluctance to part with him?

The circumstances that force these decisions will vary greatly in each situation, but generally they involve assessing the following: (1) the amount of suffering the horse is going through; (2) the horse's chances of recovery, if any; (3) the possibilities of the horse being useful to someone else, even if his age or impairment have made him useless to you; and last but not least, (4) the cost of treating him or maintaining him in his present debilitated condition. Sometimes a medical "cure" is available for what ails your horse, but the cost is completely out of your financial reach, no matter how much you love him.

Ending His Life

Once the decision has been made, the next step is to determine how and where. With an old horse you may have the option of taking him to a pre-arranged location to put him down, but with an injury victim or colic case you'll likely have no choice but to end his suffering wherever he happens to be, dealing with removal of the body later. In some areas, a rendering plant will take the body. Some landfill dumps will accept large animals, but others won't. Ask your veterinarian or county health department what the local regulations are for animal disposal, and who to contact.

The most instantaneous, and therefore the most humane, method of ending his life is with a bullet or "captive bolt" into the brain. The latter device is used for humane killing of cattle in slaughterhouses and is often the easiest way to shoot a horse, but generally it is quicker and easier to locate a gun. Most horse owners have a hard time emotionally dealing with the task and prefer to have the veterinarian euthanize the horse. The veterinarian can administer an overdose of anesthetic intravenously to induce a deep sleep and stop the heart and other vital functions. The horse becomes unconscious and his life ebbs away painlessly.

Although this method is probably easiest on the horse owner, there are times when it's not feasible to wait for a veterinarian. If a horse is mortally injured in a situation far from a veterinarian (as on a back-country ride or pack trip), or if you live a long distance from the veterinarian and a horse is going through terrible suffering in the last throes of serious illness or colic (which can happen even under the influence of painkilling drugs when the situation deteriorates during his final agonies), you have no choice but to end his suffering yourself. In this situation you must know how to mercifully end a life. Even more agony can occur (for both you and the horse) if you do not know how to properly shoot a horse and he does not die instantly.

You must aim so the bullet goes into the brain and enters the spinal cord, or at least passes through the center of the brain. The mistake most people make when shooting a horse is aiming too low, such as between the eyes or a little higher. In these locations the bullet misses the brain and merely injures the horse and makes him frantic. You must keep in mind that the horse's brain is small and high, located at the top of his head. The best way to shoot a horse is to aim for the middle of his forehead about 4 or 5 inches (10.2–12.7 cm) above the level of his eyes (in the spot where two invisible lines drawn from eyeball to ear would cross in the center of the forehead). Hold the weapon close to the forehead, almost touching it. When shot in this way the horse will drop down immediately, without a struggle, without pain. Sometimes the greatest gift you can give your horse is a merciful end, a release from the pain that binds him.

Choosing a Gun

A handgun is better than a rifle, if there is no one available to hold the horse for you; a rifle takes two hands to shoot. On a small horse, a .22 caliber gun may be adequate, but with a large horse it's better to use a larger caliber bullet; you need a 9 mm or a .38 caliber gun.

PART III

Breeding Horses

SEVENTEEN

SELECTING BREEDING STOCK

Perhaps you've decided to raise horses, breeding a mare or two for your own purposes. In that case the stallion and mares you select should have desirable qualities that will enable their offspring to excel in whatever activity you want to pursue. Or perhaps you want to raise horses to sell, breeding the type of horses you really like — for which you think you'll have a good market — and selling them to folks who have the same goals as you. No matter what your objective, it is important to make informed decisions when you select breeding stock. This chapter reviews the basic aspects of conformation and the key issues in selecting broodmares and stallions; chapter 18 gives an overview of genetics.

Goals and Purposes

The goal of any horse breeder is to raise quality animals that will be outstanding athletes in the careers they are asked to pursue, with the conformation and durability to stay sound while doing it. Animals chosen for breeding stock should be above average in the characteristics you are looking to perpetuate. The responsible breeder strives to improve his or her horses by reducing or eliminating faults. You don't want to perpetuate weaknesses or problems that will mar a horse's chances of being a useful animal. In raising horses to sell, you have an ethical responsibility to your potential customers; you should not produce horses that may cause hardship and heartache due to genetic defects or conformational faults that lead to unsoundness.

What Breed?

There are good horses and poor ones in every breed; there's no "best" kind of horse. Each breed has outstanding and versatile individuals. If you have a favorite breed, you will probably choose horses from within its ranks. If partiality for a certain breed is not a factor, look for horses of the breed or type best suited for the activities you want them to do. In raising horses for yourself, you may be interested in crossbreeding to combine the best traits of two or more breeds.

Be a Critical Judge of Your Horses

When selecting breeding stock or choosing a stallion to breed to, there are some basic rules to follow. Avoid any known genetic defects, such as night blindness in certain Appaloosas, combined immune deficiency (CID) in certain Arabians or part-Arabians that carry this defective gene, and hyperkalemic periodic paralysis (HYPP; a metabolic muscle problem) in certain Quarter Horses, Appaloosas, and Paints. Also avoid conformational faults such as crooked legs that will lead to lameness and unsoundness with hard use. Never let emotion — fondness for a certain animal, bloodline, or color — overcome good judgment. A favorite animal should be judged as critically as any other when it comes to faults or virtues that could be passed on to offspring.

Don't Be "Barn Blind"

A good breeder is ruthless when judging a horse, seeing faults as clearly as good attributes. Too many breeders are "barn blind," unable to see weaknesses in their own horses. Learning to look objectively at horses (your own as well as someone else's) and analyze their good and bad traits is crucial to breeding horses.

Visualize Perfection

Knowing in what areas your own horses are weak is the first step in becoming a good breeder, along with having a picture in mind of what "perfection" would be. No horse is perfect, but a breeder should always be striving toward a goal, applying a working knowledge of how to balance faults and strong points in choosing a stallion that will improve on the attributes of a given mare. Visualize what could result from a prospective mating, choosing animals that complement each other to produce offspring that are even better than the parents.

Look Farther Than the Pedigree

Many breeders are obsessed with pedigree, seeking popular bloodlines or horses that carry the blood of a favorite stallion or stallions. Yet no matter how perfect the match seems on paper, good offspring are not produced by pedigree alone. A foal is no better than his sire and dam's physical or temperamental compatability, and the conformation and disposition that could result. However, the study of pedigrees can be a useful tool if a breeder knows how to determine the ways in which different lines cross with each other and what they are likely to produce. A study of pedigrees is a science and an art. It should never be merely a collection of names. Pedigree without good conformation and disposition is worthless. (For more information on genetics and heritability, see chapter 18.)

Conformation

Conformation relates to body proportions and angles — how the horse is built. The way his bones, tendons, muscles, and other parts are put together plays an important role in his ability to perform well and remain sound with steady work. A sound horse has no weakness or injury to impair his abilities or make him lame. It is very important to critically evaluate the conformation of breeding stock.

A horse should be well balanced, with a smooth and pleasing overall appearance. He's an athlete; his main purpose is motion. So judge him while he's moving as well as when standing still, to see if he is clumsy or agile. He should travel "straight and clean," with straight feet and legs that move forward in straight, not crooked, lines.

Judging Conformation

Stand directly in front of (and then behind) the horse as he's being led toward you and away from you, at both a walk and a trot. Lameness, certain types of unsoundness and bad conformation are more evident when he's moving. If his front feet paddle outward or wing inward, or if any of his feet come too close together when in motion, the horse does not have good conformation.

A base-narrow horse (feet too close together) or one that is pigeon-toed (front or hind) will generally paddle outward as he travels.

A base-wide horse (feet wide apart) or with feet pointing outward (front or hind) will wing inward as he travels.

The Feet

A good judge of horses starts from the ground up because a horse is only as good as his feet and legs. The feet should be well shaped and of appropriate size. A large or heavy horse needs feet proportionally large to support his weight. A small horse needs smaller feet. Some horses have feet too small for their bodies. Small feet may look neat and dainty, but a heavy horse must have sufficient foot structure to support his weight or else he'll become lame; the small foot is subjected to too much concussion. Indeed, small feet on a large horse can lead to navicular disease. On the other hand, feet too large for a horse will make him clumsy.

Hooves should be wide at the heel, not contracted. Wide heels have some give, springing apart when weight is placed on the foot and absorbing some of the concussion when the foot hits the ground. Otherwise all the shock and jarring would be transmitted directly to the bones of feet and legs and to the joints.

If a horse has good conformation, his feet wear evenly. He doesn't wear one side lower than the other or "dub off" one side of the toe more than the other. Conformation of legs often determines the shape of the feet, the way they wear, and their flight — how they are picked up, swing through the air, and are put down. When the horse moves, the foot should break over (i.e., leave the ground) squarely — directly over the center of the toe, not off to one side.

When a horse moves, his feet should break over squarely at the center of the toes and travel forward in a straight line.

Pastern Angle

Ideally the angle of foot and pastern should be approximately 50 degrees. Hind leg angles can be a little steeper. Foot and pastern angle should not be too sloping or too upright. If it slopes too much the pastern will be weak, with the fetlock joint sometimes hitting the ground when weight is placed on the leg. If it does not slope enough the foot won't have enough give; the horse will have a choppy, jarring gait, which is uncomfortable to the rider and likely to damage the horse's feet and legs with excessive concussion.

*Pastern angles. **A**. Normal. **B**. Too much angle (weak construction). **C**. Too steep, resulting in choppy gate and excessive concussion.*

The Body

When you judge the body of a horse, look for a deep heart girth and well-sprung ribs (i.e., a wide barrel). A horse needs plenty of room for heart and lungs. Horses that are wide through the ribs and deep in the chest usually have more endurance than narrow, shallow-bodied horses. When viewed from the front, the horse's ribcage should be wider than his shoulders. There should be sufficient width between his front legs (about 6–12 inches [15.2–30.5 cm], depending on the size of the horse), yet the chest and front end should never be too wide in a riding horse because that will reduce his speed and agility. Wide shoulders are more desirable in a draft horse, for power in pulling and fitting a collar.

Muscling

Good muscling between the front legs looks like an inverted V. All muscles in a riding horse should be long and lean. Extreme heaviness of muscle on thighs, arms, and forearms hinders free action. A heavily muscled horse may be able to give a quick burst of speed but will tire much sooner than the trim horse and may have problems with muscles tying up if he is worked hard. Speed and stamina come from long, lean muscles, not bulky ones.

Head and Neck

The horse uses his head and neck for balance and to collect and extend. A long neck is more helpful to athletic ability than a short, thick one. The horse must be able to swing his head up and down like a pendulum to shift his weight and balance at each stride. The head should be set onto the neck at such an angle that the horse can flex at the poll.

Also, the upper and lower jaw should match; the horse will have dental problems if the upper and lower teeth do not meet properly. You may have to open his mouth to see how his teeth line up.

Neck conformation. **A.** *Normal.* **B.** *Too short and thick (cresty neck).* **C.** *Thin and weak (ewe neck).*

Body Proportions

For best athletic ability, a horse should be the same height at the withers as at the highest point of his hips (croup), and his height should be proportionate to his length. A long-bodied, short-legged horse or a short-bodied, long-legged horse is not desirable. If a short-backed horse has legs too long he will overreach when he moves, striking his front feet with his hinds, unless he has a long hip and shoulder, which gives him a long underline. The shorter the back, the better, provided it's due to a long hip and shoulder. A horse that is too long-backed may go lame because he cannot "round" his back to carry weight; a long back is generally a weak back.

In young horses up to 3 years of age, the croup may be higher than the withers; the latter do not attain full height as quickly. Yearlings and 2-year-olds often have no withers, but by the time they are 3 or 4 the withers catch up with the croup. By then the horse looks balanced and holds a saddle better. In some horses, the croup remains higher than the withers.

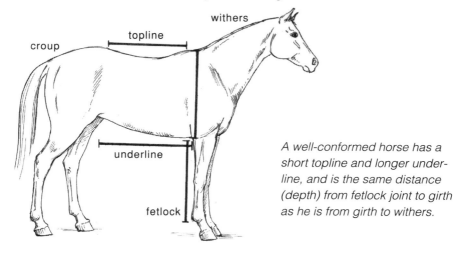

A well-conformed horse has a short topline and longer underline, and is the same distance (depth) from fetlock joint to girth as he is from girth to withers.

Shoulders and Withers

A horse with good withers and a long, sloped shoulder holds a saddle well, without needing a tight cinch and breast collar to keep it in place. A shoulder too short and straight (i.e., up and down instead of sloping back) limits free action and gives a choppy stride. A horse with short, upright shoulders usually has a short neck and short pasterns, creating more jar and concussion.

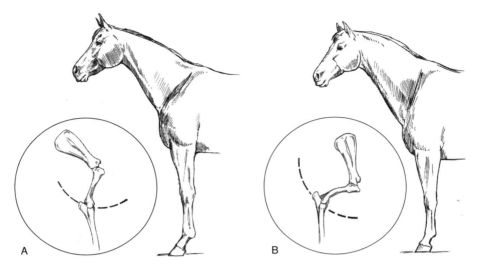

A. *A well-sloped shoulder is usually associated with proper pastern angle and good withers. The sloping shoulder allows greater freedom of leg movement and longer swing of stride.* **B.** *An upright shoulder is associated with upright pasterns, which result in more jar and concussion. The upright shoulder produces a short, choppy stride and less swing of leg.*

A good horse is square. This means he is roughly the same height at the withers as the distance from the point of his shoulder to the point of his buttocks. The length of his body is thus the same as his height at the withers. Except for his head and neck, his body fits into a perfect square because the withers and croup are the same height.

A good horse is the same distance from point of shoulder (A) to buttocks (B) as he is from ground (C) to withers (D). He's also the same distance from withers (X) to girth (Y) as from girth (Y) to fetlock (Z).

Front Legs

Front legs should be perfectly straight when viewed from the front. A line dropped from the point of the shoulder should go straight down the center of the leg — through forearm, knee, cannon, fetlock joint, pastern, and hoof. Toes should point directly to the front; feet should be exactly the same distance apart as the distance between forearms where they come out of the chest. Knees should be large, flat, and well proportioned — as large as possible without looking clumsy. A small, pinched-in knee crowds bone and tendons and hinders smooth action. The front of the knee should be flat with no roundness, and the outer edges should look square.

FRONT VIEW

Bowlegged   Knock-kneed   Bench-kneed (cannon bones set too far apart)

SIDE VIEW

Back at the knee (calf knees)   Over at the knee (buck knee)   Tied in knee   Cut in at knee

Examples of front leg conformation

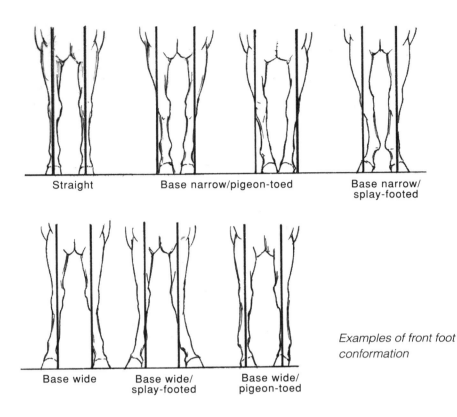

Straight   Base narrow/pigeon-toed   Base narrow/splay-footed

Base wide   Base wide/splay-footed   Base wide/pigeon-toed

Examples of front foot conformation

Cannon Bone

The cannon bone should be centered under the knee. If it is offset it will put more strain on the splint bones (the small bones that sit on each side of the cannon bone) and make the horse likely to develop splints (i.e., bony lumps on the cannon). If the cannon bone is too far back the horse is *calf-kneed*, a very weak conformation that leads to knee fractures. The opposite problem is called "over at the knee" or "buck knees."

Flat Bone

Cannon bones should be wide when viewed from the side. This is called *flat bone*; it describes the combination of bone

A   B

*"Flat bone" (desirable; **A**) versus "round bone" (**B**) in which the cannon bone and tendon are too close together.*

and tendon that gives the lower leg the appearance of being wide from front to back. The tendon should be set well back of the cannon bone for best action and strength. A common fault is being "tied in" close to the knee or hock; this means the tendon is too close to the bone, causing friction, wear, and eventual unsoundness. Undesirable "round bone" occurs when cannon and tendon are too close together all the way down. This type of leg does not hold up well.

Good Bone Is Proportionate to the Horse's Size

Coarse bone means a heavy, clumsy cannon. *Fine bone* means insufficient support under the horse — cannon bones too small in diameter. A horse should have about an 8-inch (20.3 cm) circumference below the knee per 1,000 pounds (453.6 kg) of body weight. Thus a 750-pound (340.2 kg) Arabian has sufficient bone at 6-inch (15.2 cm) circumference, whereas a 1,250-pound (567 kg) Quarter Horse needs a 10-inch (25.4 cm) circumference to be equivalent in strength.

Base Narrow or Wide

The base narrow horse has feet too close together — a condition found most often in horses with large pectoral muscles and wide front. Base narrow conformation is usually accompanied by pigeon toes, which point toward one another instead of straight to the front. This puts strain on pastern and fetlock joints, often leading to windpuffs in the joint or ringbone or sidebone on the outside part of the foot. (Windpuffs are soft enlargements in the fetlock joint area, ringbone is a bony enlargement at the upper or pastern section joint, and sidebone is bony enlargement at the cartilage just above the hoof.) The hoof wears too much on the outside edge because it tends to land on the outside wall instead of squarely flat. Base narrow, pigeon-toed horses usually paddle with their front feet, swinging them outward as they travel, which cuts down on speed and agility. The pigeon-toed horse usually paddles whether he is base wide or base narrow.

The base wide horse has legs farther apart at the feet than at the forearm. This is usually found in horses with narrow chests. Most base wide horses are also splay footed, which causes winging inward as the legs move. The inside of the limb is under great stress; windpuffs are common, and ringbone or sidebone can occur on the inside of the foot. Splay-footed horses wing to the inside, whether they are base wide or base narrow, and wear the feet excessively on the inside wall.

Hind Legs

Hind legs play a major role in how the horse travels — whether or not he is fast or slow, clumsy or agile. Stifle and hock joints work together in unison. The angle of the stifle and hock will be the same in any given individual; a horse with a too-straight stifle will also have a too-straight hock.

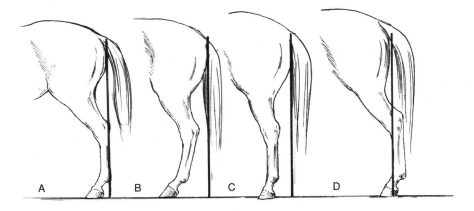

Hind leg conformation viewed from the side. **A.** *Proper hock and stifle angle.* **B.** *Sickle hocked (too angled).* **C.** *Post legged (too straight).* **D.** *Out in the country or camped out (too far back).*

The Hock

Because this is the hardest-worked joint in the body, it must be large and sturdy. It should be flat on the outside and only slightly rounded on the inside edge. Proper angle is very important for speed, agility, and soundness. Viewed from the rear when the horse is standing squarely, the hind leg should be perfectly straight. Viewed from the side, the back of the hock and back of the cannon should be perfectly straight and perpendicular to the ground. If the hock joint is too straight (i.e., set too far underneath the buttock) or too angled (i.e., hock and cannon not perpendicular to the ground — hind feet too far under the body and cannon bone angled), there is too much strain on the joints.

A horse with too straight a leg (i.e., too straight at stifle and hock) is likely to suffer dislocation and locking of the stifle joint, that is, upward fixation of the patella. Too much angle in the hock, called "sickle hock," puts strain on tendons and joint and makes a horse likely to develop curb or spavin. (Curb is strain and enlargement of the tendon at the back of the hock; spavin is enlargement of certain bones in the hock joint due to excessive strain and arthritis, causing hind leg lameness.)

Base Narrow or Wide

As in the front legs, hinds can be base wide or base narrow. These deviations from straight (when viewed from behind) put more strain on hock and fetlock joints, leg bones, tendons, and ligaments, leading to possible breakdown and unsoundness as well as loss of agility and coordination.

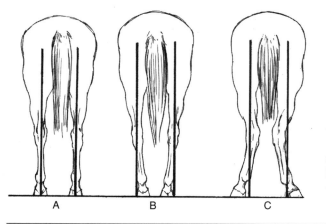

Hind leg conformation viewed from the rear. **A.** *Straight (normal).* **B.** *Base narrow.* **C.** *Base wide (cow hocked).*

Unsoundness

An unsoundness is any physical defect that interferes with a horse's usefulness, making him lame, unsafe to ride, or unable to perform as he should.

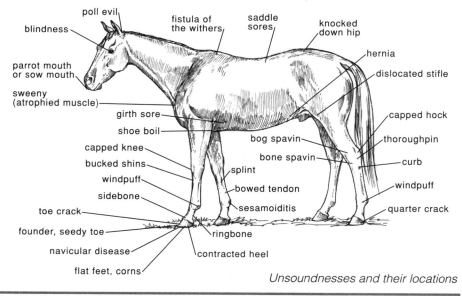

Unsoundnesses and their locations

Disposition and Temperament

Do not overlook disposition when selecting a horse. Even if a horse has excellent conformation and athletic ability, he will be difficult to work with if he has a bad disposition — and so might the foals. Temperament (e.g., bold or timid, friendly or independent, calm or nervous) is heritable, so choose animals with a pleasant basic nature. This applies to both stallions and mares.

Selecting Broodmares

Find mares that embody attributes you want in the offspring — good conformation and disposition, with the athletic ability to perform as you want. If you are raising horses to sell, also pay attention to pedigree and bloodlines; popular bloodlines are more sought after than little-known individuals, even if the latter are excellent animals. However, never sacrifice conformation, disposition, or athletic ability for name alone. A few "top" horses got to the top mainly because their owners had a lot of money to publicize them; such horses don't always produce top offspring. When buying breeding stock, look at a lot of horses before making any final decisions. Buy the best you can afford, and then try to improve your animals by selective breeding. Your mares will determine the success of your breeding program. Only good mares produce good foals, no matter how much you spend on stud fees.

Age and Price

The highest priced females are fillies with good parents and young to middle-aged (6- to 12- or 13-year-old) broodmares that have produced good foals. The most economical age groups are the very young (unproven) and older matrons. Beware of buying an older mare that hasn't had a foal for several years. Older mares can become infertile; one that has consistently produced foals may be more likely to keep producing than an old mare that hasn't had one lately.

Stallions

Many breeders like to have a stallion of their own, but this is not necessary. You can send a mare to be bred to someone else's stallion, or use shipped semen and have your mare bred at home by artificial insemination. More and more breeders are now using the latter option.

Buying a Stallion

Buy the best stallion you can afford; you are counting on him to improve your stock. He will produce many foals, whereas a mare will only produce a dozen or so in her lifetime. If your stallion is very good, you may be able to raise foals that are better than their mothers. You may decide to gamble on a weanling colt. This is probably the cheapest way to buy a stallion, but also uncertain — you can't always tell how he will look as a mature horse. Judge him closely. If you like what you see, look at his parents and siblings to get a glimpse of traits he might display when he grows up. Whenever you buy breeding stock (male or female), try to see as many of the horse's relatives as possible. If there are any characteristics you don't like, don't gamble.

Age and Price

A stallion that is 2 to 4 years old may cost more than a weanling or yearling because he's starting to mature, you can see what he'll look like, and he has had more training. The mature stallion age 5 to about 12 or 15 will probably be the most expensive because he is at his peak, easily judged, and has already produced offspring. It's always an advantage to see his foals. Sometimes the best-looking stallions don't sire the best foals, and plain animals may sire exceptional ones.

Before you buy any animal for breeding purposes, mare or stallion, have it checked for breeding soundness and fertility by a veterinarian.

Choosing a Stallion for Your Mare

If you only have a few mares, it is easier to breed them to an outside stallion than to own one yourself. This eliminates the expense and work of keeping a stallion, and it allows you to choose a different stallion for each mare to more perfectly complement her individual attributes. You can select a stallion that is strong where she is weak, with correctness where she has faults. This increases the likelihood that the foal will have better conformation or disposition than she does. You can probably produce better foals by choosing a different stallion for each mare than by breeding them all to the same one, unless your mares are a uniform group with similar genetics (i.e., closely related).

Shipped Semen

If you don't want to send a mare to be bred, shipped semen is an alternative. Many breeders are now offering this option. It greatly increases your choice of stallions and eliminates the expense of transportation and care of the mare as well as a lot of risk — especially if the mare has a foal at her side. The mare must be palpated daily by a veterinarian to determine her time of ovulation, after which the stallion owner will collect the semen and ship it by overnight air to be deposited in the mare the next day. The veterinary expense is cheap compared to the money saved in other ways. Check to see if your breed registry allows this type of breeding.

Crossbreeding

If you are breeding for excellence in a certain sport, consider crossbreeding (unless it's a sport that requires a purebred, such as Thoroughbred racing). For performance activities or any sport that demands a combination of agility, speed, and endurance, a good crossbred can do well. For versatility in English or Western pleasure riding, competitive trail riding, or speed events at a horse show, a good crossbred can be outstanding.

When crossbreeding, choose outstanding individuals that will enhance your goals with the attributes you want. For example, breeders of sport horses for jumping, dressage, and eventing often enhance their goals through careful crossbreeding to retain the athletic ability of the Thoroughbred while adding more substance and the mellow nature of a heavier framed breed such as a draft breed or warmblood. A *warmblood* is any European breed of large horse that includes some Arabian blood in its ancestry. The term is also used to describe the result of crossing a heavy (draft) horse with an Arabian or Thoroughbred.

Fertility Record

In buying a stallion or selecting one to breed to, be sure to check his fertility record. Some stallions have less ability to settle mares than others. When sending a mare to be bred, even if you get a live-foal guarantee (or when using shipped semen), don't close your eyes to a possible fertility problem. If the mare comes home empty you can always send her back, but that costs extra in transportation, board, and veterinary bills. You may also have the expense of keeping her another year with no foal if she doesn't settle by the end of the breeding season.

When selecting a stallion for your mare, weigh as many factors as possible and use your best judgment. Breeding is a gamble, but you can increase the odds of success with careful homework.

EIGHTEEN

GENETICS

When breeding horses, the mating matchups determine the quality of the off-spring. You need to have a knowledge of genetics to predict what traits the foal will inherit from its parents.

When selecting the mare and stallion to be mated, you are shooting in the dark if you only look at them as individuals; you must also know something about their genetic inheritance. Some recessive traits are not evident in the parents but may appear in offspring if the foal inherits a recessive gene from each parent. Some perfectly normal and outstanding horses are carriers of undesirable genes and may produce an abnormal foal, so it is important to know as much as possible about the genetics of the horses you wish to mate.

Inheritance of Traits

An individual's characteristics are determined by genes and chromosomes. *Genes* are chemical codes that transmit various traits. They are located on *chromosomes*, which are strands of genetic material carried in every cell of the body. Chromosomes occur in pairs. As cells divide, half the genetic material goes with the new cell; it is a perfect replica of the old one (except when chromosomes become twisted, damaged, or misplaced, resulting in mutations).

Each cell contains chromosome pairs that carry the code of inheritance. Egg and sperm cells have only one chromosome from each pair, so when they unite, the newly formed pairs are a joining of one from the male and one from the female — the offspring gets half his genetic material from each parent. Because there's such a variety of genetic material in the many genes and chromosomes, the possibilities for different matchups are great. No two foals (even full brothers or sisters) are exactly alike unless they are identical twins.

Look at Ancestors and Relatives

The ancestors of a horse contribute to his makeup. Try to see as many as you can, in photos if not in the flesh. They'll give a clue as to what the foal might look like. The first fourteen ancestors (sire and dam; grandparents and great-grandparents on each side) are the most important to consider; any ancestor farther back won't have as much influence on the characteristics the foal might inherit, unless the ancestor is present several times. Study the entire pedigree as far back as you can. If most of the individuals in a horse's pedigree have desirable traits, that horse will likely produce a good foal when mated with an individual of equally good pedigree — provided both are typical representatives of the horses in their pedigrees. The sample pedigree chart below goes back four generations.

Pedigree Chart of Arabian Stallion "Surrabu"

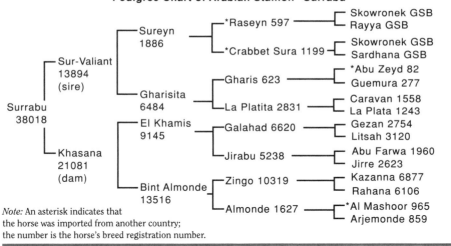

Note: An asterisk indicates that the horse was imported from another country; the number is the horse's breed registration number.

Dominant and Recessive Traits

Most characteristics are determined by several sets of genes. It's impossible to predict what the foal will be like, except in traits affected by only one pair of genes. Some genes are dominant and others are recessive. When chromosomes join to form the new individual, two genes (one from each chromosome — one from each parent) control the trait. If one is *dominant*, the trait it expresses will appear in the foal. The trait represented by the *recessive* gene won't be expressed with the dominant gene present but might be passed to future offspring if it is not masked by a dominant gene in that next union.

Recessive Traits May or May Not Appear

A recessive trait might appear in the next generation if it is carried as a hidden recessive by both parents. Their mating might produce a foal whose genes line up to combine the two recessives.

Dominant and recessive genes can be combined in three different ways:

1. The two dominants can come together, producing an animal that is *homozygous dominant* for that trait (*homo* means "same"). In this case the only genes it carries for that trait are dominant; therefore it not only expresses that trait but can pass on no other trait to its future offspring.
2. The two recessives can come together, producing a *homozygous recessive* individual that expresses the recessive trait and can only pass on this recessive trait to its future offspring.
3. The foal can inherit a mixed pair of genes — dominant and recessive — and be *heterozygous*. In this case the foal itself shows the dominant trait, because any dominant gene always masks the presence of a recessive one, but this animal can pass on either gene (dominant or recessive) to its future offspring.

Genetics of Color

Traits controlled by dominant and recessive genes, such as those for color, are readily seen. An example of color inheritance can help explain how traits are passed from parent to offspring. For instance, let's say G is the gene for gray and g is the one for non-gray. The only time a horse can be a color other than gray is when a gg combination (two recessives) exists in the gene matchup because gray is the dominant trait. The foal may inherit from one parent a recessive chestnut gene or a bay gene (which is dominant over chestnut) but will always turn gray if he has the recessive gg combination.

Gray, Bay, and Chestnut

Gray, bay, and chestnut are easy to predict because they are controlled by fairly simple dominants and recessives. In these three colors gray is always dominant. A gray horse is born bay or chestnut and becomes gray later because of the dominance of the graying gene. The gray horse must have at least one gray parent; gray can never skip a generation like chestnut can.

Bay is dominant over chestnut but still recessive to gray. Chestnut is always recessive; a chestnut bred to a chestnut can produce no other color but chestnut. Chestnut may appear from a mating of two bays or two grays, or a bay and a gray, if each parent carries a recessive chestnut gene.

Sample Crosses

In the first cross (below), the "pure" dominant parent will decide the color of the foal when mated to a recessive. The foal will then carry a recessive color gene that can be passed on to its offspring.

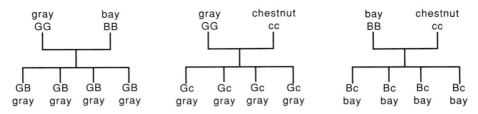

A gray horse that carries a bay gene will produce mostly gray foals but still *can* produce a bay foal. When mated with another gray horse that carries a chestnut gene, the foal has these four possibilities (below): pure gray, gray but carrying a chestnut gene, gray but carrying a bay gene, and bay carrying a chestnut gene.

If a gray horse that carries a recessive chestnut gene is mated to a bay horse that carries one also, the foal has these four color possibilities (below): gray carrying a bay gene, gray carrying a chestnut gene, bay carrying a chestnut gene, and chestnut — if he inherits both recessive genes.

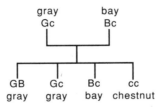

Note: Gray = dominant over all colors; bay = dominant over chestnut; chestnut = recessive; capital letters denote dominant gene; lowercase letters denote recessive gene.

Color Inheritance: Gray. A pure gray mated to an impure gray (carrying a bay gene) will produce nothing but gray foals, but some of them may carry a bay gene and pass the bay color to their offspring if mated with another horse that carries a bay gene. A pure gray mated to a bay carrying a chestnut gene will produce nothing but gray foals, but some will carry bay genes and some will carry chestnut genes. Two gray horses carrying bay genes can produce a bay foal, but the chances are one out of four.

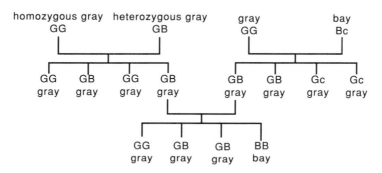

Color Inheritance: Bay and Chestnut. A pure bay mated with a bay that carries a chestnut gene will produce bay foals, but some of the foals may carry the recessive chestnut gene. Two bay horses carrying a chestnut gene can produce a chestnut foal.

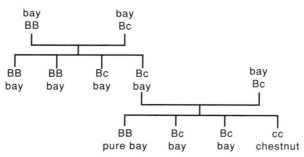

Color Inheritance: Gray, Chestnut, and Bay. Gray is dominant and chestnut is recessive, but two gray parents *can* produce a chestnut foal if each carries a recessive chestnut gene.

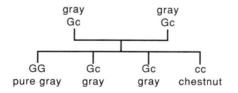

Chestnut to chestnut *always* produces chestnut.

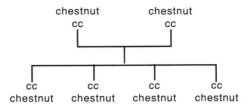

Chestnut to any other color (provided the other color doesn't carry a chestnut gene) will always produce the other color or its genetic possibilities. In this case, a chestnut was bred to a gray that carried a bay gene. The foal will be either gray or bay and will carry a chestnut gene.

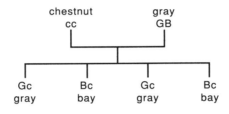

Colors and Breeds

Some colors are common to all breeds, whereas others are found only in certain breeds. For instance, there are no palomino, buckskin, or dun Arabians, but these colors can be found in Quarter Horses.

Other Colors

Breeding for black, palomino, dun, Paint, or Appaloosa colors is more complicated. Roan is dominant over all colors except gray and white. Some colors appear because of recessive factors that "dilute" another color gene. When one of these factors is present, bay becomes buckskin, brown becomes red dun, and chestnut becomes palomino. When both recessive dilution factors are present, bay becomes *perlino* (nearly white, with rust-colored mane and tail) and chestnut becomes *cremello* (creamy white with light mane and tail and blue eyes). Dilution factors can also produce grulla from black or brown.

In order to get a desired color in a foal, you have to know if that color is dominant, recessive, or heterozygous (mixed). You also must know the mare's and stallion's genotypes. This is easy if they have already produced offspring because you can see the color of their foals. Checking the colors of ancestors (in their pedigrees) also gives clues as to the presence of recessive genes.

Dominant colors are bay, gray, tobiano paint, blanket Appaloosa, and line-backed dun. Black must have multiple traits present; it can occur without black parents. Recessive colors are chestnut, perlino, and cremello. Mixed (heterozygous) colors are white, roan, buckskin, and palomino. Breeding for these is complicated because heterozygous colors cannot breed true; there is a 50-50 chance of producing the same color as the parent. White and roan exist only in heterozygous form because they are lethal to the embryo when the genes are doubled up in dominant form.

The Genetics of Color

In some ways color is a less important trait than conformation, athletic ability, soundness, disposition, or intelligence. However, understanding the genetics of color makes it easier to understand the inheritance of other traits because the expressions of color genes are so visible. If you are raising horses in a "color" breed such as Appaloosa or Paint, or for customers who want a certain color in their favorite breed, then having a working knowledge of color genetics is extremely important.

Types of Breeding

There are several genetic "tools" that a breeder can use to reach certain goals and produce outstanding individuals. A good knowledge of genetics helps a breeder use these techniques properly.

Inbreeding

Inbreeding involves the mating of closely related individuals, and it should be avoided except in special cases because it increases the chance of doubling up undesirable recessive traits or genetic defects that are hidden in the parents. Most undesirable traits are recessive homozygous, so breeding closely related animals (ones that may carry the same recessive trait) brings them out. Inbred animals are also generally less hardy; inbreeding decreases the genetic variations within a bloodline. If undesirable or lethal genes are present, inbreeding presents a greater chance of birth defects and problems. Inbreeding concentrates the genes available, both good and bad.

Many breeders use inbreeding as a tool to intensify and "fix" certain good traits in a family line. If a horse is outstanding and has no known faulty genes, a breeder may double up the good genes by mating sons and daughters (half-siblings) of the outstanding horse. This offers the chance that the offspring will be exceptional, especially in genetics that can be passed on. Inbreeding is always risky, but with luck a horse that's a product of inbreeding from outstanding parents will carry only doubled-up good attributes in his genetic makeup. He'll be genetically pure for these characteristics, serving as an outstanding sire (or dam) and passing on those traits consistently to the foals.

Linebreeding

Linebreeding, a form of inbreeding, involves the mating of distant relatives (the descendants of one particular outstanding horse) in an effort to double up and "fix" the traits of a superior ancestor so the animals will breed true and pass on the traits of that ancestor. Because the total number of gene combinations in a linebred animal are limited (owing to less outcrossing to bring in different genes), the linebred (or inbred) animal is fairly predictable in traits it will pass to offspring. Linebreeding, like inbreeding, must be done cautiously, with careful examination of all known traits of the bloodline to make sure there are no undesirable traits that might be doubled up as well.

Nicking

Nicking is a term that refers to the production of superior offspring from the mating of two individuals or family lines. If the two nick well, they produce offspring that are better than would be expected from those parents. The mating results in more favorable gene combinations than if they were mated to other individuals. Nicking is a lucky phenomenon — an unexplainable good combination of genes.

Outcrossing (Outbreeding)

Outbreeding refers to the mating of unrelated individuals within a breed. This often produces superior offspring, but their genetic makeup is more varied than that of the inbred or linebred horse, and there may be more variance among their offspring.

Crossbreeding

Crossbreeding involves breeding individuals of different breeds to obtain offspring with traits superior to those of both parents through hybrid vigor (heterosis). When the best characteristics of each parent are combined, the foals have more versatility and hardiness. The key to successful crossbreeding is to choose animals with traits you want; outstanding parents usually produce a crossbred foal that is even more exceptional — partly due to hybrid vigor and partly due to combining the traits you desired from each breed.

Selectivity in Breeding

Work toward genetic improvement by selectively keeping as breeding stock the animals that inherit traits you want, and culling animals that don't. You can do this by outbreeding or linebreeding; linebreeding concentrates the desired traits more quickly and still leaves enough genetic variation to minimize the chances of genetic defects becoming evident.

Genetic Defects

Some individuals and family lines carry genetic defects; be wary about using these for breeding. The most common inheritance pattern involves a simple recessive trait; the defective foal inherits a recessive gene from both sire and dam — individuals that are otherwise completely normal. A few defects are caused by genes with incomplete dominance, and some are caused by two or more sets of genes, but the most common problems are caused by simple recessives.

Defects run in families. Thus inbreeding or mating of closely related animals is risky unless you know there are no bad genes in that family to cause a defect when doubled up. For example, the parents of a defective foal often have at least one ancestor in common.

Lethal Genes

A few genes, when doubled up, cause death of the fetus before or shortly after birth. Some of these genes, when doubled up as an expressed trait, result in a fetus whose organs are improperly formed (e.g., lack of anus, absence of eye sockets, or water on the brain). The most common lethal gene is the one that creates white coloring. It is inherited as a dominant trait.

Lethal White Genes

White coloring, which results from lack of hair pigment, can lead to foal mortalities. Lethal roans and lethal whites die soon after conception, and lethal overo whites (a type of pinto) die soon after birth due to faulty intestinal tracts; the foal looks normal at birth but cannot survive without a functioning gut. Not every white horse carries the lethal genes for early fetal death, and not every overo-patterned Paint or Pinto produces white foals that die young. The lethal situation occurs when two horses carrying the gene are mated — such as two true roans.

Lethal Roan

This is a horse with a uniformly roaned body and dark head and legs. If two of these roans are mated the offspring would be expected to have a 50 percent chance of being heterozygous roans like their parents; a 25 percent chance of inheriting the two non-roan genes and being a solid color; and a 25 percent chance of being homozygous roan (with two dominant roan genes), but this group never survives to be born. The exact cause or time of fetal death is not known, but it occurs very early in pregnancy; the mare usually returns to heat after being bred.

True White

These horses are usually entirely white with pink skin and brown eyes. Albinos (which do not exist in equines) would have pink eyes, but none are ever born. Lethal white is a dominant gene; but because all true whites are heterozygous, they produce 50 percent white and 50 percent colored offspring when mated with colored horses. When true whites are bred to other true whites, the result is 25 percent colored foals that inherit the recessive color gene, 50 percent heterozygous true whites, and 25 percent homozygous white that do not survive — just as in lethal roan conceptions.

Lethal Overo

Paint breeders have learned to beware of mating extremely white overos to each other. A rare few white horses from overo breedings survive, but most do not due to non-functioning guts.

Combined Immunodeficiency (CID)

CID is a fatal disorder of Arabian foals (and sometimes part Arabians if both parents have the defective gene); it is a defect of the immune system inherited from two carriers that otherwise appear normal. CID foals are normal at birth, but a portion of their immune system fails to develop. As soon as their temporary protection from maternal antibodies in colostrum begins to wane (a few months after birth), they die from infections.

When two CID carrier horses are mated, there is a 25 percent chance that the foal will inherit the recessive gene from each parent and be affected with CID, a 50 percent chance the foal will be normal but carry the recessive gene, and a 25 percent chance of not inheriting the gene at all. If a carrier horse is bred to a non-carrier, none of the offspring will develop CID but half will be carriers.

There is no treatment for CID; the only way to avoid this problem is not to breed two horses that are known carriers. You can test for this defect in a horse's DNA (genetic material) via a blood sample or cheek swab (taken with a plastic handle that has bristles on one end, twirled between the horse's cheek and gum to pick up surface cells). This cheek swab can be taken by you or the veterinarian and sent to VetGen laboratory to be analyzed. The test kit can be ordered through your own veterinarian or from VetGen (3728 Plaza Drive, Suite 1; Ann Arbor, MI 48108; 1-800-483-8436). A carrier horse can still be used for breeding if the potential mate is free of CID because they will never produce a CID foal. Half of all offspring from such a mating would be carriers but the other half would not have the recessive gene. Thus you would not have to sacrifice the bloodline from an outstanding individual that happens to be a carrier.

Origin of CID

CID may be the result of a genetic mutation that occurred in a single horse in the early 1900s. Because of its rapid increase in incidence, it is thought that the mutation appeared in a relatively popular horse. Today CID may be present in almost all Arabian bloodlines and may occur to a lesser extent in other breeds created during this century that were cross-bred with Arabians during their formation.

Hyperkalemic Periodic Paralysis (HYPP)

HYPP is a muscle disorder that occurs in certain family lines of Quarter Horses, Appaloosas, and Paints. Affected horses usually have heavy muscling, as is popular in halter classes. The problem is characterized by sporadic attacks of muscle tremors (shaking), weakness, and/or collapse. Attacks may be accompanied by noisy breathing due to paralysis of muscles in the upper air passages. Sometimes sudden death follows an attack, with the horse dying from heart failure or paralysis of the respiratory muscles. The problem is sometimes mistaken for colic, choke, muscles tying up, or a respiratory ailment.

HYPP is inherited as an incomplete dominant trait. Homozygous individuals are severely affected (most do not survive to adulthood); heterozygous individuals are more moderately affected and may survive if they have proper temperament and metabolism. With careful management many of them survive and function, but this can be a headache for the horse owner.

The HYPP-affected horse must be kept on a low potassium diet (little alfalfa hay, if any). Some owners give the horse frequent doses of a diuretic (acetazolamide), which helps the body eliminate potassium. Treatment of an attack will vary with its severity. Give the horse glucose (such as Karo Syrup) in a mild attack, or seek immediate veterinary treatment for a more severe attack in which the horse collapses.

Origin of HYPP

HYPP appears to have originated as a mutation in the popular Quarter Horse stallion Impressive. During the 1970s, 1980s, and early 1990s, this famous halter champion produced many offspring, being bred to Appaloosas and Paints as well as Quarter Horses — so this genetic defect is widespread. By 1992 Impressive had more than 100,000 descendants, and that number has mushroomed since then. Breeders who wish to avoid HYPP can use a blood test to make sure they breed only animals that do not carry the defective gene.

Know Your Horses' Genetics

To be a successful breeder you must not only be a good judge of horses when assessing conformation and ability but also know their genetics. You will be mating not only two individuals but also two separate ancestries, combining their genes. You must be able to predict with reasonable accuracy what the product of each mating will be, complementing the sire and dam's characteristics as well as creating foals that inherit the best of both ancestries. There are no perfect horses, so you must try to mate sound individuals that will produce even better offspring. If a horse has faults, mate it with one that has strengths in that area. If you mate two horses with the same faults, you will likely produce a foal that is worse than its parents. Never double up bad points of conformation or any known defects.

Breeding good horses is a time-consuming job that involves much work and expense. It costs no more to keep and feed a good broodmare than a poor one, so make every effort to have the best you can afford. If you have a stallion, a good one will help sell your foals. If you consistently have good foals, people will come back for more; but if you sell a few poor ones, the word will get around. If you are conscientious about breeding good horses, you won't let fads overrule your best judgment regarding what is best for the breed. If you strive to attain consistently high goals, there will always be a market for your horses even if you aren't chasing the latest fad.

Further Reading

Genetics is a complex subject. These books are helpful:

◆ Ann T. Bowling, *Horse Genetics* (New York: CAB International, 1996).
 A look at genetics and heritable characteristics.
◆ Philip Sponenberg, *Equine Color Genetics* (Ames, IA: Iowa State University Press, 1996).
 A simplified review that is easy to understand.

Nineteen

Keeping a Stallion

You may never need a stallion; if you only have a few mares it is easier to send them to be bred or use shipped semen. If you plan to raise horses, however, you may want to have your own stallion. Before you try to raise a stallion or buy one, spend some time at someone else's breeding farm observing and learning. This will help you immeasurably when handling your own stallion.

In breeding horses as a business, the stallion is in some ways the crucial part of the operation. He is the horse that comes most before the public eye, giving your enterprise its reputation. When people come to see your breeding program, they are most interested in the stallion — his looks, ability, and condition and the foals he sires. If favorably impressed they may want to buy a son or daughter of your stallion or breed mares to him. Good mares are equally important; but because it's the stallion that many people come to see, he carries more than his share of responsibility for making a good impression.

If you have a stallion, extra attention must be given to his training, care, and handling — and to his living quarters. The stallion is more aggressive and exuberant than other horses, requiring increased responsibility in care and handling.

Facilities for a Stallion

The stallion's enclosure should be high and well built with strong, smooth materials. If he spends part of his time in a box stall, it should be large, at least 16 by 16 feet (4.9 x 4.9 m). The stall should have an adjoining paddock or pasture where he can be turned out to exercise himself. If you want him to have room to graze, the pasture should be large enough that he won't tramp out most of the grass by constantly walking the fence. Most stallions expend a lot of energy worrying about what goes on around them and spend hours pacing or cavorting along the borders of their enclosure.

The fence around a stallion pen should be high and strong.

The stallion's fence should be 5 to 6 feet (1.5–1.8 m) high or higher, depending on the individual — high enough that he won't try to jump out. It can be made of poles, board, or pipe — anything safe, smooth, and durable to withstand kicking and leaning. Spacing should be such that he can't get his head or foot between the poles (3-inch [7.6 cm] space) or wide enough for him to get a foot back out again easily if he puts it between the poles. If the enclosure is made of wood he might chew it; in that case an electric wire will keep him away from the fence. It will also prevent him from kicking or pawing it or trying to get over or through it.

Shelter

On cold or hot days the stallion should have access to shelter or shade. Standing all day long in the heat will sap his energy. In a hot climate (temperatures above 90°F [32°C]) you should run a fan in his stall or shed during the heat of the day or give him a bath after his exercise to refresh him.

King of His Own Turf

A stallion is very sensitive about his surroundings. He considers all the mares on the place to be his and is very concerned about them. He notices more things than most mares and geldings do; all the goings-on around the place interest him, for he feels they are his personal concern. If you bring a mare in from the pasture, he will notice; if you take a horse to ride, he'll be watching and probably will be the first to become aware of your return. He knows where each horse on the place belongs. Changes upset him, such as when strange horses are brought in or when his daily routine is changed. If the stallion is in a paddock next to another enclosure, make sure the other place is always occupied by the same horse — one he gets along with.

If you have more than one stallion it's best to have them widely separated, secure in their own "turf." Horses are creatures of habit; they like the security of sameness in their surroundings, daily routine, and neighbors. If you can, keep the mares a certain stallion will be breeding near him. Some stallions feel possessive about every mare on the place, but many come to develop a sense of "herd" and territoriality about the mares that live near them. A stallion may fuss and call to each of "his" mares when they are worked or turned out to pasture, only half-heartedly showing interest in the "other stallion's" mares.

Don't Isolate a Stallion

A stallion can be stabled or penned near other horses without problems, but if his stall is next to another horse, there should be a solid wall between them. It is not wise to socially isolate a stallion; he will be emotionally healthier and easier to manage if he has company, even just horses in the next field or some he can watch from his stall door. Young stallions should be raised with other horses as much as possible to keep them well socialized; they are less problem to handle later and can usually be housed in fairly conventional facilities. However, if you buy a mature stallion that already has certain habits, pay close attention to how he was housed — and make his new home similar to the old one if that type of arrangement was working and he was happy there.

Handling a Stallion

Stallions must have special care in handling because they have a very different attitude from that of mares or geldings. The stallion is proud, strong, and bold. Because of his highly exuberant and sensitive nature he should always be treated with firmness and respect but also with sympathy and intelligence. To abuse a stallion is to invite trouble — potentially very dangerous trouble.

Proper Training Is Essential

Proper handling and training of a young stallion are essential for developing and keeping a good disposition; from the time he is a foal you must teach and insist on good manners. If a stallion has a good temperament to start with and is handled properly — with tact, understanding, and respect as well as firmness — he won't be a problem to handle, especially if he is well trained while young and gets plenty of exercise as an outlet for all his energy. The confined stallion with no way to burn off energy may take out his frustrations on his surroundings or the people who handle him.

Understand Him and Gain His Respect

The stallion handler must understand him and maintain a good rapport, being able to demand and get his obedience with a mere word or touch. You should rarely have to discipline a well-trained stallion; but if so, you and he both know he deserves the reprimand and he does not forget the punishment or the reason for it. Any disobedience such as charging, rearing, striking, or biting must be halted immediately while the young horse is first experimenting with the limits of his privileges. In fact, most young stallions will try to see who is really the boss; it's their nature to establish pecking order and try to gain dominance. If the handler early on establishes who is in control and lays out the limits of behavior for the horse, a good working relationship can be developed.

The person handling a stallion must never lose his or her temper and punish the horse in anger. When the horse responds and behaves, the reprimand should stop immediately and the handler should resume what he or she was doing, with no continued anger over the misbehavior. The sooner the relationship again becomes friendly and relaxed, the less resentment the horse will carry.

A good stallion handler is tactful and kind but also constantly mindful — in everything done with that horse — that the horse is a stallion. The handler must be able to command respect and instant obedience whenever necessary.

Demand Obedience without Roughness

The stallion handler must demand obedience but should never beat on the animal to demonstrate superiority. This happens all too often when a person is afraid of the stallion. The reason for punishment should always be clear to the horse; he should never be goaded unnecessarily or he'll fight back. Anticipate confrontations so you can avoid them if at all possible; try to keep the upper hand psychologically. If you establish a good foundation of respect during the young horse's training, you can maintain good rapport and "mind control" because he knows the limits of what he can and cannot do. If you are consistent in your handling, leaving no room for doubt or trying to get the best of you, you'll rarely have to discipline him harshly.

Rough handling is a bad experience for any horse, but especially for the stallion because he is so proud and bold. When treated with roughness he will fight back. Abusive handling makes him irritable, bad tempered, and more dangerous. It is better to be gentle but firm, communicating to the stallion that you have respect for and confidence in him. Most stallions enjoy human companionship and want to please you, but if you're nervous or afraid a stallion will sense this lack of confidence and become nervous also, reacting adversely if you resort to force.

No one who is afraid of a stallion should ever handle one, for fear is usually accompanied by forceful, rough treatment and a loud voice in an attempt to cover it up. A self-respecting stallion will resent that nonsense and retaliate. Use of force just leads to more force. If there is a certain stallion you are afraid of, he should be handled by someone else.

Who Should Handle a Stallion?

Some people have the ability to handle a stallion and others don't; some horsemen may be able to handle a certain stallion (and have good rapport) but not another. It's a matter of personal aptitude and the individual horse. Some horses need a strong and steady hand, whereas others require tact, patience, and sympathetic understanding. Not every horseman or trainer gets along well with every horse. Individual personalities of horse and handler sometimes clash. If you find you cannot get along with a certain stallion or become afraid of him, someone else should handle that horse.

Build Trust

The relationship between stallion and handler should be one of mutual admiration, trust, and respect; a spirit of teamwork and good communication is extremely important. The more you treat your stallion just as you would any other horse — riding him, giving him regular exercise, using good judgment (and always remembering his male characteristics) — the easier it will be to keep him gentle, good tempered, trustworthy, and easy to communicate with.

Extra Training

Because he'll be used for breeding you should prepare the stallion ahead of time for this role, training him as a young horse so he can be handled later with good control and no problems. Being a creature of habit, the stallion can be taught to respond to certain patterns with expected behavior. If patterns are changed he becomes insecure, reacting unpredictably. It's important to establish certain ways of doing things early on and then be consistent. The young stallion must learn to respond to you instead of his environment. Even if he'd rather charge after a mare in heat, he knows he must obey you.

Verbal Commands

Proper response to verbal commands should be ingrained early, from the time of halter and leading lessons as a foal and during early groundwork. The stallion must know the meaning of "Whoa." You can also teach him "Back up," "Easy" (slow down), and so on. Proper response to commands should be taught long before he is ever used for breeding.

Have a Different Halter for Breeding

Do not use the same halter for breeding and regular work. The stallion should know that when his working halter or bridle is put on, he is to keep his mind on being exercised or ridden and forget about the mares. It's usually not wise to use the same headgear for regular handling and for breeding. If you can keep the tack and tasks separate in his mind, he'll be more apt to behave himself. If you use a halter with a nose chain for breeding, it should be different from the halter and lead you use for ordinary purposes. If you use a breeding bridle for more control during breeding, it should be different from the bridle you use for riding. The most important factor is consistency of routine. Remember that the stallion will not automatically associate the different headgear with the proper job unless *you* keep the process consistent.

Use Strong Equipment

Equipment used on the stallion should be well made and strong. There should always be enough lead shank or rope to allow extra length to work with (as when breeding mares and giving the stallion room to do the job), but not so much that there is danger of you becoming tangled in coils of rope. You may want a section of chain in the lead — to go over the bridge of his nose — to keep him under control while going to the mare, teasing, and so on. In working with the young, excitable stallion during early training you may need a chain over the nose to make him mind his manners. Many stallions like to rear and strike when they are being led near other horses, and the young stallion must learn this is unacceptable behavior.

Discourage Rearing

If a stallion wants to rear you should nip this bad habit in the bud, being prepared for it while leading him. A sharp jerk with nose chain before he actually gets off the ground is best. If he rears before you have a chance to halt him, allow him to complete the rear — a sharp jerk while he is going up or at the peak of his rear may make him throw himself over backward or sideways in his attempt to avoid more pain. It could also cause him to strike at you. It's better to wait until he's starting back down and then punish him with a jerk on the nose chain, with you well off to the side for safety. Split-second timing is best for good control and punishment.

Let an Older Horse Educate the Young Stallion

Some young stallions are so aggressive and difficult to handle that they need more "education" than a human being can physically accomplish. The best solution for this problem is *not* to use more force — which merely creates a confrontational situation and precludes a good working relationship — but to turn the young stallion out for a while to live with an older, domineering gelding or a few pregnant mares. They'll quickly put him in his place and teach him humility. Once he learns about being at the bottom of the pecking order instead of the top, he'll be a lot more manageable and responsive to his human handlers as well. He knows that someone else can boss him.

Grooming Tip

One aspect of training that you shouldn't neglect is getting the young stallion accustomed to having his scrotal and sheath area washed. Some individuals resent this at first, but you can avoid problems by making it part of the normal grooming routine from the time he is a foal.

Provide Daily Exercise

Daily exercise is important. Because the stallion is by nature active and aggressive, if he is cooped up you will have trouble. He needs room to exercise himself or have regular supervised exercise. Once he is old enough to be ridden, the best exercise is just to ride him regularly. This will do him good both mentally and physically, keeping him fit and also constantly handled and disciplined; he'll be a lot happier, less bored, and more apt to behave like a regular horse than a monster stallion. Work out some method of daily exercise, especially if your stallion does not have room to exercise himself.

Lead a stallion for exercise on a regular basis.

Boredom Is Your Biggest Enemy

Boredom and lack of exercise, especially while confined in a stall or small paddock, can make a stallion sluggish or ready to explode — and bad tempered. Any high-strung horse that is well fed and confined needs an outlet to burn up extra energy. If that horse is an underexercised, undisciplined stallion there will be serious trouble when you try to handle him. Keep in mind that regular exercise and handling will do more to mellow a horse than anything else — and will keep the stallion more physically fit for the breeding season. He'll also be less bored, doing less cribbing and masturbating.

Pasture Breeding

Most valuable stallions are never allowed loose with mares; breeding occurs in a closely supervised situation. Pasture breeding has many advantages, however, if you have a small group of mares and aren't using a stallion to breed visiting mares. Pasture breeding generally results in higher conception rates than hand breeding because the stallion courts and breeds the mares at the proper time. There are no "shy" mares that get missed due to silent heats; the stallion knows when each mare is in heat. Pasture breeding also gives the stallion plenty of exercise, eliminating that daily chore from your management tasks.

A stallion that grows up interacting with other horses and learning about mares at an early age (and how to stay away from kicking feet) can do very well in this situation. He never rushes into anything and is wise enough to never approach a mare from the rear, taking his time and teasing her properly until she is ready to breed. It is difficult, however, for an inexperienced older stallion that has only been used for hand breeding to suddenly be put out with a group of mares; he will invariably get hurt because he has not learned the patience of courting.

Some Stallions Will Not Pasture Breed and Hand Breed

Not every stallion will pasture breed and hand breed. You are fortunate if you have an easygoing individual that lives with his mares but also lets you catch him and take him away from them to hand breed a visiting mare now and then. Some stallions that have been used only for pasture breeding do not cooperate for hand breeding; they are accustomed to courting and breeding a mare at their own leisure, as nature intended, and may be hesitant to breed a stranger that is suddenly presented for breeding. Being handled and confined at breeding time rather than being able to do it their own way is distasteful to some stallions.

Records

Complete and accurate records are an essential part of any breeding program. Keep individual records on every mare: results of teasing (how the mare reacted to the stallion), visits by the veterinarian, breeding dates, vaccinations, dewormings, special considerations in handling the mare, and so on. A good set of records is one of the best management tools you'll have when breeding horses.

The Pasture

If you pasture breed, a large, clean, grassy pasture with safe fencing and no hazards for horses to run into is best. (See chapter 1 for fencing information.) There should be no horses through the fence in adjoining pastures unless they are young ones (weanlings or yearlings) or pregnant mares. There should absolutely not be another stallion through the fence or any mares that might come into heat. Geldings are also taboo, for a stallion may hate a gelding as much as he would another stallion. If you have an adequate and well-fenced pasture with no adjacent troublemakers, and if you select a group of mares for your stallion to be with during the breeding season, pasture breeding is an ideal way to settle the mares. It saves time and labor and usually ensures getting the mares in foal.

Be careful when pasture breeding to avoid problems caused by personality clashes. You may not be able to add mares to the stallion's group without upsetting the pecking order. It's best to give him a small group at the start of the season and make no changes to disrupt the social order.

Age and Fertility

The stallion is not mature until age 4. Some people start using a stallion for breeding at age 2 or 3, but it's better to wait until he's 4 because overuse of a young stallion can be physically and mentally damaging; he may lose his desire for the job. Even at age 4 he should have a limited number of mares, usually less than twenty.

How Many Mares?

A mature stallion can breed many mares, the number depending on the individual, his physical fitness, semen quality, and situation. Forty mares is usually considered an adequate number for any one stallion during a breeding season, but with careful hand breeding (i.e., trying to breed each mare at the proper time and not having to breed her more than once or twice at most), a stallion can successfully breed many more than that. For best conception rates, one a day is plenty. If he must service two, the breedings should be spaced out, one in the morning and another in the evening. If your stallion has good semen quality, he can service even more mares through semen collection and artificial insemination instead.

Be aware that if a stallion is used too often you can't count on him settling every mare; he's more fertile if used sparingly. Also, an older stallion should not be used hard. Like a young one, his number of mares should be limited.

At the start of the breeding season the stallion shouldn't be expected to be perfectly ready for his job. It takes a little time for him to become fit and fertile; moreover, sperm production is lower in early spring than later in the season. Have his semen checked by your veterinarian before breeding season starts. If there are any problems it's better to find out beforehand rather than after several mares have been bred fruitlessly.

Sperm Production

Many factors affect sperm production, including age of the stallion, size and condition of his testes, disease, fever, trauma to his reproductive organs, and time of year. For example, daily sperm production is directly related to testicle size. As the stallion matures and his testicles become larger, he produces more sperm. The testicles "shrink" during winter (when stallions as well as mares are not as fertile), and also may suffer degeneration with advancing age.

Sperm production and sex drive peak in May and June. If you are trying to breed mares early to bear early foals, you may find fertility problems not only in the mares — because they are still in winter anestrus — but also in the stallions. Most stallions produce only 50 to 75 percent of possible sperm early in the year as compared to their output in June.

Sperm Life

Different stallions have different sperm life — meaning how long the sperm remain alive within the mare's reproductive tract. Average sperm life is at least 40 to 60 hours for most stallions, but a few have sperm that are viable only up to 24 hours. By contrast, other stallions can breed a mare as many as 7 or 8 days prior to her ovulation and still get her in foal.

Illness and Injury

Infertility in stallions can be caused by fever from illness or inflammation from injury. Anything that raises the temperature of the testicles for any length of time interferes with sperm production. The higher and longer the fever, the more severe the effect — but it may not be evident for several weeks because of the large amount of sperm already produced and maturing. Those sperm may still be viable, enabling the stallion to settle mares for a while until he runs out of mature sperm and has no more coming on due to the gap in production while he was sick or injured. Formation and maturation of sperm cells take 60 to 70 days. Thus it is important to do a periodic semen check following any illness or injury to determine whether the stallion will have an infertile period and how long it will last.

Contracts

If you breed mares owned by other people, have a contract specifying the responsibilities of both parties. A breeding contract protects both mare owner and stallion owner. It ensures that the mare owner knows the breed fee, terms of payment, daily charges for care and feeding while the mare is at your farm, and whether the mare will have a return privilege to be rebred if she doesn't settle or produce a live foal. A *live foal guarantee* means the mare must produce a live foal before all or part of payment is due. *Return-in-season* means the breed fee is due at end of the breeding season regardless of results. Other subjects to be clarified include liability for veterinary services and any other possible points of contention between mare owner and stallion owner.

A veterinary exam for breeding soundness of the mare is beneficial to both owners, for if any problems are found they can be cleared up before the mare is bred. The exam also ensures that only healthy mares are bred to your stallion.

Breeding Fee

A stallion's breeding fee should be reasonable, especially if he is young, unproven, or not well known. Some breeders offer free breedings to good mares during a stallion's first season to attract mare owners and get some offspring on the ground for people to see. If a stallion can produce good foals, they are his best advertising.

You can start a stallion's fee low and raise it later if he becomes desirable as a good sire. If you start too high and later want to lower the fee, this can be embarrassing. If the fee is reasonable, more people will be able to afford and use your stallion. In trying to promote your stallion and make him well known, start him at a reasonable fee, then as his foals begin to make a good name for him, you can go on up from there.

TWENTY

BREEDING THE MARE

Whether you want to breed a favorite mare to raise a foal that might be like her or raise horses as a hobby or a business, to be successful you must understand equine reproduction. It is also important to give the broodmare proper care for conceiving, carrying, and raising a foal.

Age to Breed

A mare should not be bred until she is mature and fully grown. Breeding too young is hard on her; if she's still growing she can't maintain herself and the foal she's carrying. Some mares mature earlier than others and can be bred at age 3 to foal at age 4, but for others it is better to wait until age 4 for foaling at age 5. In most cases 2-year-olds should not be bred.

If a mare is old she may be hard to settle (get pregnant), but it is not impossible to get older mares in foal unless they have a fertility problem. If a mare is over 10 years old and has never had a foal she may be difficult to settle, but even some 20-year-olds have been known to raise foals successfully when bred for the first time. Regardless of age, every mare should first be checked by a veterinarian to make sure there are no problems that might interfere with her ability to conceive or carry a foal.

The Reproductive Tract

The mare's genital tract begins at the vulva, the external opening just below the rectum. Inside the lips of the vulva is the clitoris, a walnut-size protrusion that is visible when the mare urinates or "winks" when showing signs of heat. Beyond the vulva is the vagina, extending to the brim of the pelvis and leading to the cervix — the opening into the uterus. The cervix remains tightly closed except when the mare is in heat or ready to foal. When she is in heat and ready to breed, the cervix opens up to enable sperm from the stallion to enter the uterus. There are two horns to the uterus that extend to either side from the main body of the organ; at the tip of each horn the oviducts extend to the ovaries, which are bean-shaped and roughly 1 to 2 inches (2.5–5.1 cm) wide by 2 to 3 inches (5.1–7.6 cm) long.

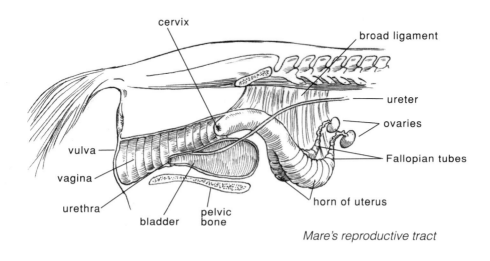

Mare's reproductive tract

Make Sure She's Healthy and Sound

The mare should be in good physical condition before she is bred — not too thin or too fat — otherwise she may be hard to settle. She should be on a regular deworming schedule and have vaccinations up-to-date. After all, several diseases cause abortion in mares (e.g., rhinopneumonitis, leptospirosis, equine viral arteritis), some of which can be prevented by vaccination, and you do not want her losing the fetus just because you neglected to vaccinate her.

Before a mare is bred, have her teeth checked and make sure she has no lameness or chronic problem that might cause her discomfort. Any serious leg problems could become worse with increased body weight during pregnancy. The mare should be in proper body condition as well — if she is too fat she may have unpredictable heat cycles or none at all, and if she is too thin she may not cycle. A mare with a hormone problem also may not have normal cycles. Suspect a thyroid problem if a mare stays fat on reduced rations.

Breeding Soundness Exam

The best mare in the world cannot be a broodmare if she has a physical problem that interferes with conception or carrying a foal to term. Have your veterinarian examine the mare before she is bred. Then if there's a problem you'll know about it before taking the time and expense to have her bred, and if it can be corrected you'll have time to take care of it before she is bred.

Rectal Palpation

When examining a mare for breeding soundness the veterinarian checks her uterus and ovaries by means of rectal palpation, reaching gently into the rectum with a well-lubricated long-sleeved plastic glove. The veterinarian can tell a great deal about the health of the uterus and ovaries from this examination. If the ovaries are hard, fibrous, or excessively large, the mare may have problems with reproduction. Sometimes a cyst or tumor in the ovary interferes with fertility. If one ovary is greatly enlarged, the other may be small and atrophied. A single hard ovary is usually not a problem in terms of fertility if the other is normal. While palpating the ovaries the veterinarian also checks the oviducts for adhesions that might interfere with the passage of eggs.

Safety Tip

Most veterinarians prefer to have the mare confined in stocks for rectal palpation, but if this is not possible be assured that many mares will stand quietly with no restraint. Some may need a twitch or Stableizer or a front leg held up (see chapter 5). If a mare is quite nervous and upset, she may need a tranquilizer. A mare that is prone to kick may still do so even when tranquilized.

The Uterus

The veterinarian checks the cervix and uterus by feeling them through the rectal wall. Enlargement of the uterus or thickening of its walls may mean the mare has had an infection that left scar tissue. A uterus that lacks proper tone, feeling flabby to the touch, is abnormal and may reduce the chances of a successful pregnancy. Any adhesions of the uterus to the surrounding structures, or hematomas on the wall of the uterus or near the oviducts, also limit the mare's breeding ability. These blood-filled swellings may be due to injury or rupture of a blood vessel during a previous foaling. A thickened cervix from injury at foaling or from infection also makes a mare unsound for reproduction. The veterinarian may perform an ultrasound examination along with the rectal palpation to clarify any uterine or ovarian abnormalities.

The Vagina

The veterinarian may take a closer look at the vagina and cervix through a speculum. This examination should be done while the mare is in heat. A sterile, well-lubricated speculum is inserted through the vulva, expanding it so the inner tissues can be viewed with a flashlight. This enables the veterinarian to check for adhesions that might indicate previous tears and injuries. Color and health of the vaginal tissues can indicate the presence of infection. Any abnormal discharge or drainage should be noted.

Infection

To determine if there's an infection, a cervical culture should be taken during the heat period. At that time the cervix is open and the veterinarian has access to uterine fluids. These should be cultured for 48 hours; if bacteria grow they should be tested with various antibiotics to determine which ones would be most effective in clearing up the infection. It also helps to examine uterine cells from a swab under a microscope because the presence of white blood cells may indicate infection. Any genital tract infections should be treated as soon as they are discovered. If they are allowed to persist, they may cause permanent infertility due to uterine damage.

Abnormalities

Occasionally during a breeding soundness exam it is discovered that a mare does not have a properly developed reproductive tract or has an inherited or congenital defect that will prevent reproduction. These cases are rare, but if your mare has such a problem it's best to find out before you try to breed her. The veterinarian should also look at the mare's udder to see if there are abnormalities or scar tissue from previous mastitis.

Imperforate Hymen

More common is a maiden mare (i.e., one that hasn't been bred before) with imperforate hymen or an obstructing membrane partition. These problems can be corrected during the breeding soundness examination. Otherwise, the membranes would be torn at breeding, which could lead to infection. It is better to correct them with sterile techniques, allowing the mare at least 2 or 3 weeks to heal before being bred so there's no soreness.

Caslick Repair

Proper conformation of the mare's vulva is important. A mare with a tipped vulva (sloped inward rather than up and down) is likely to develop vaginal infection from constant contamination due to feces falling through the opening. Correcting this problem involves suturing the lips of the vulva. The veterinarian uses a local anesthetic so the mare won't feel pain while he or she trims the edges of the vulva and sutures them so they will grow together. This simple Caslick repair can prevent the most common type of infection that causes infertility in older mares. The grown-together vulva lips must be reopened for breeding (and then restitched) and again for foaling so the vulva does not get torn. In some cases artificial insemination can be accomplished with a high Caslick in place; this may be the preferred breeding method for such mares because there's less chance of introduced infection.

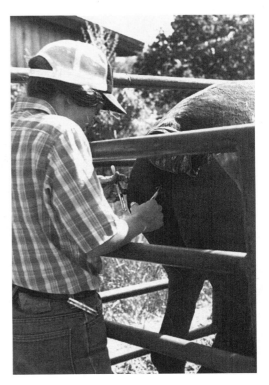

To perform the simple Caslick repair (suturing the vulva) on a mare, the veterinarian will have you put the mare in stocks or have you use a Stableizer or hold her next to a gate or two bales of stacked hay. The veterinarian stands behind them while doing the repair to avoid being kicked.

Check for a Caslick Repair

Before a mare is bred, check to see if she's been sutured and if there's enough room to permit easy access by the stallion. If there's not enough room, her vulva must be opened in time to allow the area to heal so breeding won't be painful. If she's sutured and there is room to breed her, have the veterinarian put a breeding stitch (cross stitch) at the end of the suture seam to avoid tearing during breeding.

Sending a Mare to Be Bred

For best success, sending a mare to be bred should take place well ahead of when she's due to come into heat. She'll be more likely to come into heat properly and conceive if she has a chance to get used to the unfamiliar place and the stallion before coming into heat. If she is pregnant and you are sending her to be foaled and rebred, send her at least 2 months before she is due to foal.

Have her well groomed and clean with feet properly trimmed and no hind shoes (for the stallion's safety). Also send her with a good halter. Remember to include any records concerning her wormings, vaccinations, and heat cycles.

Breeding to Your Own Stallion

Mares can be bred at pasture, hand bred (i.e., with mare and stallion both under supervision), or artificially inseminated. The method you choose will depend on the mare and your circumstances. Pasture breeding is discussed briefly in chapter 19. Most breeders prefer to hand breed because there is less risk of injury to the animals and it is possible to take precautions against infection.

Conscientious breeding practices and techniques are crucial to success. If hand breeding is overseen in a careless manner by inexperienced people, it can result in distress and injury to the horses or injury to people; it may also distort the animals' attitudes to such an extent that they don't behave normally. For example, a stallion may develop impotence or aggression, and a mare may become overly fearful and unwilling to be bred; she may also become so aggressive that she injures her foal. A mare deprived of normal social interaction before breeding can become uncontrollably violent when forced to accept the stallion's unwanted advances, even if she is biologically ready to be bred. Proper teasing and determination of heat cycles is important to breeding success.

Understanding Heat Cycles

When breeding by hand you must determine when the mare is in heat and when she is most likely to accept the stallion. A mare may be in heat from 1 to 10 days, but there is a certain time in that period when she's ready to breed. Breeding too early will not be successful. Ovulation usually occurs about 24 to 48 hours before the end of heat. It is traditional to breed on the third day of heat and every other day thereafter as long as the heat lasts, even though one good breeding is enough if it is done at the proper time. The mare's ova live only about 12 hours, but sperm can survive in her reproductive tract for 60 hours or more; thus breeding every other day keeps viable sperm present for fertilization of the egg.

Knowing how long a certain mare tends to be in heat helps you determine when to breed her. On farms where there are many mares to be bred and the owners don't want to overuse their stallions — or irritate a mare's tract with repeated breedings, which could lead to infection — the mares are palpated by a veterinarian to determine the time of ovulation so they need only be bred once.

The Cycle

To breed a mare successfully, you must be familiar with her heat cycles. No two mares are alike in their reproductive cycles, and any given mare may show different characteristics from one year to the next. For example, she may have winter *anestrus* (lack of cycles) one year and cycle all through winter the next. Mares are unpredictable.

The cycle is divided into two parts, *estrus* (the period of heat), and *diestrus* (the time when she is not in heat). The average length of the combined cycles for most mares is 21 to 23 days. During estrus a few follicles (containing immature eggs) in the ovaries grow rapidly and secrete estrogen, causing the mare to come into heat and display signs of sexual responsiveness. Near the end of heat one of the follicles ruptures and releases a mature egg into the oviduct, where it can be fertilized by sperm. Half of all mares go out of heat within 24 hours of ovulation.

Diestrus

The diestrus period begins immediately after ovulation and varies in length from 5 to 33 days, but for most mares it lasts 14 to 19 days. The length of diestrus is actually more predictable than the length of estrus; most mares will come back into heat 15 to 17 days after they go out. Estrus periods tend to be longer in early spring, however, and get progressively shorter throughout the summer. Most mares are fairly consistent in their cycles through the breeding season, and the pattern and timing can be determined with a good teasing program.

Silent Heat

Some mares are obvious about their heat cycles, but those that are not may have to be teased regularly by a stallion to show any changes. A mare who never sees a stallion may not show obvious heats. If she has a diestrus longer than 33 days, she may be having

Signs of Heat

- ◆ Relaxed vulva
- ◆ Spasmodic winking (opening and closing of the clitoris right inside the vulva)
- ◆ Mucous discharge
- ◆ Slightly raised tail
- ◆ Frequent urination
- ◆ Gregarious attitude (wanting to be near other horses)

regular heats but not showing. This is called *silent heat*. Sometimes an observant person who knows the mare well can detect it. A mare suspected of having silent heats should be regularly palpated by a veterinarian to determine her time of ovulation — then she can be bred artificially if she will not accept the stallion. At times silent heats are a sign that something else is wrong in the reproductive cycle or that the mare has a uterine infection. Silent heats may also be intermingled with normal heats.

Some mares with foals at side do not cycle at all until the foal is weaned, and others fail to show heat because they are preoccupied with their foals. Estrus in these mares can often be detected by means of a careful and diligent teasing program. Shy mares may become more obvious about their cycles if a stallion is nearby and some breeding activity is going on. Horses are very social animals, and often a little more "social stimulation" causes a mare to become more obvious about her cycles.

Mares Are Seasonal Breeders

The most important aspect of equine reproduction is the fact that mares are seasonal breeders; unlike cattle, they usually cannot be bred at any time of year. Fertility varies greatly during different seasons. About 75 percent of mares stop showing signs of heat during winter when days are short. This period of winter rest, called *anestrus,* occurs in most mares in the northern hemisphere from November through February, although in some mares it may begin as early as September.

Erratic First Cycles

As spring days lengthen, mares that were in anestrus start cycling again. Breeding season for most mares begins soon after the vernal equinox (March 21, when there are equal hours of daylight and dark) and reaches its peak by summer solstice (June 21, the longest day of the year). The first cycles in spring are usually erratic and the mare is not particularly fertile, because follicles at this early season may not ovulate. Heat periods during this transitional period are often long because the follicles persist. Mares may come into heat for a day or two, go back out for a week, then come in again for 2 weeks — or some other erratic pattern. Try not to breed mares early in their heat cycles. If a mare is going to be in heat for 20 days, it's a waste of time. Have her palpated first, then breed only if a follicle is close to ovulation. Transitional heats are usually not fertile but sometimes produce a viable follicle at the end.

Breeding in Late Spring and Summer

As heat cycles become shorter and more normal toward summer, conception rates improve. By late April nearly 80 percent of mares are cycling, and by June almost all mares are coming into heat regularly with properly functioning cycles. Heat periods in June are short (3 to 5 days); getting mares pregnant during May, June, and July is quite easy compared to early spring. Some breeders initiate teasing in early spring to help stimulate the mares' reproductive system.

Teasing Mares

If a mare is not obvious about her heats, she's more likely to show if you take her near the stallion every day (or bring him to her) to discover her reaction. This is called "teasing." You can put the stallion in a box stall with top door open so he can see and talk to the mare as she's led up to him, but he is protected if she suddenly decides to strike or kick. Introduce the mare to the stallion several times before she is actually ready to breed, so she'll be acquainted with him and accept his services when the time comes. Many mares do not submit to breeding unless they undergo a lengthy teasing and courtship first.

Most mares show stronger estrus the more they are teased, and some won't accept a stallion without this teasing process. The act of teasing stimulates the release of oxytocin in the mare. This hormone has many purposes, such as stimulating uterine contractions and milk let-down in the broodmare. It also initiates several processes that enhance ovulation and the transport of egg and sperm to their destinations. If you tease problem mares often, and long enough to stimulate oxytocin release, they will ultimately show signs of heat.

Keep Records

Keeping a teasing record throughout the breeding season is very valuable; it shows when the mare's cycles become normal and indicates the most favorable time to breed her. It helps you determine the number of days per heat so you can keep the number of services on each mare to a minimum, with maximum likelihood of conception. If the mare is to be bred at foal heat (i.e., her first heat after foaling), teasing should begin on the fourth day after foaling.

A good teasing stallion is one who simulates the actions of a stallion under natural conditions, being patiently aggressive. A stallion living with mares is interested but controlled; the mares are not restrained (i.e., haltered or hobbled), so he has a healthy respect for their view of the situation. He's not about to barge right up and get kicked. He has learned how to be patient, knowing that when a mare does come into heat he will be the one to breed her.

Teasing Routines

Large farms usually identify one stallion to do the teasing so their breeding stallions do not have to spend so much time and energy at this job. In contrast, the breeder who has only one stallion generally uses him for teasing as well as breeding. There are advantages to this, because some mares object to being bred by a stallion different from the one that did the courting.

When using your stallion for teasing, establish certain teasing routines that are distinctly separate from breeding routines. Use different headgear on the stallion for the different jobs. He should have a specific piece of tack for each job he does, so he knows what to expect — whether he is to be ridden or taken to a show, or is going to tease mares, or is going to breed. For instance, some horsemen tease with a chain over the nose and breed with a headstall and rubber-covered bit (i.e., a bridle and bit never used for riding). Whatever you choose, use it consistently with that particular stallion. With his teasing headgear on he knows he is only going to make "small talk," which is a large part of his springtime work. Having one site for teasing and another for breeding also helps the stallion keep his jobs straight.

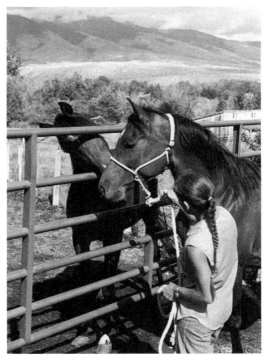

Always use the same headgear on the stallion when teasing a mare; this helps establish a teasing routine.

Teasing Methods

Some breeders tease with one of the horses in a stall. By bringing the stallion to the mares, a manageable stallion and one handler can tease a barnful of mares in a short time, making this an efficient method for reliable mares. However, some mares feel threatened when another horse invades their space; they show aggression when approached in their stall by an inquisitive stallion at the door. This is often true of mares with foals. Be aware that the foal might be inadvertently injured by his overprotective mother charging at the stallion. In this case it's better to take the mare to the stallion, with the foal left behind in the stall out of harm's way. Then the mare can focus on the stallion.

Some mares need to meet the stallion on neutral turf (neither her stall nor his) and require more stimulation than just nose to nose at a stall door. Use trial and error to determine if a certain mare does best being teased with her foal beside her, within sight but out of reach of the stallion, or completely out of sight. The site and method of teasing should be tailored to the individual mare and stallion, not the convenience of the handlers. What's easiest for you may be inadequate for a particular mare.

Teasing Chute

Some breeders use teasing bars or a teasing chute to check a mare at close quarters. The chute has a solid partition about chest high between mare and stallion. Once the mare is in the chute, the stallion is led up to her. She can reach over the top to sniff noses with him and show her inclination or objections, but the solid chute prevents her from kicking or striking out. Another method is to use adjacent stalls with a window open between them.

Teasing Mares at Pasture

If mares are kept outside, the stallion can be walked past their paddocks or pastures. This reveals a great deal about the status of the mares' heat cycles. A mare that is not in heat will usually go on about her business after glancing up or coming over to say hello, whereas a mare that is in heat and ready to be bred will come over and show interest. A shy mare or one that shows no obvious signs of heat may have to be brought to the stallion every day for more intensive teasing at closer quarters.

Teasing Cage

An effective way to tease mares at pasture is to build a small pen or teasing cage (like a portable stall) in the field or pasture where the mares are kept. The stallion is put into the pen for a while each day, and the mares approach him at their leisure. This is very similar to pasturing a stallion with the mares — but without risk of him being hurt. Many mares are more likely to show signs of heat while in a natural and relaxed herd situation. Even the timid ones can approach the stallion if they wish, when the more dominant mares have grown tired of socializing and wandered off. An alert observer can tell a lot from the mares' actions. If there is a stallion nearby, even the shiest mare may give a clue that she is in heat, even while calmly grazing. The constant presence of the stallion is helpful in getting timid mares to cycle and show heat.

Be Observant

Experience will enhance your ability to interpret the actions of the stallion and mares. If a good teasing stallion loses interest in courting a certain mare, trust his judgment. The veterinarian will probably confirm she isn't ready to breed. If the stallion continues to insist that she ought to find him attractive despite her vigorous protests, the veterinarian will probably confirm that her reproductive tract has all signals "go." If the stallion doesn't give any clues, you will have to read the signals from the mare. A mare that is in heat may demonstrate obvious willingness or disinterest (the strongest sign of heat in some shy mares), or less enthusiasm about efforts to kick him. Certain mares require palpation by a veterinarian to determine the true status of their heat cycles.

When teasing or breeding, the stallion should always be under control and have excellent manners. He should be allowed to approach the mare and make her acquaintance by nibbling and making sweet talk ("chuckling"), but never allow him to bite. If the mare is in heat and ready to breed, she will usually show it and be willing to stand for him. If she isn't, she will reject him firmly.

Preparing the Mare for Breeding

Tease the mare again just before breeding, to make sure she is still ready to accept the stallion and to ensure her cooperation — and to encourage her to urinate and defecate. Then wrap her tail in a clean bandage or enclose it in a clean nylon stocking or pantyhose. This keeps the tail hairs from cutting the stallion's penis and also keeps the penis from passing through the tail hair and picking up dirt.

Washing the Mare and Stallion

Wash the mare's genital area and buttocks with a mild soap and rinse them thoroughly, then wipe dry with paper towels. Chlorhexidine (Nolvasan) is a good disinfectant that cleans without irritating. The stallion's penis and sheath should be washed with clean water (no soap) once he drops it down — which he'll do if he is near enough to see the mare. Repeated washings with soap or disinfectant kill normal skin bacteria, opening the way for infection with resistant pathogens; repeated washings

This tail is wrapped for breeding and enclosed in a nylon stocking.

also kill sperm. For good sanitation, use a plastic garbage bag to line a bucket for wash water (have a separate bucket for mare and stallion), to be thrown away after each use. The water should be body temperature. Both mare and stallion should be well rinsed and dried because any remaining water tends to kill sperm. The mare can be briefly teased again so that she and the stallion are both ready.

Safety Tip

Breeding should take place in a safe, open area with no hazards for the animals to run into, no dust that might introduce infection, and plenty of room for the horses and handlers to maneuver. Footing should be solid, dry ground, with good traction for both mare and stallion. If the mare has a foal, it should be confined in a pen close by where the mare can see it, preferably in front of her.

Use of Breeding Hobbles

Some breeders use kicking hobbles and other restraints on the mare as a matter of course, but this is not always a good idea. Most mares will submit willingly to the stallion when they are ready to breed. A mare that kicks and fusses a lot is probably not ready. Hobbles are unnecessary on most mares and can be a hazard to both mare and stallion if he gets tangled up in them. A maiden mare (i.e., one that hasn't been bred before) may be nervous and need quite a bit of steadying to keep her from moving around too much, but hobbles and excessive restraint can do more harm than good, making the breeding a bad experience for her. Reserve the use of hobbles for the few mares that truly need them, mainly to protect the stallion.

Hobbles should be properly designed for the job, with a quick-release feature. With young mares, hobble them a few times before you expect to breed them. You may even need to twitch a mare or use a Stableizer prior to hobbling her.

Breeding hobbles and tail rope should be used only when necessary, mainly to protect the stallion.

Approaching the Mare

Lead the stallion up behind the mare, about three horse lengths behind her and to one side so she can see him. She may try to turn and face him if she is nervous; the person holding her (see below) must keep her calm and steady. If she's quite upset, lead the stallion around her at a safe distance for a few times so she can see him better and become less afraid. If she won't stand still yet you are sure this is the right time to breed her, use a twitch or Stableizer to make her stand. (See chapter 5 for proper use of these restraints.)

Handling the Stallion

The stallion should be led up only when he is fully ready to breed her, with penis erect but not so large, or "belled," that he cannot enter her. The approach should be a little to one side, at a 45° angle, in case the mare tries to kick. He should approach calmly, not at a run or on his hind legs. Let him mount from the mare's hip and work his way around to the proper position. Give him enough slack at the proper time so he won't be distracted by a pull on his head. Both handlers should be very alert at this time to try to prevent and be out of the way of any striking or kicking on the part of the mare, and to avoid flailing forelegs of the stallion.

If the stallion has been pasture bred he may have developed his own pre-mating ritual; it's usually best to let him approach the mare in his own way as long as he does not rush and frighten her.

As he mounts, one person should hold the mare's tail to the side and make sure the stallion enters properly. If the stallion is an individual that likes to bite the mare while he is breeding, you should muzzle him or protect the mare's neck and withers with a pad.

Holding the Mare

The best method is to have a solid fence or panel with the mare's head over it and her chest up to the fence so she can't go forward much. The mare's handler stands on the opposite side of the fence (with the release rope for her hobbles — if hobbles are used — in his hands at all times, long enough for him to release it from his side). The mare will move forward a step or two as the stallion mounts, but the solid fence keeps her from lunging forward and gives her something to brace against so both she and the stallion can have solid footing. Moving forward a bit helps prevent injury to the mare and allows the stallion to find his proper position, but the mare should not rush forward. The mare's handler should keep her head high as the stallion comes up to her; this will inhibit kicking if she feels so inclined. If she starts to kick, the handler should pull her head around toward the stallion to get her hind feet away from him.

The Service

Copulation usually takes 30 seconds to a couple of minutes, followed by ejaculation. Most stallions show a rhythmic flagging of the tail while ejaculating. A few stallions do not ejaculate during the first mount and must do it again. After the stallion has finished his service let him dismount, or move the mare a step to the side to help him dismount. If you are collecting semen for checking, it is easy to obtain at this point because some usually runs out of the mare's vulva as the stallion dismounts. Catch it in a clean paper cup to check under a microscope.

The mare's handler should be alert as the stallion dismounts, turning her to keep her from kicking him. Lead her for a few moments to keep her from standing and straining, which would discharge much of the semen, then put her back in her stall or paddock.

Washing the Stallion After Breeding

After the stallion dismounts and is led away from the mare, his penis should be rinsed. You don't need to disinfect his skin; just rinse him off to make sure there is no dirt. It is more important to have the mare and her hindquarters clean so that no contamination is taken in with the breeding.

Artificial Insemination

Some breeders use artificial insemination to impregnate a mare that is difficult to breed or settle by natural service, or to extend the semen of a valuable or older stallion (one ejaculate can be divided and used to inseminate a number of mares). Some breeders use artificial insemination for *all* mares being bred to their stallions. The easiest way to collect semen is to let the stallion mount a mare in heat or a "phantom mare" (a breeding "dummy" designed for this purpose), and divert his penis into a specially designed artificial vagina.

Some breeders collect and ship semen so mares can be bred without having to go to the stallion. The semen must be shipped by overnight or same-day air freight so the mare can be inseminated at the right time. This requires advance planning, letting the stallion owner know exactly when the mare will be ready to breed.

Check Mares Closely After Breeding

If a mare does not conceive after being bred, she will come into heat again about 15 to 17 days after going out. Because mares are often irregular, it is wise to watch every bred mare closely for several weeks after service to see whether or not she has settled. (Up to 10 percent of pregnant mares show one or more heat cycles after being bred and settled, and this can be deceptive.) Check her by teasing about 20 days after service and again at what would be her next heat cycle. If she shows no signs of heat for the next two possible heat periods she is probably pregnant, but you should have her checked by a veterinarian to be sure.

Fertility Problems in Mares

Many things cause infertility in mares. Most can be detected ahead of time with a breeding soundness exam, but sometimes mares do not settle for unknown reasons and return to heat. A fairly large number of embryos are lost in the first 45 days, usually due to resorption. That is, the embryo dies for some reason and is resorbed by the uterus. Losses between 45 and 100 days have been estimated at up to 16 percent.

Infection

Uterine infection is a common cause of infertility. Antibiotics and suturing (Caslick repair) to prevent contamination can clear up some of these. Good breeding hygiene also prevents the transmitting of infection from one mare to another via the stallion. Almost all mares pick up infection at foaling time, for the open genital tract inevitably comes in contact with some contamination, but most mares throw off this infection within 1 or 2 weeks without treatment. If the mare is in poor condition or for some other reason has lowered resistance at that time, she may not be able to overcome the infection.

Be aware that the genital tracts of some mares are more resistant to infection than others. The older the mare, the more susceptible she is to infection; the healthy young uterus is more able to ward off infection because it is not scarred and damaged.

It is unwise to breed a mare that has any infection; if she does settle, chances are she will abort later. It is better to spend a few weeks treating the infection and clear it up than to breed the mare and have her abort at 3 to 6 months' gestation. If a mare has an infection or any kind of fertility problem, work with your veterinarian to try to correct it before attempting to breed the mare.

Breeding at Foal Heat

Because most mares have minor infections for a short time after foaling, it often is unwise to rebreed a mare at *foal heat*, which occurs 3 to 22 days after foaling. Some mares (especially those that return to heat soon after foaling) are not fully recovered at this time; breeding may not be successful or may lead to further problems if the mare does conceive. Breeding before the infection has cleared up seals it in and may cause abortion. Breeding at foal heat is not recommended unless the mare had a normal delivery with no complications, has an exceptionally healthy reproductive tract, and has her first heat late enough after foaling (2 weeks or more) to be fully recovered from the temporary infection.

TWENTY-ONE

CARE OF THE BROODMARE

The arrival of a new foal is a long-awaited event that requires careful planning and the best possible care of the mare. You wait approximately 11 months for the new arrival after the mare is bred, so it is important to make sure that all goes well; a lot of time and effort are invested in the coming foal. Proper care of the broodmare begins even before she is bred.

Care of the Mare Before Breeding

If a mare is to conceive, carry, and produce a healthy foal she herself must be in excellent health — free of disease, parasites, and physical impairments that would jeopardize the pregnancy. She should be well fed, but not to the point of being fat. Mares that are overfed and underexercised are often hard to settle.

Nutritional Needs Vary

Broodmares are categorized as *maiden mares* (those that have never been bred), *barren mares* (those that have had a foal but are not in foal presently), *mares in foal,* and *lactating mares.* Because these four groups have vastly different nutritional needs, they should not be fed as a single group. Maiden mares that are still young and growing need extra protein and calories. Mares in foal need extra protein and nutrients, especially during the last trimester of gestation. Lactating mares need the most nutrition due to the heavy demands of producing milk.

Is She Pregnant?

After a mare is bred you hope that she has settled. If she doesn't show signs of coming into heat again at the expected time, she may be with foal. Because many mares take a long time to show definite signs of pregnancy, it may be a good idea to have your veterinarian check. In some mares the pregnancy is fairly obvious by 5 or 6 months; the belly begins to drop and widen toward the lower part of the flanks. This drooping belly looks different from a round belly that is merely fat. When you look at the mare from the rear as she is standing still or traveling, if she is pregnant one side of her belly toward the flank may appear a little larger than the other side, and as she moves there may be a slight swinging motion.

Other mares, especially with first foals, do not look pregnant even right up to foaling time. In contrast, certain fat mares look deceptively pregnant even though they aren't. After about the eighth month there may be signs of movement toward the mare's right flank as the fetus kicks and squirms, but you may not necessarily observe this. If you see movement, you know the mare is carrying a foal; however, not being able to see it *does not* mean she isn't.

The pregnant mare's belly will begin to droop and widen toward the lower flanks.

Feeding the Pregnant Mare

While she is carrying a foal she needs the best care possible, being kept in good condition but not overfed. Nutritional requirements increase significantly only during the final third of pregnancy. If you increase the mare's feed as soon as she becomes pregnant, she may be too fat when she foals. Proper amount and quality of feed are important. The broodmare needs a well-balanced diet containing necessary nutrients for the growing fetus. Also, the feed must be clean and free of mold and spoilage to eliminate the risk of abortion.

Exercise

Exercise is important because the mare needs good muscle tone for foaling; she should not be soft and flabby from inactivity. Don't limit her exercise in fear of hurting her or the unborn foal. She is not fragile.

If you were riding her before she was bred, continue riding her afterward. Just use good judgment in what you ask of her. A mare that participates in stressful athletic events before pregnancy can probably continue the same level of exercise during the first 6 months of pregnancy while the fetus is still quite small. Many pregnant mares have successfully completed endurance rides, for instance, during the first part of pregnancy.

During the last month or two many mares become quite heavy with foal, which cuts down on their agility and stamina. Usually after 300 days of gestation it's best to slow down. The mare still needs regular exercise, so keep riding her if you wish, but make the rides shorter and slower. Stick to walking and jogging; avoid fast or strenuous workouts. Older mares should not be exercised as extensively as younger ones. Be aware that fatigue, strain, or overexertion can be a cause of abortion.

Pasture Is Ideal

A good pasture with room to move around can provide needed exercise as well as proper nutrition. Pasture is an ideal environment for pregnant mares and those with foals. If it contains good types of grasses and fertile soil, it may provide all the nutrition a pregnant mare needs. However, during winter and whenever pastures are dry, the mare will need supplemental feed.

When a mare is pregnant, keep her with a few other horses she gets along with — preferably other mares. Several pregnant individuals at pasture by themselves get along the best. Don't keep a pregnant mare in the same pasture with a group of young horses or geldings; this reduces the risk of injury to the mare from being kicked or slipping and falling while running. Geldings may tease and fuss with a mare and have been known to make them abort. If a mare foals at pasture while she is with other horses, a more dominant mare may steal the foal. Some geldings will even try to kill a newborn foal.

Preventing Edema

In late pregnancy some mares develop edema, a swelling around the udder that extends forward along the flanks and is caused by fluid in the tissues. It results from disruption of the normal interchange of fluid among blood vessels, tissue spaces, and lymph vessels. Occasionally the area along the mammary vein in the lower abdomen is swollen too, almost as far forward as the girth. This happens when a large fetus and the weight of its fluids distend the uterus, compressing the mammary veins and causing poor blood return. Inactivity (leading to poor circulation) makes the problem worse.

As the mare gets closer to term and heavier with foal, she becomes more clumsy and sluggish and moves around less. It's a vicious cycle; the heavier she gets, the less she exercises, the more swelling she develops, and the less she feels like moving. She may have trouble getting up and down. Some mares get so clumsy they rarely lie down during the last weeks or days of pregnancy, because the effort of getting up is so great. The inactive mare's legs may also fill with fluid due to her sluggish circulation and capillary leakage.

A mare at pasture moves around during the normal activities of grazing and going to water, so her edema may not become extreme. But a mare confined in a stall or paddock may not move around at all, especially as she gets uncomfortably heavy during late gestation.

In late pregnancy, some mares develop swelling around the udder that extends forward along the flanks and belly.

Daily Exercise

To prevent or alleviate edema in the heavily pregnant mare, make sure she gets mild exercise. This will help her circulation, clear away fluid from the tissues, and enable her to feel better (perhaps prompting her to exercise more on her own). She doesn't need vigorous exercise because she's clumsy and might stumble and fall, but brisk walking can be very helpful. If she is confined, give her a good walk at least twice a day at morning and evening (more often, if you can). Leading her at a brisk walk for at least 20 to 30 minutes at a time will help relieve the swelling and prevents more edema. It also keeps her muscle tone in better condition for delivering the foal.

Keep Her in a Hazard-Free Place

If the broodmare is on good pasture she'll need little extra care as long as she has good feed and water as well as shelter from flies and hot sun. Be sure to remove any obstacles or hazards from her pasture or paddock. Do not allow her in a pasture with ditches or gullies where she may get into trouble if she lies next to the ditch and rolls into it. Avoid narrow gateways and doorways when moving her out of the pasture or into a barn or stall. In the stall, remove projecting edges. During winter make sure she has access to shelter if she is at pasture. If kept in a box stall she should be turned out or exercised every day.

Abortion

Abortion, or failure to carry a foal to term, happens all too often with mares. It has been estimated that 20 to 30 percent of all equine conceptions end in abortion. It is a larger problem than many people realize because few aborted fetuses are ever found; the most common abortions occur while the fetus is small, and many times it is simply resorbed rather than expelled. It is interesting to note that before the use of ultrasound it was difficult to accurately diagnose pregnancy until 30 days or more after conception, so pregnancies lost before 30 days were usually considered a failure of the mare to settle.

There are many causes of abortion. Among these are bacterial or viral infections such as equine viral arteritis, rhinopneumonitis, leptospirosis, bacterial contamination of the uterus, and — occasionally — mold or fungus. Non-infectious causes are not always as easy to detect and diagnose, but they are responsible for a large number of abortions.

Non-Infectious Causes of Abortion

These include genetic defects of the embryo or fetus, lethal genes that cause a defective embryo to die early, twins, hormone problems (in particular, progesterone deficiency) that interfere with the continuation of pregnancy, and nutritional factors. In addition, some types of grasses and feeds, such as plants containing high levels of estrogen, can make a mare more susceptible to losing the pregnancy. Even a defect in the umbilical cord or placenta can interfere with the normal exchange of blood between mare and fetus; twisting of the cord is sometimes severe enough to cut off circulation to the fetus. (See under Feed Problems for more natural causes of abortion.)

Placental Insufficiency

A common cause of abortion is placental insufficiency due to degeneration of the uterine lining, especially in older mares. The uterine tissue becomes less efficient with age. Scarring from earlier pregnancies may not leave enough normal lining for adequate attachment of the placenta; thus, blood supply to the lining is diminished. The developing embryo cannot make a good attachment and dies early, or the fetus dies later on, due to inadequate nutrition from the poorly attached placenta. Less severe cases of uterine insufficiency may lead to the birth of small, weak foals, "windswept foals" with curved leg bones, and those with other abnormalities.

Twins

Twinning is another leading cause of lost pregnancy. Twins are often aborted because there is insufficient nutrition for both, due to the limited surface area of the placental attachments. On the mare's placenta there are no specialized areas that attach to the lining of the uterus (unlike the placenta of a cow or dog). It must have contact with the entire uterine wall. When twins share this space, neither one has quite enough contact. Occasionally one dies and the other takes over its space and continues to develop. The surviving twin may still be small at birth as a result of nutrient deprivation during early gestation.

Loss of a twin is fairly common during the first 30 days of pregnancy, occurring in 3 to 5 percent of all pregnancies. Complications arise only when the loss occurs later; then the death of one twin causes the mare to abort them both. Abortion of twins can take place at any stage of gestation but is most common from about the time the fetuses are 10 to 12 pounds (4.5–5.4 kg) on up to full term.

Twin Pregnancies Are Fairly Common

Mares often produce multiple eggs. When both fetuses die early and are resorbed or aborted, the owner usually doesn't know why the pregnancy ended unless the mare is checked by palpation or ultrasound and twin fetuses detected. If a veterinarian detects twins early (and one fetus doesn't resorb on its own), many breeders have the veterinarian terminate one in order to enhance the survival of the other. If the mare is less than 35 days pregnant, some breeders abort the mare with prostaglandin and bring her back into heat again to be rebred.

Lack of Progesterone

The hormone progesterone stops the mare's heat cycles and keeps her pregnant. It also changes the uterine environment and enables it to support the fetus. Abortion partway through pregnancy sometimes occurs because a mare fails to make enough progesterone. Some mares habitually abort between 30 and 140 days of pregnancy because the level of progesterone falls too low. If you have a mare with this problem, your veterinarian can advise you on treatment with additional progesterone.

Stress

Actual physical trauma, such as being kicked, falling down, or hitting a doorway, rarely causes abortion. The uterus, placenta, and fluids in which the fetus floats usually provide adequate insulation against bumps from the outside world. A mare hardly ever aborts from the trauma of an injury itself, but the stress may be another matter.

Stress can trigger labor contractions and cause abortion, with the uninjured fetus dying because it is born before it can survive outside the uterus. The stress of any major problem — illness, severe colic, high fever — may lead to abortion. In fact, any drastic change in the mare's environment or handling can upset her, and in some cases it causes the release of too much histamine and prostaglandins, which can lead to premature birth. If the fetus is almost at full-term when this happens, it may survive the premature birth; otherwise, the fetus is lost.

Drugs

Certain drugs may cause abortion. Always read labels or consult a veterinarian before giving medications to pregnant mares. Most medications and drugs are safe to give during pregnancy, but it is always wise to err on the side of caution.

Feed Problems

Some feeds can cause a mare to abort, including fescue grass. The grass itself is blameless; the villain is a toxin-producing fungus in the bud of the plant. Not all fescue is infected with this fungus, but many fescue pastures are. The toxin can cause diminished blood supply to the placenta. It thickens, preventing nutrients from reaching the fetus. The affected mare either aborts or carries the foal to term, only to have problems during foaling. The placenta may become so thickened that the foal cannot break through the membranes at birth and suffocates.

Overfeeding of high-quality feeds occasionally leads to abortion. Rich alfalfa hay, high-protein supplements, or lush legume pasture sometimes cause a "flushing" effect that triggers estrogen production and premature opening of the cervix.

Lush Pasture Can Cause Abortion. In years with above-average rainfall or wet fall weather that promotes prolific regrowth of patures, lush green pastures can cause a higher incidence of fetal loss or abortion. Bluegrass pastures, for instance, can have as much as 30 percent protein in early spring when growing rapidly (this percentage drops to about 15 percent in summer when grass is not so lush). The 30 percent level is too high for a pregnant mare. Moreover, new spring grass and lush green feed contain estrogens, and the combination of protein and estrogens can stimulate a mare to cycle. This is nature's way of helping a mare come out of winter anestrus and start cycling again, getting her ready for the breeding season. However, if the mare is already pregnant it causes trouble.

The pregnant mare's cervix should remain tightly closed throughout gestation. If she is bred early in spring the hormone levels and high protein in pasture grass may cause her to cycle (even though she is already pregnant) and her cervix to relax. This allows bacteria to enter the uterus and infect the placenta and fetus; the mare then loses the pregnancy.

Limit Her Time on Lush Green Feed. To prevent this situation, you can limit a mare's time at pasture under lush growing conditions and put her on good hay instead — a mix of grass and legume hay in which total protein level does not exceed her nutrient requirements. You can safely feed a mix of good grass hay and legume hay, or first cutting alfalfa containing a high proportion of grass. Second and third cutting may be too rich unless it was cut very mature — in which case the protein level will be lower. If you limit a mare's time on green pasture when it's growing the fastest and select hay that is not too rich, you will eliminate the risk of abortion caused by too much protein and estrogen. This will also keep the mare from foundering.

Health Tip

To reduce the risk of abortion, make sure the mare has wholesome feed (no mold or toxic plants) and water and no access to poisons or toxic chemicals. Make feed changes gradually; a sudden change of diet may upset her system.

False Pregnancy

Sometimes a mare loses the embryo or fetus but the ovary or the uterus continues to secrete progesterone, the hormone that maintains pregnancy and keeps the mare from coming back into heat. Because she may not return to heat for several months, you might think she is still pregnant. Up to 15 percent of bred mares that have been checked and determined to be pregnant turn up empty before the hundredth day of gestation; some of these fail to return to heat for a while, giving the owner false hopes of a foal.

Infectious Causes of Abortion

Infections in the reproductive tract are a serious threat to pregnancy and should be cleared up before breeding. Certain infections are introduced at the time of breeding if good hygiene is lacking or if the stallion passes an infection to the mare. All mares bred by natural service are infected to some extent at breeding, but a healthy mare is able to overcome this infection unless it is a particularly bad pathogen or the mare's uterus does not have normal immune defenses. Infection can also be introduced by unsterilized equipment during a breeding examination or artificial insemination.

If the mare becomes ill with equine viral arteritis (EVA), rhinopneumonitis, leptospirosis, or other diseases that affect the fetus, infection may invade the uterus via the bloodstream. Ask your veterinarian about vaccinating the mare for diseases that might be a problem in your locale and situation. She may need several vaccinations during her pregnancy to prevent certain diseases, such as rhinopneumonitis.

If a mare in late gestation starts bagging up too soon or acting like she's in early labor, watch her closely and have your veterinarian examine her. Udder filling is the main sign of an impending abortion, but if they are caught in time some of these cases can be saved with proper treatment. If ultrasound shows that the placenta is still intact, the mare can be treated and the abortion prevented. Mares that have prefoaling colic act like they are in labor, so be sure to investigate any signs of abdominal pain. If a mare in late pregnancy is aborting, she may have a difficult labor. In any of these situations you should not hesitate to call your veterinarian.

Udder Problems

Check the pregnant mare's udder periodically. This not only gives you a clue as to when her body is preparing for foaling (the udder will start to enlarge) but also alerts you to problems or abnormalities. Problems that occasionally occur in broodmares include inverted teats, small nipples, or even extra teats. Whereas extra teats are generally not a problem, inverted or small teats make it difficult or impossible for the newborn foal to nurse, especially if the udder is large and swollen. If a mare has this problem, watch closely when the foal is born to make sure it is able to nurse. If it cannot, be prepared to bottle feed by milking out the mare by hand until the swelling goes down and the foal can nurse more easily.

Dirty Udder

Sometimes dirt, oil, and sweat build up between the teats and cause discomfort and itching. The affected mare rubs her tail a lot (you might think she has pinworms) just because she's trying to get at the itch. If you look at her udder you will see the flaky debris and accumulated material between the teats. Cleaning her udder periodically will resolve the problem.

Get Her Used to Having Her Udder Touched

The pregnant mare should have her udder touched regularly — especially if she's never had a foal before — so she'll be used to the sensation. Then she won't be so upset when the foal tries to nurse. Also, if you have to milk her out or try to help the foal nurse, she'll be cooperative and not resent your efforts. When handling the udder, be sure your hands are clean. Do not touch the ends of the teats. You can gently touch and massage the rest of the udder with your hands; or use a clean cloth, or a cloth and warm water, to wipe her udder and remove debris between the teats.

Mastitis

Mastitis is infection and inflammation of the udder. Mares do not get mastitis as often as cows do, but the condition can be very serious nonetheless. A mare's udder is divided in halves with each half containing two quarters, like the udder of a cow. The two quarters on each side feed into a single teat with two openings. If you ever milk a mare, you'll notice two streams of milk from the nipple when you squeeze it. Mastitis usually affects just one or two quarters on the same side. Only rarely are quarters on opposite sides affected. Infection from bacterial invasion causes the affected quarter to become swollen, hard, and warm. The milk — if the mare is lactating — may become lumpy, thick, or watery.

Mastitis usually occurs in lactating mares, especially at weaning when the full udder is painful and slow to dry up. Infection may develop at any time up to 8 weeks after the foal is weaned. A mare doesn't have to be milking to develop mastitis. Even pregnant, barren, and maiden mares occasionally get udder infections due to trauma, bruising, or insect bites. Mastitis can also occur in young fillies.

Many cases of mastitis occur in summer from insect bites. The inflammation is triggered by fly bites and other external irritations of the teat opening that introduce infection. Mastitis in a non-lactating mare may be due to abnormally high estrogen levels from eating lush spring pasture or legume hay — causing mammary development and secretion of milk, even in mares that aren't pregnant and haven't foaled. Hypothyroidism can also cause false lactation and make a mare prone to mastitis. Moreover, udder injuries such as a kick, or contamination from bacteria getting into the teat openings from dirty bedding, can lead to mastitis.

Symptoms

Mastitis may be quite obvious or hard to detect, depending on the severity of infection and amount of pain involved. A mare with mastitis generally shows discomfort; she may stand off balance and rest one hind leg continually while trying to ease the swollen side of her udder. Or she may hold the stifle out from her body, so that leg does not touch the swollen udder. She may show slight lameness in the hind leg next to the affected quarter, moving stiffly in an effort not to bump the udder. Because of her discomfort, a nursing mare may kick at her foal when it tries to nurse.

Signs of mastitis include swelling, heat, pain, and edema in surrounding areas. Usually only one teat is affected. Some mares become depressed and feverish or go off feed. If she has a foal, it may avoid nursing the infected teat because the milk tastes different; as a result that side becomes even larger, adding to the mare's discomfort. In some cases of acute mastitis abscesses form in the udder and nearby lymph glands, and these require surgical drainage. Occasionally mastitis completely destroys the udder tissue and results in death of the mare. Highly pathogenic bacteria can lead to endotoxic shock and cause the death of both the mare and the nursing foal.

If you ever suspect mastitis, have your veterinarian examine and treat the mare. The veterinarian may take tissue smears or samples of milk from the affected quarter to identify the pathogen and determine which antibiotics will be most effective.

Treatment

A mare with mastitis needs antibiotics, some of which should be inserted directly into the affected quarter through the teat canal. Treatment includes frequent milking of that side of the udder. Hot packs and nonsteroidal anti-inflammatory medications (such as phenylbutazone or flunixin meglumine [Banamine]) help reduce the pain, swelling, and fever. Your veterinarian can instruct you on what to do. Some cases of mastitis occur simultaneously with a uterine infection; you may need to treat both to resolve the problem.

Mastitis Can Cause Permanent Damage

If a mare has a serious case of mastitis you won't know until her next foaling whether the affected quarter will produce milk again, even if the infection was treated immediately and cleared up. It depends on whether the mammary tissue was damaged. As she nears her next foaling date and the udder fills, you'll be able to tell. A damaged quarter that's not producing milk won't fill, and the udder will look a bit lopsided. If more than one quarter is damaged, you may have to find another source of milk for the foal.

Countdown to Foaling

When the mare is 6 to 7 months along in pregnancy you may want to give her a booster shot for flu and tetanus if she did not have them earlier. She will need rhinopneumonitis shots at 5, 7, and 9 months of gestation to make sure she does not become susceptible to this viral disease that could cause her to abort. Repeat the flu and tetanus shots at 2 to 4 weeks before foaling so she will have antibodies in her colostrum — to give protection to the newborn foal. In some regions it is a good idea to vaccinate against botulism as well. Discuss this with your veterinarian. As the mare gets close to foaling, also reduce or eliminate her grain ration if you've been feeding grain. Otherwise she may have too much milk at first, which would give the foal diarrhea.

TWENTY-TWO

FOALING

The birth of a foal is always exciting. If the mare is healthy and foals in a clean place, most of the time there are no problems. If something goes wrong, however, it's best if you are there to give assistance or to summon the veterinarian for help.

The Foaling Place

Well before the mare's foaling time approaches, decide where she will foal. If she must give birth in a small enclosure such as a shed, corral, or box stall, it must first be thoroughly cleaned. Take out old bedding and scrub the entire stall with disinfectant — floor, walls, and all. Sprinkle the clean floor with lime before putting in new bedding.

A foaling stall should be at least 14 x 16 feet (4.3 x 4.9 m), preferably larger, to minimize the risk of the mare lying too close to the wall during her labor. Identify a safe place in the stall where you could set up a heat lamp if the foal is born during cold weather.

Clean straw makes the best bedding for foaling.

Use Straw for Bedding

Shavings or sawdust make a satisfactory bedding while the mare is pregnant, but when she foals you need good clean straw instead. It won't stick to the foal or get into his nostrils as much as other bedding would; sawdust or shavings are sometimes sucked into a foal's air passages or swallowed by the mare when she tries to lick him. Also, wood products may carry *Klebsiella* bacteria, which cause uterine infection. The straw should be comfortably deep for the mare but not so deep that the foal will have a hard time getting up and moving around. Remove all obstacles from the stall — feed tubs, buckets, or anything else that might get in the mare's way during labor.

Supplies to Have on Hand at Foaling

At least a month before the mare is due to foal, you should gather supplies and keep them in a clean, convenient place.

◆ A new, clean, 3- to 5-gallon (11.4–19 L) bucket
◆ Clean container to milk into if needed (an open pan works well because of the divergent flow from the teats — they don't "shoot straight")
◆ Tail wraps or nylon stocking to bandage the mare's tail
◆ Pint or quart (0.47 or 0.95 L) of disinfectant such as chlorhexidine (Nolvasan) (for adding to wash water to clean the mare's hindquarters)
◆ Small bottle of tincture of iodine (for disinfecting the foal's navel stump)
◆ Plastic shoulder-length obstetric gloves
◆ Clean bath towels, three or four
◆ Paper towels
◆ Roll of cotton
◆ Plastic garbage bag to put the placenta in
◆ Prepackaged enemas (human adult size in plastic squeeze bottles, fitted with a prelubricated rectal tube) or foal enema kit
◆ Flashlight and new batteries
◆ Heat lamp
◆ Obstetric chains and handles, or nylon pull straps and handle
◆ Sterile syringes and needles
◆ Suction bulb, for clearing mucus out of the foal's nostrils
◆ Injectable antibiotic (refrigerated)
◆ Obstetric lubricant
◆ 10 mL of oxytocin (refrigerated) to be given if advised by your veterinarian if the mare does not clean (shed her placenta)

The Stall and Mare Must Be Clean

For foaling in a stall (especially if other mares have foaled there) wash and dry the mare's udder, belly, buttocks, and upper legs (any place the foal will nuzzle while trying to find the udder) with warm water and chlorhexidine. This will greatly reduce contamination to the foal when mouthing around trying to nurse. It is especially important if there have been any cases of *E. coli* and other foalhood diseases on the farm. Also, pick up all manure in the foaling stall several times a day so the mare has no chance to get dirty. Washing her just before foaling reduces the risks of diarrhea and septicemia, which are the leading causes of death in newborns.

If you think the foal may be at risk because other foals on the place have been sick, milk the mare and get a good amount of colostrum into the foal *before* he starts nosing around. It's a race between the bacteria and the colostral antibodies to be absorbed into the bloodstream before the gut lining closes.

Consider Foaling at Pasture

The best place for a healthy mare to foal, especially if she has a history of easy deliveries, is a level grassy pasture where she can be by herself. It should be a clean, dry place with safe, smooth fencing and no brush, rocky ground, gullies, or other hazards — no loose nails sticking out of fences, no machinery or junk that might injure her or the foal, no ditches or puddles. Foals have been known to drown in tiny pools of water before having a chance to get to their feet. Even a dry ditch is dangerous if a foal rolls into it in his efforts to get up, because he cannot stay on his back very long without suffocating. Foals have died because they got on their backs under a fence. If a mare lies next to the fence when she foals, you may have a problem.

Advantages

If the pasture is safe and weather is good, there are many advantages to foaling at pasture rather than in a confined box stall. A good, clean pasture offers less risk of infection than a stall or corral. Also, it's a more natural environment where the mare can seek her own privacy and be less upset and nervous. As a source of feed and exercise, it prepares the mare in advance as well.

If a mare is living at pasture, on green grass, she will have exercise for better muscle tone and looser, more natural bowel movements than a confined mare that is fed grass hay and grain. Green grass or alfalfa hay are more laxative than grass hay; the mare on pasture will have an easier time foaling than a hay-fed mare that is constipated.

Disadvantages

It is hard to keep constant watch on the mare if she's in a large pasture and not visible from your house. It helps to have a spotlight of some kind for checking on her at night. You may want the mare closer at hand so you can check on her often and be there in case she needs help, especially if it's her first foal. Foaling at pasture is not advisable in bad weather. Consider letting her out on nice days and keeping her in a stall at night where you can more readily observe her.

When Will She Foal?

Normal gestation for mares is 305 to 395 days, with an average of 340 days. Mares are notoriously unpredictable; veterinarians don't even agree as to what the "average" should be. Many factors influence length of gestation, including the age of mare, time of year, genetics of the foal, stress, infection, and nutrition.

Genetics of the Foal Affects Gestation Length

The individual foal's gestation length may or may not correspond with the due date. His genetic makeup, nutritional status, and other factors within the mare influence his timetable for maturation. Some stallions sire foals that are carried longer or shorter than average. Also, light breeds tend to have longer gestation than draft breeds.

It is not unusual for gestation to vary between 320 and 360 days, but most mares foal somewhere in between. That still leaves quite a period for watching and waiting up nights checking on a mare, so horsemen try to predict the actual event more closely.

First Foals Sometimes Come Early

Certain factors affect the likely time of foaling. Keep in mind that these are generalities — averages or trends — and that mares don't always follow the rules. Mares with first foals often (but not always) have shorter gestations than individuals that have delivered before. Midlife mares generally carry foals longer than younger mares, but the first foal tends to have a shorter gestation regardless of the age of the mare. In other words, a 10- or 12-year-old mare bearing her first foal may have shorter gestation than a 7-year-old delivering her third. Because first-time mothers are likely to foal ahead of schedule, they need extra watching. They also may not show as much change in the udder or other areas as the veteran broodmares that often display signs of foaling several weeks ahead of the event. Older mares (age 15 and up) may have short gestations, especially if they have uterine insufficiencies.

Time of Year Makes a Difference

Mares foaling in January, February, or March usually have longer gestation length (10 days longer, on average) than those foaling during April, May, and June. Perhaps the lengthening days, warmer weather, green grass, and other beneficial changes in the environment help the mare bring a late spring or summer foal to term early. Indeed, some veterinarians believe the biggest factor affecting gestation length is the time of year the mares are due to foal.

Individual Patterns

Some mares consistently foal early, late, or on time. Often a mare's own pattern is more reliable than her age, number of pregnancies, or foaling time of year. Always keep in mind that mares foal when they are ready (or more accurately, when the fetus is ready) rather than according to due date.

Sometime a low-grade uterine infection causes the mare to have a longer or shorter gestation than normal, if the fetus has been nutritionally deprived due to poor placental attachment. If a mare goes more than 3 or 4 weeks past her anticipated date, have your veterinarian check her to make sure nothing is wrong.

The Mare Can Delay Foaling If She's Upset

The foal triggers the onset of birth, but the mare determines when it actually happens. The final stages of labor and delivery depend on voluntary contractions of the mare's abdominal muscles as well as involuntary uterine contractions. If she is nervous or upset, she can delay foaling for a day or more. Many mares put off foaling when they are constantly watched. They are ready but prefer to do it in privacy. (Mares are notorious about putting off foaling, doing it suddenly when bleary-eyed owners head to the house for a cup of coffee.)

It is a mare's instinct to leave the herd and go off by herself to foal. This is nature's way of protecting the newborn from other horses that might try to steal it, injure it, or interfere with the bonding of foal and dam.

The Foal Triggers the Birth Process

When the unborn foal is full-term and ready, its hormones stimulate a change in the mare's hormones. Throughout the pregnancy her own hormones have been inhibiting uterine contractions, but toward the end of gestation this delicate balance shifts. The foal's hormones start triggering changes that cause her cervix to dilate, uterus to contract, and milk to let down.

Pinpointing Foaling

You can purchase foaling prediction kits with test strips for checking the mare's milk before foaling (the strip has four zones that change color as the concentration of calcium increases in the milk being tested). Only a small amount of milk (about 1 mL) is needed for the test. It is recommended that milk be tested daily once the mare starts to develop an udder and twice daily after a color change occurs in two or more zones of the strip, with one of those tests taken late at night. The test is helpful when interpreted in conjunction with a mare's expected foaling date and signs of imminent foaling, but it is not 100 percent reliable.

Other methods involve installing television cameras in barn stalls (viewed through a TV screen or special monitor in the house) and acquiring alarm systems that signal the start of foaling. For example, the Baby Buzzer (Locust Farms, Kirkland, OH) fits around the broodmare's girth like a surcingle, sounding off in the house if she lies down prone. The Birth Alarm (Kincardine, Ontario) is a small transmitter that is sensitive to changes in temperature; implanted in the vagina several days before foaling, it stays there until expelled by early contractions — then the change in temperature causes the sensor to trigger an alarm. The Foal Alert (Foalert, Inc., Acworth, GA) is a transmitter that is sutured across the vaginal lips when the mare's Caslick repair is opened before foaling. It comes apart and gives a signal when pressure from the amniotic sac causes the vulva to open.

Signs of Approaching Labor

About a month before the mare is to foal her udder begins to look fuller and larger, especially at night when she is at rest. Initially, this enlargement may recede during daytime. About 2 weeks before foaling the udder will remain larger, fill with milk, and look shiny. A few days before foaling the muscles on each side of the tail around the pelvic bones become relaxed and droop away from the root of the tail. The mare's vulva may appear relaxed and swollen. There may also be a secretion from the teats.

Waxing

Secretion from the teats is one of the most common signals that foaling is imminent. The "wax" is formed by the congealing of secretions forced out of the end of the teat. Most mares wax within 24 to 36 hours of labor, but there are always exceptions. Some don't wax at all. Others wax for as long as 10 days before foaling. Still others may make up a large udder and leak streams of milk before they foal, with the milk dripping down their hind legs. The mare may leak milk for just a few hours before foaling or for several days. A few mares foal without waxing or making much udder at all. Some have no milk when they foal; they come to their milk within a few hours after delivery. A mare may bag up, wax, then stop waxing. The most predictable thing about mares is that they are unpredictable. If you are worried about a mare that is not "going by the book," have your veterinarian check her.

Collect Leaking Milk

If milk streams from the udder for several hours and the broodmare has not foaled yet, collect as much of it as you can and freeze it. This is the colostrum that is crucial to the foal, giving him important antibodies for temporary protection from disease. Don't milk her, however. That would stimulate her to let down her milk even more and lose precious colostrum. If a mare leaks milk for more than 24 hours, have your veterinarian check her. Also check the foal after its first few hours of life to make sure it has sufficient antibodies from the milk.

Caslick Repair

If the mare has been sutured, her vaginal opening (vulva) will be very small; she'll have to be opened before foaling to prevent tearing and damage to these tissues. If you're not sure whether she's had a Caslick repair (if she's a mare you bought, for instance), you can check by gently spreading apart the lips of the vulva, beginning at the bottom. In a mare that hasn't had a repair, the opening will extend up to within about 2 inches (5.1 cm) of the anus. In a mare that has been repaired, the top part of the vulva will be sealed shut with no opening and there will be just a small opening at the bottom. Most mares do not object to gentle lifting of the tail to check. If you are in doubt as to her reaction, have someone hold her for you next to a gate or stable door that you can stand behind while you check.

If the mare has a Caslick repair, it should be opened a few days in advance of the anticipated foaling so she will not tear herself if she foals unexpectedly and unobserved. Have the veterinarian do this for you, and watch how it is done so that you can do it yourself in the future if you wish.

Give Her Privacy

As the mare gets ready to foal, give her privacy. If she is at pasture she should be by herself, with no other horses that might pester her or injure the new foal. If she's in a stall, observe her quietly from a place where she cannot see or hear you. If people are constantly hovering around, she may put off foaling. Remember that mares often foal at night in the quiet seclusion of darkness.

Health Tip

Keep the mare clean and well groomed. The cleaner she is, the less chance of infection when the genital tract is exposed during foaling, and the fewer bacteria to be picked up by the foal as he noses around to begin nursing. Wrap the mare's tail in a clean bandage when you wash her for foaling, to prevent dirt from getting into the open tract as she foals and to keep her tail free of blood and mucus.

Early Labor

The mare may be in early labor for just a few hours or for as long as 72 hours. Mild uterine contractions prepare her for delivery. During this first stage of labor the uterus contracts and the foal is positioned for birth so his head and front legs are aimed toward the birth canal.

The mare may not show much outward sign, but if you are observant you'll notice subtle changes in her attitude or behavior. She may go to the far corner of the pasture or stand with a faraway look in her eye. She may become restless, eating a few bites, then pacing a bit before eating again. If you suspect she is in early labor, check her every 10 or 15 minutes.

Alert Your Veterinarian

If you made arrangements for your veterinarian to be present at the birth, the time to call is when the mare is in early labor — when she is merely restless and periodically uncomfortable as the foal is shifting position inside her. Keep the veterinarian posted on progress so he or she can be there when the mare goes into second stage (active) labor. Usually everything goes according to plan and the mare foals with no problems at all. However, if anything does go wrong, there may not be much time to correct it because a mare usually delivers within 15 to 30 minutes of beginning second stage labor. Thus, if you want your veterinarian to be there, contact him or her when the mare is still in early labor.

Signs of Early Labor

During early labor the mare is restless and nervous. Her pelvic muscles relax more fully, her tail rises, and her sides appear caved in just in front of the hip bones. The foal is shifting position, heading into the birth canal. As contractions occur the mare shows distress, switching her tail and nervously pacing, pawing, or nosing at her flank. This is the time to bandage her tail, supply fresh bedding if she's in a stall, and then leave her alone (checking on her without her seeing you). If there is time you can gently wash her teats and udder, and legs and buttocks if they are dirty, with soap and water or a mild disinfectant such as chlorhexidine

Early Labor Resembles Colic

The mare in early labor may sweat and paw, getting up and down and showing other symptoms similar to colic. She will also keep grabbing bites to eat, whereas the colicky horse generally is not interested in food.

Second Stage (Active) Labor

Active labor occurs when the foal is expelled from the uterus in a series of hard contractions. This stage does not last long. During early labor the mare moves around a lot and is fairly comfortable between contractions, but after she starts active labor she is more uncomfortable because the contractions are coming faster. When she shows more constant signs of distress and gets up and down, she is starting the second stage and the foal will come soon. She may go down only once and start to strain, delivering the foal in a very short time, or she may get up and down several times.

Be sure she has soft bedding, for she may throw herself down rather violently as the pains hit. Mares have been known to rupture themselves internally by going down hard on a firm stall floor.

Rupture of Water Bag

During active labor the water sac ruptures. You may see the dark-colored sac protrude from the vulva or just the stream of amber-colored fluid rushing out. When this happens you'll know the foal should come within 5 to 30 minutes. Cows can safely be in active labor for several hours and still deliver a live calf, but if a mare takes longer than 20 to 30 minutes she needs help — *immediately.* She is so strong that if she keeps straining with the foal in an abnormal position she will kill it or push part of it through her uterus. A first-time mother may take a little longer than other mares, but if nothing has happened within 30 minutes after the water breaks, the mare is in trouble. Either the foal is being presented wrong and cannot be born, or it is too large and not coming through the birth canal readily.

After the water sac breaks, the clear white amnion sac should soon appear, with the foal's feet within it.

Normal Position

The normal position for the foal is head first with front feet extended (see page 384). A clear whitish membrane filled with fluid will appear, then in a moment the feet should be visible within it. As the feet protrude farther, the foal's nose should appear tucked between them. Usually the feet are not even;

one front foot may be a little in advance of the other as the foal moves through the birth canal one shoulder at a time. Any deviation from front-feet-first position is abnormal; professional help will be needed at once, the sooner the better. A foal in an abnormal position cannot be born, and it and the mare may both die.

Passing the Shoulders

The foal's shoulders are the widest part to pass through the birth canal, and they usually come through one at a time. This is why one front foot may be extended farther than the other. If both are protruding evenly and the mare seems to be having difficulty moving him out, you can pull on one of the foal's front legs as she strains, to advance that shoulder so he can move along more easily. In most cases, however, you should not try to assist in the birth because the mare will manage fine on her own and too much pulling and tugging at the foal might injure the mare.

Pull on the foal's legs, one at a time, to bring the shoulders through the birth canal one at a time; the head is now appearing.

If you need to pull on a leg, it will be slippery and hard to get a grip with your bare hands. For a hard pull you should use obstetric chains or straps, but for a little pull you can do fine with just a small towel around the leg to give you more traction to hang onto it.

The mare will alternately strain and rest, often grunting as she strains, moving the foal gradually but steadily through the birth canal. After his shoulders pass through he should slide out easily and quickly, with perhaps another hard strain or two from the mare while passing the hips. Then he should lie there with hind feet still in the birth canal while the mare rests for a few moments after her labor.

Large Foal or Hard Birth

Occasionally the birth may be hard even if the foal is presented normally. This is rare, but sometimes a mare takes longer than usual and needs some help. Once in active labor she must deliver the foal quickly or its blood supply will be compromised and it will die. If the foal is partway out, the umbilical cord may be pinched off; if there is pressure on the foal's ribs and body, he will not be able to survive long.

If for any reason the mare is taking too long, grasp the foal's forelegs (using chains if you have them) and pull with each contraction. If it is a difficult pull you may need one person on each leg. Exert a downward pull toward the mare's hocks, directly in line with the birth canal. Pull strongly with each contraction, but *do not pull when the mare is not straining*. If you pull too hard when she is not pushing, you may damage the birth canal. If you tear the mare too much, she may hemorrhage.

If the mare is having trouble pushing the foal out, pull on the foal's legs each time she strains.

Malpresentations

If the foal is not coming properly, his position must be corrected immediately. As soon as you see there is a problem, call your veterinarian or have someone else call while you get the mare up (if possible) to keep her from straining until the veterinarian arrives. You may have to put a halter on her and pull on her head or tap her hindquarters to encourage her to get up. If it is a simple problem you may be able to correct it, and even if it's serious you may have to try to help the mare while you wait; unless the veterinarian is able to get there quickly, most serious malpresentations will result in a dead foal.

Be Clean When Helping a Mare

Don't reach into the birth canal without first taking precautions to prevent contamination. Bandage the mare's tail and wash her buttocks and genital area with a mild disinfectant such as chlorhexidine mixed with warm water. Be sure your hands are clean and scrubbed, and use a plastic obstetric glove if one is available. Lubricate the glove with plenty of obstetric lubricant or K-Y jelly (available at any drugstore). Carefully reach into the birth canal to see how the foal is positioned.

One Foot Turned Back

If only one front foot appears, reach in and feel to determine if the other foot is close enough that you can pull it out. If it is close but you cannot pull it out even with the first foot, get the mare up and push the first leg back in a little while you pull on the missing one. If you

normal presentation

cannot find the missing leg — just the foal's nose instead — it is probably turned back at the shoulder. This will require veterinary assistance.

If the Veterinarian Cannot Get There Soon Enough. If you cannot get help quickly, wash up again and apply plenty of lubrication to your gloved arm. Push the foal's nose back into the uterus and try to reach farther in and find the missing leg. The mare's strains will make it difficult so you'll need her to be on her feet with someone holding her, trying to keep her from straining so much.

Be careful when trying to manipulate the foal; the uterus can be easily torn by a foot or even a nose pushing hard in the wrong place. If you can get hold of the missing foot, cup it in your hand so it won't tear the uterus, then bring it inward and upward — under the foal's neck and over the pelvic brim. Once both legs are in the birth

one foot turned back

canal and the head is in proper position, you can start pulling on the foal's legs. Make sure the foal's nose is in the right place and coming through the birth canal, doublechecking that by bringing the missing leg up you have not pushed the head back too far.

Safety Tip

If you pull on a foal with chains or straps, place them above the fetlock joints. Otherwise you may injure the joints or feet.

Nose But No Feet

Sometimes only the nose of the foal appears, with both legs turned back. This often means the foal is already dead; a live foal usually wiggles and squirms and gets his legs aimed into the birth canal during early contractions, but a dead one is limp and may not get positioned properly. The head first, legs back position is often typical of a dead twin. The second one might still be alive, so get help immediately. If the veterinarian cannot come right away, try to get help from an experienced horse breeder, if possible. If you cannot get help, you must try to push the dead foal far enough back into the uterus to make room to maneuver the legs into the birth canal.

Head Turned Back

If both feet protrude but no head is coming, it is serious trouble. Attempts to correct this malpresentation should only be made if your veterinarian is unable to get there within the next 20 minutes. If you must try to correct it yourself, scrub up and lubricate both arms to your shoulders. The mare must be on her feet, so have someone hold her. Push the foal back into the uterus with one hand and try to find the head with the other.

head turned back,
over body

head turned back,
down along one side

The head may be down between the legs or off to one side. If you can get hold of his lower jaw you may be able to pull the head up into position. Even if you are eventually able to get the head up and the foal out, it may be dead by the time it's delivered. Any birth that takes much more than 30 minutes can result in a dead foal because the placenta begins to detach.

Unlike the cow, which has "button" attachments between the placenta and uterus that detach gradually, the mare's placenta comes loose from the uterine wall quickly. If the placenta begins detaching, the foal may die before you can get him out even if you correct the malpresentation. This is why it's so urgent to have professional help immediately when there's a problem. Even if the foal dies, you must still get him out to save the mare. If you try to stay as clean as possible and are careful not to damage the mare, she will probably be able to have another foal even if you lose this one.

Three or Four Feet at Once

A rare but extremely difficult situation is when a foal comes front feet first with one or both hind feet entering into the birth canal as well. The hind foot or feet may not be discovered until you've already started to pull on the foal (the foal does not progress through the birth canal, and you may think he is just

all four feet coming; hind legs must be pushed back

a large foal that needs help). It may be impossible to push the extra leg or legs back in. If the foal is still alive your veterinarian may try to do a caesarean, but if it is already dead he or she will probably choose to cut the foal into pieces so it can be brought out of the mare through the birth canal, a procedure that is not without risk to the mare but safer than a caesarean.

Rotated Transverse Foal

A rare but frustrating situation is when a foal presents crossways, with part of its body in each horn of the uterus. This problem is usually seen only in large draft mares. The uterus is tipped so the foal cannot be reached from the birth canal at all. A caesarean is the only way to get the foal out.

foal upside down, with head turned back

Make Sure You Have Front Feet, Not Hinds

Even if there are two feet coming, if you do not soon see the nose, you should always check to make sure they are front legs and not hind. Front feet have the soles down, whereas hinds point upward. Even if they are pointed upward and you think they are hinds, it is wise to check. Reach in alongside the leg to see if there are knees or hocks. The foal may be coming out frontward but upside down. If the foal is upside down and the head is turned back, you have a serious problem; but if the head is there, the foal just needs to be turned over. This may happen naturally if the mare gets up and down a few times; often as she finishes early labor she gets up and down a lot, helping the foal find the proper position. A small foal can be turned over without too much difficulty, but for a large one you'll need help. To correct this position, push the foal back into the uterus, then take hold of each leg at the cannon bone and try to twist him into normal position. Once you get him rotated, check his head to make sure it's coming into the birth canal.

Backward Foal

If the hind feet are presented the foal can still be born, but the birth must be accomplished quickly so the foal will not suffocate or breathe in some of the surrounding fluid before he is completely out of the birth canal. If when you reach in you discover hind feet instead of fronts, and the veterinarian is not yet there, keep the mare on her feet and find someone to help you pull. The mare could deliver the backward foal by herself, but it will die without extra help because she will stop to rest during the birth (as when the hips are coming through). As the umbilical cord is broken off or compressed by the foal's ribcage, the foal will have to start breathing — and will drown in amniotic fluid.

If there are at least two people to help, you can probably get the foal out alive. Put obstetric chains or straps above the fetlock joints, and when the mare starts to strain, pull with all your strength, one person on each leg. As the hips emerge, keep pulling,

backward foal

even when the mare stops to rest. The foal must come out immediately or he will die. Next you must get him breathing.

The cord will break as the foal comes out, so there is no point in letting the foal lie quietly by his mother as you would with a normal birth — you must try to get him breathing. He may be limp and unconscious from being short of oxygen. Clear away any mucus from his nostrils with a suction bulb. If there is a lot of fluid in his air passages, have someone help you hold him upside down just briefly to let it drain out. If he doesn't start breathing, tickle his nostril with a piece of clean straw to make him cough and sneeze. Slap his ribcage if that doesn't work. If he still won't start breathing, put him back down on the ground and give artificial respiration. (See "Giving Artificial Respiration" on page 391.)

Breech

A breech presentation is much worse than a normal backward foal. Because the hind legs are forward in a sitting position, nothing appears. The mare is in labor but nothing protrudes. When you reach in to check, all you can feel is the tail and rump. Seek veterinary help immediately!

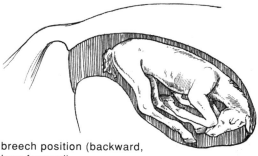

breech position (backward, legs forward)

If you absolutely cannot get help, scrub up and have someone hold the mare (she should be standing), while you reach in with both hands, pushing the foal's buttocks forward with one and reaching for a hock with the other. If it is a long-legged foal, it may be impossible to reach the hock. Still, if you can get hold of the hock and bend it, you may be able to reach the cannon bone. If you do, slide your hand to the foot. Cup the hoof in your hand to keep it from tearing the uterus, and bring it up over the pelvis and into the birth canal. If you can get one foot, you can usually get the other. Then follow the procedure above for delivering a backward foal. If the feet cannot be reached (the veterinarian may not be able to reach them either), the foal may have to be delivered by caesarean. If the foal is still alive after a breech position is corrected and pulled — a rare occurrence — he will need artificial respiration. (See page 391.)

Placenta Previa

One situation that requires immediate action to save the foal is when the placenta detaches too soon and part of it precedes the foal through the birth canal. If you see a mass of dark red membrane coming out when the mare starts labor, instead of the clear white amnion sac and feet of the foal, you should immediately scrub up and break through the thick membrane with your hands, reaching into the birth canal to get hold of the foal's legs and pulling him out swiftly. Placenta previa signals that his lifeline is detaching from the uterus, and the sooner you can get him out, the better.

Feces

If a birth is difficult or the foal is severely stressed from shortage of oxygen, it's not unusual for him to pass feces during birth. A foal covered with yellow excretia is in stress and may be in worse trouble if any of this material is drawn into his lungs when he starts to breathe. Wipe off his nose immediately, and clear his nostrils with a suction bulb.

Uterine Torsion

Occasionally a mare has torsion of the uterus in which the heavy uterus and its contents flip over, making a twist in the cervix. The foal cannot be delivered if the uterus stays that way. A mild torsion that occurs at foaling may correct itself when the mare rolls. Indeed, it is quite common for a mare in early labor to roll. If the foal is coming out a little crooked or the uterus has a mild twist, rolling may help.

In contrast, a complete twist prevents the foal from coming through. If this is the case, the veterinarian must try to correct the twist. Unless this is done promptly, the blood supply will be compromised, causing swelling and shock, death of the foal, and possible rupture of the uterus and death of the mare. Fortunately, this situation is rare.

The Importance of Breathing Soon After Birth

While the foal is developing in the uterus, he gets oxygen from the mother's bloodstream via the placenta and umbilical cord. At birth he must begin to breathe for himself. Once the umbilical cord pinches off or breaks, he's on his own. At this point the drop in oxygen level in his blood stimulates him to try to breathe (this is why the backward foal tries to breathe as soon as the umbilical cord crosses the pelvic brim and is compressed — and why you must get him out quickly). In a normal delivery, the birth process begins the transition. As the foal comes through the birth canal, his ribcage is compressed as it squeezes through the mare's pelvic opening. This forces fluid out of his air passages. As the foal's body emerges from the mare the ribcage expands again, which creates a vacuum in the lungs, drawing in the foal's first breath.

Usually he makes a smooth transition from depending on oxygen in his mother's blood supply to his own breathing, but this process can be hindered by his position immediately after birth. To breathe effectively he must be able to expand his ribcage and inflate his lungs, and his circulatory system must be working properly so blood reaches the lungs to pick up oxygen and carry it to the body tissues.

After a normal birth the foal positions himself upright, resting on his chest with forelegs extended in front as he begins the process of trying to get to his feet. Because of the way his chest and ribcage are constructed he breathes best standing up or lying in an upright position (head up) for his lungs can fully expand, increasing the levels of oxygen in his blood.

A foal lying flat has lungs partially collapsed. If he is limp or weak after birth, gasping for breath, or lying flat instead of trying to raise his head and roll up onto his chest, you must help him get upright. It is crucial that he begin to breathe properly.

Clearing the Foal's Airways

Make sure his nostrils are clear of mucus. A squeeze-bulb suctioning device works well; keep one handy in your coat pocket when attending a foaling mare. If you don't have one and it is obvious the foal's airways are obstructed, position him briefly with head downward to help the fluid drain out, then gently squeeze out excess fluid by pressing thumb and forefinger along the top of the nostrils toward the muzzle — like squeezing a toothpaste tube.

Stimulating the Blood Circulation

After making sure his airways are not obstructed, position the foal upright on his chest. You can also stimulate his circulation — which will aid his breathing reflex — by lightly rubbing and drying him with a rough towel and by gently moving his legs.

Checking the Foal's Heartbeat

Sometimes a foal is limp and lifeless (as after a difficult delivery and prolonged birth, leaving him short on oxygen), but if he's still alive there's a chance to save him. Check for a heartbeat with your hand on the lower left side of the ribcage, just behind and above the elbow. A young foal's heartbeat is easy to feel because there is very little tissue between it and the outside chest wall. If there's a heartbeat, try to get him breathing immediately.

Measuring heart rate is a good way to tell if the foal is in trouble; it drops as body tissues are depleted of oxygen. If the foal fails to start breathing within 20 to 30 seconds of birth and has a heart rate lower than normal, he is in trouble (normal is about 60 to 100 beats per minute for the newborn foal in the first few minutes, then rising as high as 200 per minute in the first hour, before tapering off at about 70 to 130 per minute). If the heartbeat has dropped to as low as 30, the foal's condition is critical. Also check his gums. They should be pink. If he's short on oxygen they will be grey and colorless or blue. He needs to start breathing.

Giving Artificial Respiration

If the foal does not respond to a piece of straw tickled up his nose, clear his airways with suction, then roll him up onto his breastbone with head and chin resting on the ground, nose as low as possible — to allow more fluid to drain out the nose. Then lie him on his side, head and neck extended. Cover one nostril tightly with your hand, holding the mouth shut. Gently blow into the other nostril. Don't blow rapidly or forcefully or you might rupture a lung.

Blow until you see the chest wall rise. Then let the air come back out on its own. Blow in another breath until the chest rises again as the lungs fill. Continue filling the lungs and letting them empty until the foal starts breathing on his own. Once his tissues become less oxygen starved his heart rate should rise, and he will regain consciousness and start breathing. His breathing may be erratic at first, but if everything else is normal he will develop a more regular pattern within a few minutes.

Normal Foaling — After the Birth

Let's look at what happens after a normal, unassisted birth. When the foal is born, he may be still wrapped in membrane but should quickly free himself by shaking his head. If he doesn't, and the sac has not broken, you should break it and make sure he begins breathing — being sure your hands are clean.

Don't Break the Cord Too Soon

He'll be lying with hind legs still in the birth canal. Don't try to pull him out or get the mare up. As soon as you are sure the sac is off his head and he's breathing, leave the mare and foal alone. The mare will generally lie there for 10 to 20 minutes before she gets up, regaining her strength after the effort of labor. Also, this is nature's way of making sure the foal gets his full blood supply. The placenta is full of blood that is pumped into the foal through the umbilical cord as the uterus starts to contract.

If the cord is broken too soon by the mare jumping up or you pulling the foal out, he'll be deprived of part of this blood and be somewhat weak and anemic for a few days. As much as 25 to 35 percent of the foal's blood is in the placenta at birth, but it drains into the foal if the mare stays down for a few minutes. If she gets up too soon not only will the cord break prematurely (which can cause it to hemorrhage) but the placenta that is detaching and starting to work its way through the birth canal may fall back into the uterus. Because the uterus and cervix are contracting, it becomes harder for the mare to expel the placenta if it all goes back into the uterus as she gets to her feet. Try not to disturb her as she lies there resting.

Keep the Cord from Breaking

If a mare foals standing up, as some do when nervous or when having help with a difficult birth, it's best if two people catch the foal as it is delivered and hold him up for a few minutes so the cord doesn't break immediately.

Disinfecting the Navel Stump

After the cord breaks — which usually happens as the foal struggles while trying to get to his feet, or when the mare gets up — thoroughly soak it in tincture of iodine. This disinfects it, prevents entrance by bacteria, and acts as an astringent to help it dry. A good way to iodine the navel stump is to immerse it completely in a small jar, making sure it is thoroughly dipped, without spilling any on the foal. If the cord does not break naturally, you can pull it apart with clean, well-washed hands (while you hold the cord close to the foal's body with one hand so there's no pull on his abdomen, which could injure him internally). It should be broken about 2 inches (5.1 cm) from his body; there is a natural stricture there, where it would normally break when the mare gets up.

If your veterinarian has suggested any medications for the foal (such as selenium, if you live in a deficient area), give these when you iodine the navel while the foal is still down.

If the Cord Bleeds

The cord rarely bleeds when broken naturally. The blood vessels of the cord have a greater elasticity than the outer covering, and if the cord is broken rather than cut these vessels draw back into the stump and seal off the opening. If the cord does bleed, or if for some reason you have to cut it and it bleeds, sit with the foal and pinch the navel stump firmly with your fingers for a few minutes. Then apply iodine to the stump. Never tie it, as there will be a certain amount of natural drainage; tying could cause swelling and increase the risk of infection. Also, if you were to tie it, the tie would act as a wick for infection to enter the stump.

Final Stage of Labor

The third and final stage of labor is when the mare expels the placenta. This should occur soon after foaling; the placenta can usually be seen hanging down after the mare gets up. The placental membranes surrounded the foal in the uterus, attached directly to the uterine lining except at the cervix. During labor the cervix relaxed and uterine contractions put pressure on the fluid around the foal in its placental envelope. The only direction it could go was through the cervix. The placenta ruptured at that spot and allowed the inner envelope — the amnion — and fetus inside it to start through the birth canal. The fluids between the inner and outer sacs preceded the foal (i.e., breaking of the water).

The placenta, the thick outer envelope, begins detaching from the uterine lining during labor, finishes detaching soon after the birth, and follows the foal out the birth canal 15 to 45 minutes later, except in abnormal situations when the placenta is retained. As soon as the placenta is shed (usually about the same time the mare gets up, or soon thereafter), remove it from the stall and spread it out on a clean surface to see if it is intact. A fresh, healthy placenta is bright red to mahogany in color, with a velvety appearance. An off-color placenta may be infected.

During third stage labor when the uterus is shedding the placenta and begins shrinking, the contractions may be so severe that they cause the mare discomfort and colic. In severe cases of abdominal pain she will need medication to halt the cramping and ease the discomfort. Consult your veterinarian if this is the case.

Checking the Afterbirth

If any part of the placenta is retained, it will cause serious uterine infection. When the membranes are spread out for checking, they will look like a pair of trousers (one leg smaller than the other) with the ends of the legs closed. There should only be one tear in the membrane — at the "waist" of the trousers where the foal came through. If any part of the placenta is missing, it will usually be the tip of one of the closed trouser legs. If you think there might be a problem, save the placenta in a clean garbage bag for your veterinarian to examine. A placenta that weighs more than 15 pounds (6.8 kg) is abnormal and may be diseased.

Is the Mare's Vulva Torn?

Soon after the mare *cleans* (sheds her placenta), check to see if her vulva ripped during foaling. If she tore very much your veterinarian should repair her as soon as the swelling goes down. If she needs a Caslick, she should be sutured up again soon after foaling.

Retained Placenta

The placenta attaches to the uterine wall during gestation with tiny finger-like projections that plug in to glandular openings in the uterine lining. This hookup connects the mare's bloodstream with the placental membranes, which send nutrients and oxygen to the fetus through the umbilical cord. Once labor starts the placenta starts to detach, especially during second stage labor when the foal is being delivered. The placenta comes completely free after the birth when the uterus goes through more contractions and begins shrinking back to normal size. If for some reason it does not completely detach it will hang out of the vulva, dragging on the ground.

If it takes the mare more than 2 hours to shed the placenta, this is abnormal. After 3 hours you should suspect problems and consult your veterinarian. Anything longer than 8 hours means the mare may be in grave danger. If the afterbirth is hanging down and dragging, tie it in a ball above the level of the mare's hocks so neither she nor the foal will step on it and pull it out forcibly. Use very clean hands. The weight of the balled-up placenta will keep a gentle traction on the membranes to help ease them on out. Never try to remove the afterbirth; it must come away by itself. If you pull on it you may cause hemorrhage or tear it, leaving part of it inside the mare.

Risk of Infection

The placenta starts to decompose rapidly after the birth of the foal. The birth process opens the way for contamination, and the membranes hanging out of the mare serve as a wick by which bacteria can gain entrance. Most mares that retain the placenta longer than a few hours develop uterine infection. It may be temporary, local, and easy to eradicate, or it may become systemic and life threatening. Even a localized infection has serious consequences if it leads to pus in the uterus. The infection may reduce the mare's ability to have another foal.

Retained Placenta

Retained placenta is fairly common after a difficult birth, twins, a premature foal, or induced labor. Sometimes retained placenta is due to dietary imbalance (e.g., a deficiency of phosphorus or selenium, or too much calcium from a diet too rich in legumes). Hypothyroid mares are also prone to retained placenta.

Even though the infection may eventually clear up, the resulting fibrous scar tissue that develops during the healing process will likely damage part of the uterine wall to such an extent that it cannot support another pregnancy. Extensive scar tissue in the uterus may interfere with conception or cause the embryo or fetus to die early in pregnancy.

Treatment

Your veterinarian may give the mare an intravenous or intramuscular injection of oxytocin to stimulate contraction of the uterus and hasten expulsion of the placental membranes. Anti-inflammatory drugs such as flunixin meglumine (Banamine) are usually given also to combat inflammation and prevent founder. The mare may need systemic antibiotics to treat infection. The veterinarian may also irrigate her uterus to remove any bits of afterbirth that might still be inside and to flush away debris that could foster infection.

Indeed, retained placenta can cause acute and fatal infection if bacteria get into the bloodstream. The mare becomes obviously sick with high fever, dullness and depression, lack of appetite, and a drop in milk production. The infection may also lead to founder.

Watch the mare closely after birth, especially if the placenta was retained. Check her temperature twice daily. Any fever, lameness, colic, loss of interest in feed, or drop in milk flow should be considered a medical emergency.

Sometimes in early stages of infection following a retained placenta the mare seems normal, only to colic and founder a few days later. If this happens, only swift medical attention can save her life.

Foaling Founder

If a mare develops infection in the uterus after foaling, changes occur in the bacterial flora and endotoxins may be produced. These toxins set the stage for production of lactic acidosis, an excess of lactic acid in the bloodstream. The lactic acid and endotoxins in the blood affect the whole body, causing shock and damaging liver cells. The latter condition interferes with the normal filtering of poisons from the body. The toxins also damage tiny blood vessels, resulting in laminitis. The mare's extremities suffer from impaired circulation, and the horn-producing laminae in the feet are deprived of oxygen and nutrients. Mares with acute laminitis caused by uterine infection after foaling may be difficult to treat because all four feet are affected. Laminitis can also result just from inflammation (which causes the release of histamine and prostaglandins, substances that damage body tissues) without any infection being involved.

Founder Can Become Chronic

Some mares suffer foaling founder a day or two after foaling even if the birth was normal and the placenta shed promptly. An examination or uterine flush reveals the uterus to be infected. Most cases of severe founder follow a difficult birth or retained placenta and sometimes become chronic, giving the mare a low-grade laminitis that flares up periodically. Hypothyroid mares are also extremely susceptible to foaling founder.

Whether or not a mare will founder again with future foaling depends on the severity of the initial laminitis and the potential lingering presence of laminitis when she foals again. A mare whose founder episode is successfully treated, with full recovery, may not have further problems, but laminitis due to bacterial endotoxins may recur if she is ever exposed to a sufficient quantity of endotoxins again. Once a horse suffers from founder, it's much more susceptible to founder in the future. Thus, a mare that founders due to uterine infection may readily founder again if conditions in the uterus promote another infection.

Uterine Prolapse

A rare problem is prolapsing of the uterus. When this occurs the uterus comes out after the foal and hangs down from the vulva all the way to the mare's hocks. This serious condition is fatal unless you get prompt help from a veterinarian. While waiting for him or her to arrive, saturate a clean blanket with a solution of disinfectant and warm water and use it to support the uterus. Tie the mare so she cannot move around, and try to keep the inverted uterus supported so it can't come out any farther and put strain on its attachments. Too much trauma or strain might cause a blood vessel to break, resulting in fatal hemorrhage. It may take two people to support the prolapsed uterus, holding the blanket between them. If possible, it should be held up almost as high as the vulva.

Prolapses usually result from the mare's continual straining after the foal is born. The problem is most likely to occur after a hard foaling but can happen even under good conditions.

Hemorrhage After Foaling

Another serious complication of foaling is internal bleeding. This rare but life-threatening problem usually occurs within 24 to 48 hours of the birth. It is not easily recognized if you haven't witnessed it before, but is a condition you should be aware of. If the person attending the foaling realizes early on that something is wrong, there is a good chance of saving the mare and the foal. Some mares collapse suddenly onto the young foal, killing or injuring it.

Bleeding results from rupture of an artery in the uterine wall or one of the major vessels that run along the uterine ligaments. The mare's chances for survival depend on the size and location of the damaged artery. Bleeding from the uterine wall is not likely to be fatal because blood seeping into the uterine tissues creates a large hematoma that tends to block and slow the flow. However, if the damaged vessels are in the uterine ligaments they bleed right into the abdominal cavity, causing rapid and fatal blood loss.

Internal bleeding is usually caused by degeneration of artery walls rather than by a difficult foaling, and it is more common in mares older than 15 years. Continual stretching and contracting of the uterine arteries during pregnancy result in less elastic artery walls, and the force of foaling may break one. The mare shows no evidence of bleeding at the vulva and may die before the problem is recognized.

Hemorrhage sometimes occurs prior to foaling due to uterine torsion or trauma during violent exertion, but most cases happen soon after the mare foals or as late as the next day.

Signs of Internal Bleeding

The discomfort of foaling and colicky cramps due to placental expulsion may mask signs of hemorrhage, but certain subtle signs can lead you to suspect the colic is from bleeding. A mare with a ruptured artery is weak and may be reluctant to get up after foaling. If she doesn't get up, don't push her; the less exertion, the better. She may get up and go down again, being anxious, agitated, uncomfortable, and colicky. If she is standing she may sweat and shift from one foot to another, sometimes curling her upper lip. She may paw and roll; if this is the case you should keep the foal away from her so he won't be injured. If bleeding continues the mare becomes dull, her mucous membranes become pale, and her temperature drops and pulse weakens. Sweating is profuse and her skin clammy as she goes into shock; she starts to quiver, ready to collapse.

If you suspect hemorrhage, keep the mare as quiet as possible and get veterinary help immediately. Blanket her to conserve body heat. Protect the foal (in a cage of hay bales just outside her stall door, so she can see him and not be too upset by his absence). The veterinarian will use tranquilizers to calm her and lower her blood pressure, as well as intravenous fluids or transfusions to keep up her blood volume. He may give oxytocin to hasten uterine contractions and seal broken blood vessels if the bleeding is in the uterine wall. Coagulants and painkillers also help. If the mare recovers she will resorb the hematoma and may foal normally the next time.

Most veterinarians recommend that mares who survive a bleeding episode take a year off from breeding. If the mare cannot be saved, she should be milked out immediately after death to provide colostrum for her foal.

Most Foalings Are Normal

There are many, many things that can go wrong at foaling, but do not be discouraged; most mares foal with no problems. It is important, however, to know about possible problems, because many can be prevented with good care — such as cleanliness to avoid infection — or corrected with prompt assistance. If you know what to do in an emergency, you have a much greater chance of being able to save the foal — and the mare.

Care of the Mare After Foaling

After foaling the mare will be tired, dehydrated, and somewhat constipated. Offer her plenty of clean, lukewarm water to drink and palatable food. Green grass or alfalfa hay is ideal. She might like a bran mash; it not only contains moisture (which she needs) but is also quite laxative. If she has been on a diet of grass hay, she may be constipated. (For a bran mash recipe, see chapter 2.)

Don't be alarmed if she passes no bowel movements for the first 12 hours after foaling. She "cleaned out" rather completely when she was nervous, restless, and pacing during early labor, and it may be a while before any more feces work through the digestive tract. Also, the mare may be sore from foaling and reluctant to pass a bowel movement because passing manure causes pain. The best remedy for constipation is a good, palatable, laxative feed such as green grass, alfalfa, or a bran mash. If the mare continues to be constipated for the first 24 hours (passing no manure or just a few firm, dry balls), consult your veterinarian. In serious cases of post-foaling constipation the veterinarian may decide to give the mare mineral oil.

Deworm the mare with ivermectin immediately after foaling to prevent parasitism in the foal by *Strongyloides*, a primary cause of so-called foal heat scours. (See the section on parasite control in chapter 24.) If the foal does not pick up worms from the mare's milk (tiny threadworms are passed to the foal from the mare's udder during the first days of life), he will scour less. He will also have less initial parasite load from eating his dam's manure — something foals do to obtain the intestinal flora necessary for digestion of roughage — since her manure will be clean.

TWENTY-THREE

CARE OF THE NEWBORN FOAL

Your mare has just foaled a beautiful long-legged baby. The foal will need immediate attention (i.e., having his navel stump disinfected) and then careful monitoring for the first few days to make sure all goes well. He must nurse soon after birth to gain protection from disease through absorption of antibodies in colostrum, and he must also start passing bowel movements. You must check to make sure he does.

Guidelines for Care Immediately After the Birth

There are several things you should attend to when the foal is born, but be sure to do these calmly and quietly. Too much fuss, noise, or activity can cause problems.

Avoid Foal Rejection

If you rush to iodine the navel, help the foal to its feet, and aim him toward the mare's udder — and your family and friends are there to admire it — you may upset the mare. Sometimes all this attention can lead to problems. Occasionally a mare will ignore, reject, or even viciously attack her new foal, or she may seem unsure or confused about motherhood — especially if it's her first baby. Some mares balk at mothering the foal because of a hormone imbalance or physical problem (e.g., pain from a sore, swollen udder when the foal tries to nurse), but often the problem is too much human interference during the critical time when mare and foal should be alone for bonding, establishing that all-important mother-baby relationship.

Wait to Begin Imprint Training

A popular procedure is imprint training — handling the foal soon after birth and getting him used to the feel of your hands all over his body, impressing on him that human handling is part of his world and nothing to fear. Done properly, imprint training can be beneficial, making the foal more at ease later when you work with him. For example, he won't refuse to let you handle his ears, put something into his mouth, or pick up his foot. His subconscious will "remember" this has all been done before.

Sometimes, however, the person who rushes to handle the new foal is focused on one thing only — imprinting the foal — and forgets how this might disrupt nature's program for the mare and her newborn. If you're going to imprint the foal, keep the mare in mind and stay tuned to her reactions. If she's an old dependable broodmare and your handling of the foal is no problem for her, there is nothing to worry about. If she's a first-time mother or a highly nervous individual, go easy. Wait until that very important bonding process is fully accomplished before you do anything with the foal.

Too much human attention and handling of a new foal may interfere with bonding, as the foal may be attracted to the human moving around him instead of seeking out the mare to nurse.

Safety Tip

Be prepared for your friendly, trusting mare to have a sudden change of personality when she foals. Many mothers become extremely protective of their babies — it's their instinct to defend the newborn against predators, and they become aggressive toward intruders for the first few hours after birth. Keep an eye on the mare the first time you go into the stall or pen after foaling, in case she decides to attack you as you tend to the foal to iodine the navel. Keep all children, pets, and extra people away from the mare and foal the first day, to avoid upsetting the mare.

Minimize Handling of the Foal

A nervous or insecure mare, or any first-time mother, may have her maternal instincts disrupted by too much human handling at foaling time. Even having people observe her during the birth may upset her and make her postpone labor for several hours or become less apt to mother the baby after she foals. Once the foal is born, don't rush right in unless his head is still encased in the amnion sac and he needs help to start breathing. Let the mare and foal lie there until the cord is broken and it's time to iodine the navel; then move in quietly, take care of it, and slip away again.

A first-time mother may be nervous and uneasy but needs to bond with her new baby.

Rushing to iodine the navel, clean up soiled bedding, dry the foal, or help him to his feet and to nurse can be too much disruption for some mares, causing them to reject the foal. If the weather is cold you will have to dry the foal; otherwise leave mare and foal alone once you have done what you need to do (e.g., start foal breathing if there's a problem, iodine navel, give selenium injection if recommended by your vet). By moving slowly and quietly you won't upset the mare. If everything is normal, leave the pair in peace for at least an hour. Watch from a distance. If the mare isn't worried or uneasy about hovering observers (which in her mind pose a threat to her baby), she'll be more likely to concentrate on the foal and begin her role as mother. Foaling time reactivates deep-seated instincts; even though your mare trusts you, too much human presence and fussing at this time will upset her.

Remember That Bonding Takes Place During the First Hour

During this time you should stay away, quietly observing from afar. Throughout the first hour the mare smells, nuzzles, and licks her foal, permanently identifying it as hers and establishing the bond of scent that locks her newborn's identity into her brain.

Human interference can also confuse the foal, because his instinct is to follow any large moving object. Under natural conditions the only nearby large moving object is the mare. With people in the stall he may be attracted to them instead and not seek the mare to nurse.

Don't Help Him Stand Up Unless He Really Needs Help

The foal's first attempts to get up are clumsy; he falls down several times before gaining his feet. It's tempting to help him, but usually unnecessary and occasionally harmful. Sometimes a foal has a fractured or cracked rib from the pressure of coming through the birth canal. These usually heal just fine (and you probably won't even know they occurred), but picking him up or helping him stand may displace the fractured pieces and cause them to puncture a lung or other vital organs.

Out in a pasture the foal generally has less trouble getting to his feet, as the footing is better and he's not encumbered by the straw of a foaling stall. He's not hindered by hitting walls or bumping against a manger or water tub. If he's born in a stall, make sure there is nothing there that might hurt him. Before the mare foals you should remove tubs and other obstacles, eye screws, or protruding nails at foal height. Monitor his progress in getting up and blun-dering toward the udder, helping him only if he really needs it.

Nursing

By the time he's an hour old, the foal should be on his feet and looking for the udder. Make sure he finds it and gets his first drink of colostrum. It's best if he can accomplish the first nursing by the time he is 2 hours old.

Waiting and Watching

You don't need to panic and assist if the foal is a little slow to find the udder. He'll get more strength and coordination with each unsuccessful try, unless he has a physical problem (e.g., born weak, oxygen deprived, or with some other abnormality). Even if he doesn't accomplish the first nursing within an hour, he is still not in immediate danger, because his ability to absorb the colostrum's antibodies remains high for 2 to 3 hours following birth. Then it begins to decline as the pores in the intestinal lining close.

If you wait at least an hour before helping him, you'll be less apt to hinder the bonding process and the mare will be less confused by your attempts at intervention. As long as the foal gets a good nursing by 2 to 3 hours of age, and an adequate amount of total colostrum within 24 hours in subsequent nursings, he will obtain the necessary immunities. Let him try to do it on his own, and allow the mare to mother him.

Some mares are confused at first, especially first-time mothers or mares that had earlier bad experiences. Some habitually reject their foals or are indifferent, but most will mother the foal if given a chance. Usually once he nurses, she accepts him. This is partly because nursing stimulates the produc-tion of hormones that prompt the so-called maternal instinct. Also, the mare's udder won't be so sore and tight after some of the milk is out of it, so she may become more cooperative. Most mares that reject their foals eventu-ally accept them if time, care, and patience are used, but you can often pre-vent potential problems if you keep a low profile at foaling time and give nature a chance.

Eventually most foals find the right place to nurse, although to an observer it may seem like an eternity as the newborn noses about here and there. You only need to help if he is weak and tired from a long or hard birth and gives up too easily, or if the mare kicks at him. If she kicks, you'll have to restrain her until he has nursed.

Her Udder May Be Sore

The mare's udder is a little swollen and tender at first, and she may be nervous about the first nursing because it hurts her. Sometimes the mare moves around a lot or squeals and kicks at the foal, but her kicks are not meant to hurt him; she's just upset because her udder hurts. Rarely does a mare kick viciously at a foal, but if this happens you'll have to intervene. Often restraining the mare is more helpful than trying to guide the foal.

A mare with a sore udder may kick at her new foal.

Helping a Foal Nurse

If the foal has not succeeded in nursing by the time he's 2 hours old, help him before he becomes tired and discouraged to make sure he gets crucial colostrum. If the mare is moving around too much, halter her and stand her against the fence or stall wall. You may need to hold up her front leg or shoulder twitch her (grasping the skin over the shoulder with your hand) to keep her still and steady if she wants to keep moving her hindquarters to avoid the foal or insists on kicking at him. Talk to her soothingly and try to keep her calm and still. If the foal is strong and vigorous, this may be all the help he needs so he can catch up with the udder. If he is clumsy and uncoordinated, or scared off by her kicking and too timid to try again, you may need someone to guide him to the udder while you restrain the mare.

If a newborn foal is weak or confused, he may need help with his first nursing. If he has not nursed on his own by the time he is 2 hours old, you should assist.

Use as little force as possible to keep an uncooperative mare calm and quiet, but use a twitch or Stableizer if necessary to hold her still (see chapter 5). Usually by the second nursing, which should be tried 20 minutes after the first, the mare is more cooperative. In most cases by the third or fourth nursing she will accept the foal. You may merely need to hold her until he starts nursing, then you can release her gently and she'll let him continue to nurse. Occasionally a mare refuses her foal for 24 to 48 hours, but usually if you are persistent in helping the foal at each nursing she will come to accept it. Whenever a mare is aggressively vicious toward the foal you should keep them separate except at nursing; it may even be necessary to find a substitute mother if she fails to accept the foal after several days.

When trying to help a foal nurse, hold him in place and steady him by hanging on to his tail. Don't handle his head much or you'll confuse him. If he can't get onto the udder himself, milk a little out of the mare into a clean container. Then put some colostrum on his nose and lips and all over the mare's udder (or use corn syrup, if necessary) to get him interested in nursing and to help him find the right place. Never force his head down to the udder. He can't nurse in that position, and it will make him resist your efforts.

If he simply will not nurse, milk more from the mare and use a sterile bottle (presterilize bottles in boiling water) and lamb nipple to get him sucking. Usually after he gets a taste he'll be more eager and try harder, and you can guide him to the udder or lead him to it with bottle and nipple. If he's confused and tries to suck on you instead of the mare, get him sucking the colostrum from your bottle and move it toward the udder until you can transfer his mouth from nipple to teat. This may be easiest if someone else is on the other side of the mare, holding the bottle underneath her and bringing it up next to her udder to "fool" the foal into getting into the proper position to nurse.

The foal's first nursing is extremely important: The mare's first milk (colostrum) contains nutrients crucial to his health.

The Importance of Colostrum

The newborn's first nursings are very important. Foals that don't nurse soon after birth are more likely to become ill and die during the first weeks of life than foals who nurse promptly. This is because the mare's first milk contains ingredients crucial to the foal's health and survival. Colostrum has more energy, protein, vitamins, and minerals than regular milk, with twice the energy and five times the protein. It also serves as a gut stimulant to help the foal pass his first bowel movements. These first evacuations are often firm and hard-packed, therefore difficult to pass. Sometimes an enema is necessary, but in most cases the laxative effect of colostrum will get things moving.

Benefits of Colostrum

Colostrum contains a creamy fat, high in energy and easily digested. It's an ideal first meal, giving high-energy calories to generate body heat and strength. This is why a foal that nurses soon after birth is more lively, strong, and vigorous (and able to stay warmer if born on a cold night) than a foal that has not yet nursed. Once the foal gets some colostrum he usually has the strength to keep looking for more, becoming stronger and more coordinated with each energy-laden meal. The newborn nurses in small amounts at frequent intervals, about every 20 minutes.

Antibodies

Especially important are the antibodies in colostrum. The foal is born with no immunity, no resistance to disease. Unlike a human baby who picks up his or her mother's immunity while still in the uterus, the foal has to acquire his through colostrum when he nurses.

Disease Protection

During late pregnancy when the mare's udder is filling she makes this special milk, with a concentration of all the antibodies she's developed, to give the foal immunity for several weeks after birth. This passive transfer usually lasts until his own immune system has encountered bacteria and viruses in his environment so it starts producing its own antibodies. Passive transfer failure can occur, however, if there is an inadequate amount of antibodies in the mare's colostrum or if the foal is unable to get enough colostrum in time to receive proper levels.

If the mare has encountered diseases and developed antibodies, she can pass this immunity to the foal. Vaccination also stimulates her body to develop immunity against specific diseases. For instance, if the mare has a tetanus or botulism shot during gestation, she develops antibodies that will be present in her colostrum to protect the foal as soon as he nurses. Keeping the broodmare's vaccinations up-to-date is the best way to make sure the new foal will have immunity against most problem diseases.

A foal that fails to nurse soon enough runs a high risk of developing scours (i.e., diarrhea), pneumonia, or other serious problems. Failure of passive immunity transfer is the greatest single factor resulting in disease and death in foals.

Timely Nursing

The foal must nurse within 2 or 3 hours to gain immune protection. At birth his intestinal lining has openings directly into the bloodstream, and the antibodies can slip through. However, these pores begin to close soon after he's born, so he must nurse promptly to get the full benefit of colostrum. Foals lose 75 percent of their ability to absorb antibodies by 4 hours of age; the large molecules can no longer slip through the gut lining into the lymph system and bloodstream. A foal that gets his first nursing after 4 hours receives only a fraction of the antibody protection he needs. The colostrum remains beneficial for protecting against some gut infections, because certain antibodies still have local activity within the gut. Even though he can no longer absorb antibodies after the first few hours, the foal needs colostrum in subsequent nursings throughout the first day because of its additional benefits. He not only needs the antibody protection but also its high energy content (to keep him warm and give him strengh) and it's laxative effect.

If he's born during the night and you don't discover him until morning, you may not know whether he nursed soon enough. If he's up and about, strong and frisky, nursing vigorously, passing bowel movements and urinating (which most foals do after nursing), these are good signs that mean he's getting sufficient milk. If he isn't passing urine or bowel movements, or you find him cold and shivering or slow and weak, make sure he nurses as soon as possible.

If there's failure of passive transfer for any reason (e.g., whether the foal failed to nurse in time, or the mare had poor colostrum), the foal can be transfused with plasma from the mare or from another horse on your farm, or with commercial plasma preparations.

Testing the Foal's Antibody Levels

If the foal is not able to nurse within the first few hours, or if the mare foals unexpectedly and when you find the newborn you don't know how old he is or if he nursed soon enough, your veterinarian can check a blood sample after the foal is 18 hours old to test for immunoglobulin levels. The peak level of antibodies in the foal's blood is attained at 18 to 24 hours of age. The veterinarian can do a simple test at the clinic in just a few minutes. There are also test kits available to analyze blood samples; in fact, on many breeding farms it is routine procedure to test foals. Test results are also required by some insurance companies as part of the criteria for determining foal insurance premiums.

Lack of Colostrum

Sometimes a mare does not have an adequate supply of colostrum in her first milk — either because the birth was premature or because most of the colostrum leaked out before foaling — or does not come to her milk until several hours after the birth. First milk should be thick, sticky, and more yellow-orange than regular milk. If it's thin and white, it is low in colostrum and may not give the foal much protection. Some mares have little or no milk — a common problem in mares grazing fungus-infested fescue pastures.

Usually a mare with no milk comes to her milk if the foal nurses often, stimulating hormones that trigger milk production. Meanwhile you must find an alternative source of colostrum. If you feed him an alternative source, make sure he gets an adequate first nursing or two of colostrum, then keep him a bit hungry so he'll work on the mare's udder between feedings to stimulate the milk to come. Colostrum can be tested for antibody content; if you suspect a problem with the milk, or if the mare doesn't seem to have enough, consult your veterinarian.

Avoid Stress

High stress at birth interferes with a foal's ability to absorb antibodies. A difficult birth or cold weather may reduce his ability to absorb antibodies. Stress seems to hasten closure of the intestinal pores.

Getting Colostrum into a Foal That Hasn't Nursed

If he can nurse a bottle, that's the easiest way to get him fed if he cannot nurse his mother. If he's weak and unable to suck, you must feed him by syringe or stomach tube. With weak foals it is often beneficial to syringe milk into the mouth, rather than use a stomach tube, just to keep the swallow reflex going. A stomach tube is generally used as a last resort on a foal that cannot swallow, although some veterinarians prefer to use a tube for any foal that has trouble sucking.

If you must feed the foal his first colostrum, he needs about 6 to 24 ounces (0.2–0.7 L) each hour for the first 3 or 4 hours, depending on his size. A small foal (less than 80 pounds [36.3 kg]) will need just less than a cup (0.2 L) per hour, whereas a large foal (more than 100 pounds [45.4 kg]) needs as much as a pint and a half (0.7 L). It's best to divide that up and feed him every 30 minutes. Your veterinarian can advise you on the proper amount for your particular foal.

Feeding by Tube

The stomach tube is a plastic tube approximately 3 feet (0.9 m) long that goes into the nostril to the back of the throat, then down the esophagus to the stomach. The veterinarian usually puts it in and leaves it, stitching it in place with a few strands of suture thread. Then you can milk out the mare periodically (with clean hands, into a sterilized container) and carefully funnel the milk down the tube at specific intervals and specified amounts as directed by your veterinarian.

This has advantages over bottle feeding if a foal is weak; you get the proper amount of milk into him without any effort on his part that might tire him, and without danger of getting milk into his windpipe — which would cause aspiration pneumonia. Also, the foal can still nurse the mare with the tube in place, and if you are not bottle feeding him he won't be so apt to think of you as his mother; he may still try to nurse the mare. Once he starts nursing (leave him a little hungry now and then so he will try to nurse his mother), the tube can be removed.

Emergency Colostrum

If the mare has no milk or poor colostrum, it may be possible to obtain colostrum from another mare or a breeder who has a supply for emergencies. Most large breeding farms keep frozen colostrum on hand, and if you have many mares you should keep some too. Store it in small plastic containers or in snap-top freezer bags in 4 to 8 ounce (0.1–0.2 L) quantities — easy to thaw quickly when you need them — or in baby bottles with plastic liners. Milk the mare into a clean non-glass container, pour the colostrum into the plastic liner or other container, seal it, label it with date and name of the mare, and put it in the freezer.

Taking Colostrum from a Mare

Colostrum is only produced during the first 24 hours; thereafter the concentration of antibodies falls sharply. If a mare is a heavy milker and her foal has several good nursings, you can safely take extra colostrum from her during that first 24 hours to freeze and save. Before milking the mare, gently wash her udder with plain warm water, or brush it clean with a dry, soft cloth to make sure you don't get any dirt in the milk you are collecting. Don't use anything but plain water for washing, because the foal finds the udder by smell and you don't want to confuse him.

Freezing and Thawing Colostrum

If colostrum is frozen quickly and kept frozen at low temperatures (approximately −5°F [−20.6°C]) it will keep well for about 2 years without losing the value of its antibodies. If it is older than that when you use it, check its potency by thawing a small amount and having it tested.

Thaw colostrum in warm water, not a microwave; even a setting on low heat will destroy part of the antibodies if it gets much above body temperature (100°F [37.8°C]). Thaw it in water in which you can comfortably place your hand. Colostrum should never be thawed and refrozen.

Keeping a Foal Warm in Cold Weather

If it's cold when the mare foals, you may have to dry the newborn with towels and put a heat lamp in a corner of the stall, placed where it cannot be accidentally bumped. Another way to keep the foal warm is to use a zippered sweatshirt as an overcoat, putting his front legs in the sleeves and zipping the shirt under his belly. If the sweatshirt has a hood, use it to cover the top of the foal's neck. When most of the foal's body is covered, he won't get too chilled.

Dressing a foal in a zippered sweatshirt will keep him warm in cold weather.

Bowel Movements

The foal's first bowel movements are dark and sometimes hard. While in the uterus he continually swallows some of the fluid in which he floats, but he never evacuates his intestinal tract until after he is born. Hard pellets accumulate in the tract, and the foal may have trouble getting rid of them. After he has nursed a few times the colostrum stimulates his gut to move, and his bowel movements become soft and yellow, but if he remains constipated he will need an enema.

Constipation is a common cause of death in newborns. If a foal does not have a bowel movement during the first few hours of life, you should gently remove the pellets from his rectum with a well-lubricated (use vasoline or K-Y jelly) finger. To soften up pellets beyond your finger's reach, give the foal an enema.

Giving an Enema

It is simplest to use an inexpensive preparation such as a human adult Fleet enema from a drugstore. If you don't have one of those, try a cup (0.2 L) of mineral oil mixed with a quart (1 L) of warm water or some slightly soapy water. Gently squirt it into the foal's rectum with a regular enema tube (inserted carefully into the rectum a few inches [cm]) and squeeze bulb, or use the nozzle of a large syringe to squirt it in. Have someone hold the foal for you with one arm around his chest and the other around his buttocks to keep him still while you give the enema.

Giving a foal an enema (mineral oil and warm water) with a large syringe can relieve constipation.

Stubborn Cases of Constipation

Some foals may need a small quantity (half cup to a cup [0.1–0.2 L]) of mineral oil orally, or magnesium hydroxide (Milk of Magnesia), to soften up a high impaction out of reach of an enema. If the foal continues to have trouble passing bowel movements or shows discomfort (e.g., restlessness, straining, rolling and colic, or nosing at his flank) even after several enemas, consult your veterinarian. If constipation is severe, the veterinarian may give the foal a sedative or mild tranquilizer to keep him from injuring himself until the laxative works.

Exercise

If the foal is born in a stall, both he and his dam should be turned out into a paddock or pasture as soon as he has nursed a few times, the pair has bonded well, the mare has cleaned, and everything is normal. The foal needs exercise; moving around will help him become stronger and more coordinated, and it will stimulate his appetite and digestive tract.

Turn the pair out in a safe enclosure with no other horses; they might pester the mare or try to steal the foal. It's not unusual for mares to swap foals if several are in the same pasture. Keep each pair by themselves for the first few days until bonding is fully established. The paddock should be away from the influence of other horses because an overly protective mare may charge at an interested onlooker peering over the fence, possibly injuring her foal in the process. If the ground is wet or the weather is cold, leave them out only a few hours the first time, bringing them back in the barn when the foal gets tired and wants to lie down. If the weather is good, the pair is better off left outside.

The mare and foal both benefit from exercise.

Stall Must Be Clean

If the mare and foal spend any time in a stall, it must be very clean and safe. Remember that ammonia from urine and manure is highly damaging to young lungs. A stall might look clean and seem pleasant but still be noxious at floor level; the best way to check is to actually lie down in the straw to see if you can smell ammonia. This is one of the leading causes of foal pneumonia in housed foals. If you can smell ammonia near the floor, where the foal spends most of his time when napping, you must clean the stall more thoroughly. The foal must have a clean stall and dust-free bedding.

Problems in the Newborn Foal

The first few days and weeks of life are a critical time for the new foal. Closely observe the pair so you can quickly detect and deal with a problem before it becomes serious or life threatening. Some problems are obvious right after birth — such as external congenital defects, limb deformities, and prematurity — and require professional care. If anything seems abnormal about the foal, don't hesitate to have your veterinarian examine him.

Birth Defects

Some foals are born with cleft palate, crooked head (from lying against the mare in an abnormal position), or crooked legs (from being cramped for space in the womb). Others have incomplete digestive systems; the intestine is closed off before it gets to the rectum or anus and the foal cannot pass bowel movements. Still other foals are born with a leaky navel (pervious urachus); the canal in the fetus that connects the bladder with part of the umbilical cord and placenta does not seal off at birth, so urine leaks from the bladder and out through the navel. Some foals may be weak, with severe heart defects. Among all these problems, some can be corrected or repaired with surgery whereas others cannot. If a foal is born with turned-in eyelids, for instance, a veterinarian can correct this condition.

Some abnormalities correct themselves with time, such as crooked legs or weak tendons. Occasionally a male foal has trouble urinating because his sheath and penis have not come down yet; he crouches in urinating position for a long time and strains and dribbles. The first time you see this condition you may think the foal is abnormal, but after a day or two the situation usually resolves and the foal is perfectly normal.

The Foal Has a Delicate Nervous System

Keep close watch on the foal and be aware of subtle signs of trouble. Time is critical when a foal is young because small problems can become serious in a hurry. For instance, the foal's nervous system is immature and easily damaged by infection, oxygen deprivation, stress, or trauma. In mammals the brain is protected from most insults (including drugs) by a system called the *blood-brain barrier* that keeps ordinary disease organisms and poisons in the bloodstream from damaging the brain. In the young foal (up to about a month of age), this barrier is not completely functional, so disease can have a devastating effect. A foal that has diarrhea or some other problem, or one that suffered stress and trauma during a hard birth and passed fecal material during birth, could experience neurologic consequences and collapse in a seizure because his brain has been adversely affected.

Ruptured Bladder

Sometimes this condition is present from birth, either as a congenital defect, a result of trauma during birth (e.g., when a big foal is squeezed through the birth canal with a full bladder) or a weakness in the bladder wall that gives way later. The most common cause is an accident during birth, but a sudden jerk on the navel cord (e.g., when the mare foals standing up or jumps to her feet immediately after foaling) can also rupture the bladder.

A tiny hole or tear in the bladder will not keep the foal from passing some urine normally, but small amounts continue to leak into the abdominal cavity. A large tear will make him unable to pass urine because it goes out the leak faster than it can collect in the bladder — although he may often squat and strain due to the uncomfortable full feeling. A large leak will make his condition obvious before he is a day old, but a small slow leak may take 2 or 3 days to accumulate enough pressure for symptoms to appear.

Symptoms

The foal may seem normal at first, but before the day is over he becomes dull and uncomfortable, losing his desire to nurse. Periodically he may raise his tail and try unsuccessfully to pass urine. His abdomen may become distended and tender as fluid seeps out of the ruptured bladder and into the abdominal and pelvic cavity. As the waste products of urine are absorbed into his body instead of being eliminated, the foal develops a toxic condition called *uremia*. The fluid build-up puts pressure on his diaphragm and lungs; he pants shallowly, unable to get enough air. If this condition is not corrected quickly by surgery, he will go into shock and die.

Treatment

Your veterinarian can check for ruptured bladder by lying the foal on his side and inserting a needle into his abdominal cavity to drain off fluid and check it. If the abdomen is full of urine, it will flow freely out the needle and smell like urine. Blood tests can confirm the presence of urine products. Prompt surgical repair results in rapid and complete recovery. If the condition is neglected too long, the foal will die from uremic poisoning or suffocation.

Neonatal Isoerythrolysis

This problem may not show up for a day or two. The foal is born normal and is lively, vigorous, and nursing the mare. Then he starts to show signs of weakness. Sometimes called *Rh foals*, these newborns have a different blood type than the mare, an incompatability that causes antibodies in her colostrom to attack the foal's red blood cells. The condition can be fatal unless it is quickly corrected.

Cause

If the foal inherits an incompatable blood type from the sire and some of the foal's blood cells pass through the placenta into the mare's bloodstream, her immune system creates antibodies against the foal's red blood cells if these components are not already part of her normal blood makeup. The antibodies are usually produced in large amounts by the eighth to tenth month of pregnancy but usually don't affect the unborn foal because they are too large to pass through the placenta from mare to foal. Thus, no reaction occurs until the foal is born and ingests the mare's colostrum, which contains a high concentration of the antibodies.

The antibodies are absorbed into the foal's blood after he nurses, where they attack his red blood cells and destroy them, inhibiting the flow of oxygen to his body tissues. This results in anemia, bloody urine (due to the breakdown of red blood cells), and jaundice; his tissues look yellow because of the hemoglobin pigments that are scattered throughout his blood from the destroyed cells.

Symptoms

In acute cases the foal becomes weak before you notice the jaundice. After the foal has nursed a few times he becomes dull and lethargic and loses interest in nursing. There is no fever, but his pulse rate rises. Respiration remains normal until the anemia takes its toll, then he has trouble breathing and starts yawning; at this point his respiratory rate goes up to 80 breaths per minute. To save the foal, you must take him away from the mare immediately or muzzle him so he cannot nurse.

Some cases are not evident until the foal is 2 to 4 days old. Mild cases may not develop symptoms until day 4 after birth, and these may recover without treatment. If a foal shows symptoms soon after birth, you must take prompt action to save him. Your veterinarian can test his blood and the mare's colostrum for diagnosis.

Treatment

Treatment consists of feeding the foal an alternative source of milk and milking out the mare every 2 hours to keep up her milk flow. This hastens the coming of her regular milk and gets rid of the last traces of colostrum that contain the deadly antibodies. Your veterinarian can check the foal's blood to see if he needs a transfusion, which may save his life if he is weak and anemic. Of course, the blood used must be from a horse with a compatible blood type.

Prevention

Blood from both mare and stallion can be checked before breeding. If the mare is already pregnant, have your veterinarian test her blood serum against the sire's blood cells a week or two before she is due to foal. If the result is positive, meaning the foal will be endangered by her colostrum, you must not let the foal nurse his mother when he is born (until his own blood has been tested; it only takes a few minutes at the clinic). If the test shows the foal is at risk, do not let him nurse the mare until he is fully 3 days old. Usually after the second day of life the foal's intestinal membranes can no longer absorb the antibodies and he is out of danger.

Be present when the foal is born, making sure he doesn't nurse. Have colostrum from another source available if you think there might be a problem. Muzzle the foal, test the milk, and if it's harmful feed a formula for the first 3 days.

Once a mare has produced an Rh foal you need to prevent it from happening again. Breed her to a stallion with a compatible blood type.

Umbilical Hernia

Some foals are born with umbilical hernia, a soft tissue swelling or protrusion at the navel. While the fetus is growing in the uterus, the navel is a natural opening through which umbilical blood vessels enter. After he's born these vessels begin a natural degeneration process; within the first month of life the skin has healed over. In some foals, however, the underlying tissues of the belly wall do not close, leaving a hole.

Symptoms

Some hernias are obvious soon after birth, but others don't appear until a few weeks later. Trauma, infection, or even straining to have a bowel movement can cause more tissue to protrude through the opening, creating a visible swelling under the skin at the navel. As the foal grows in size, the increasing weight and fluid in the abdomen will cause the protrusion to enlarge if the hole in the abdomen wall is large.

Treatment

Large hernias require surgical repair, but a small one may disappear on its own as the foal grows. If you have any questions about the young foal's navel, ask your veterinarian to check it. Though umbilical hernias often heal on their own, if left untreated a large one can have fatal complications.

Infections

Probably the most serious problems of young foals are infections. Not only is the new foal acutely susceptible to certain pathogens because of his immature immune system, but the viruses or bacteria often gain the upper hand swiftly, causing damage or death in just a few hours. Most infectious organisms that might harm the foal are common in the environment, especially on breeding farms or wherever horses have congregated. In large part these pathogenic invaders are bacteria that enter the foal through contaminated material that is ingested or chewed on (e.g., bedding), or through the respiratory system, or via the navel stump if it's not quickly sealed off.

Bacterial infections in the foal can usually be treated with antibiotics early in the course of illness, but viral infections do not respond to antibiotics. The best treatment in these instances is good supportive care — giving extra fluids to prevent or combat dehydration and electrolyte imbalances, keeping the foal warm and dry, and reducing stress as much as possible — so his body is better able to fight the infection.

Injections

In years past, most horsemen gave the mare and foal injections at birth — vitamins, antibiotics, or iron shots. Usually these are not necessary. In fact, they may even be harmful. For example, it used to be standard practice to give mares and foals tetanus antitoxin, but this sometimes leads to serum hepatitis. It's wiser to keep the mare's tetanus toxoid vaccinations up-to-date, with a booster a month before she foals. If she's on a well-balanced and adequate diet, her foal won't need supplemental vitamins. He won't need antibiotics either, if you kept the foaling area clean and disinfected his navel. The only time the mare and foal need medication is for a specific problem, in which case your veterinarian will advise you.

Navel Ill (Joint Ill)

Sometimes a foal is born with an infection already in his bloodstream — transmitted from a diseased uterus — and it settles in his joints, or he may develop a similar infection soon after birth if bacteria enter the navel before it dries and seals off. Either type of infection may produce acute septicemia, which causes high fever and makes the foal weak, dull, and too listless to nurse. His condition will deteriorate rapidly and become fatal unless it is treated immediately.

In some cases the infection localizes in the body, such as in the joints or organs, without as much general illness. There may be mild fever, lameness, and local swelling if the infection settles in the joints, in which case the disease runs a much longer course. It may or may not develop into fatal illness.

If acute septicemia is halted in time to prevent death, infection may still have invaded the joints. Joint ill can be the aftermath of any serious infection that invades the circulatory system.

Joint Infection. Leg joints in the young foal are susceptible to infection due to the stress of weight bearing. If the condition is untreated the joint surfaces become damaged, leaving the foal with enlarged and painful arthritic joints. The first sign of trouble is tenderness or lameness. The foal may have a fever and his joints may be hot, with swollen soft tissues. *Streptococcus* infection usually initiates a rapid destruction of the joint cartilage and bone, so it is essential to make an early diagnosis and begin treatment. The foal will need massive doses of antibiotics; the veterinarian may also flush the affected joint to remove the harmful enzymes.

Preventing Joint Ill. A high incidence of joint ill occurs in foals that suffer abnormalities at birth and in foals that don't nurse soon enough to get antibodies from colostrum. Attending all births and making sure all goes smoothly are important. So are making sure the foaling area is scrupulously clean and disinfecting the foal's navel stump immediately after the cord breaks. Uterine infections in the mare should be treated and cleared up before breeding, to make sure no infection is transferred from mare to unborn foal. If subsequent uterine infection is suspected prior to foaling, your veterinarian may prescribe antibiotics for the mare.

Health Tip

When a foal is ill it is often helpful to treat the symptoms, such as reducing fever and discomfort. If a foal has high fever, drugs such as flunixin meglumine (Banamine) or phenylbutazone (Butazolidin) can bring it down. Even though these types of drugs can cause ulcers in foals, they are the lesser of two evils when a foal is suffering from life-threatening high fever that can cause irreversible damage to his nervous system and other systems. Once the fever and pain are reduced, the foal may feel well enough to start nursing again, taking in urgently needed nutrients and fluids.

If a foal has to be confined to a stall for treatment, your veterinarian may prescribe antiulcer medication such as ranitidine (Zantac) or cimetidine (Tagamet). Years ago many foals were successfully treated for infectious disease or other serious conditions only to die later from gastric ulcer brought on by the stress of stall confinement and aggravated by medications for the primary illness.

IVs and Stomach Tubes

In severe infections, the veterinarian may put the foal on a round-the-clock schedule of antibiotics adminstered intravenously. For ease of treatment and to prevent repeated damage (and pain) when putting needles into a vein, it's common practice to insert a catheter into the foal's jugular vein and secure it with a small suture so the medication can be given every few hours as needed, along with intravenous fluids. Dehydration is a serious problem in sick foals because once they stop nursing the fever and diarrhea dehydrate them.

The veterinarian may choose to give fluids and medication via stomach tube if the foal can absorb fluids through the gut, if the infection is in the digestive tract and will respond to oral medications, or if the diarrhea can be eased with magnesium hydroxide, mineral oil, or other preparations to soothe the gut lining.

Crooked Legs and Weak Tendons

Sometimes a foal is born with crooked legs or weak tendons that allow the limbs to sag or bend at abnormal angles. In some cases the crookedness may result from nutritional imbalance or inadequate diet in the pregnant mare; in other cases it may be due to the way the foal was lying in the uterus or to uterine insufficiency. The leg deformity or lax tendons are not usually genetic defects that can be passed on to future offspring.

Treatment

Mild cases generally straighten on their own without treatment, leg braces, or corrective foot trimming. If the foal can get up and around, nurse his mother, and travel, the weight bearing and exercise will strengthen the legs and they will gradually straighten. It may take a few days or weeks. If the foal has trouble getting around or is causing himself injury, consult your veterinarian or farrier for advice in bracing or splinting the leg until it becomes stronger.

This newborn foal has weak pasterns; notice how the fetlock joints appear to have collapsed.

Premature Foal

If your mare gives birth to a premature foal or twins, the chances of survival depend on how fully developed the baby or twins are and what kind of care can be given. You'll need a good place for the mare and her fragile offspring — clean, warm, and dry, and well away from other horses. These tiny babies must be in isolation for a while until they gain strength and resistance to disease, and you must be careful not to introduce any disease germs yourself. Consult your veterinarian immediately after the birth for advice in trying to save and care for these newborns.

A premature foal or underdeveloped twin may need to be in an intensive care unit at the clinic. If it is strong enough it may be able to stay in a roomy box stall with the mare. When weather is cold you'll need a heat lamp in a dry corner where the foal can snuggle in but the mare cannot. A pole across a corner, chest high to the mare, will keep her out yet allow her to still see and smell the foal. If she has twins and they both live, she may or may not have enough milk for both. You might have to give supplementary bottles until they are big enough to eat more solid food.

Keep the stall absolutely clean, changing the bedding often. You can often save a premature foal or twins with diligent good care.

Orphan Foal

At some time you may be confronted with an orphan (e.g., if the mare dies at birth or shortly thereafter) or a mare that won't mother her foal or one that does not produce milk. Unless you can locate a nursemare, you'll need to feed a formula or milk replacer. The problem is more complicated if the foal is newborn and has had no colostrum; he will need at least a pint (0.5 L) of colostrum, preferably within 4 hours of birth.

Colostrum

If the dam died from foaling complications rather than disease, her colostrum can be milked out immediately after death to feed the foal. If the dam is fine but refuses to mother the foal, she can be restrained, twitched (or held still with a Stableizer), and milked of enough colostrum through that first day to feed the foal. When there is no way to obtain colostrum from the dam you must find another source, such as frozen colostrum, or as a last resort use a commercial serum containing blood plasma and antibodies to give intravenously.

Milk Replacer

After the foal has had several feedings of colostrum, you can switch to a foal milk replacer such as Foal-Lac (Pet Ag, Hampshire, IL), Foal-Mate (Manna Pro, St. Louis, MO), or Mare's Milk Plus (Buckeye Feed Mills, Dalton, OH) and mix it according to directions. All are composed of cow's milk, with additional ingredients. Prices vary, reflecting the quality and digestibility of the added ingredients. Match your foal's nutritional needs with a product's ingredients, asking your veterinarian which one would be best. Milk replacers containing a non-lactose type of carbohydrate such as maltose, a sugar found in grain, may not be tolerated well by sick foals or young foals. Foals under 2 weeks of age lack the enzymes to digest some types of carbohydrates. A foal that is premature or stressed by infection might do better on a high-fat milk replacer than on a high-carbohydrate product; he cannot digest the carbohydrates but can still get energy from the fats.

A milk replacer should not contain fats derived wholly from vegetable sources because they irritate the gut, causing diarrhea. Milk fats and animal fats are more easily digested. Milk replacer also contains vitamins and minerals, and some have probiotics (i.e., gut microbes) to aid digestion. Sick foals may do well with probiotics, but most foals don't need them.

Don't use calf milk replacer; most of these preparations don't meet the foal's nutritional needs. If you don't have access to foal milk replacer, you are better off to mix a home recipe.

Home Recipes

Don't use straight cow's milk. Mare's milk is lower in fat and protein, but higher in carbohydrates, than cow's milk. In making an emergency formula using cow's milk, you can modify it by lowering the fat and protein content and increasing the sugar. There are a number of workable formulas:

- 1 can (12 ounces [0.3 L]) evaporated cow's milk mixed with equal part water; 4 tablespoons (60 mL) limewater or tap water; 1 tablespoon (15 mL) sugar or corn syrup
- 1 pint (0.5 L) low-fat cow's milk (2%); 4 ounces (0.1 L) limewater or tap water; 1.5 teaspoons (8 mL) sugar (white sugar is harder to digest than corn syrup or lactose — it's best to use something other than white sugar)
- 1 pint (0.5 L) low-fat cow's milk; 2.5 ounces (74 mL) water; 2.5 ounces (74 mL) limewater or tap water; 2 teaspoons (10 mL) lactose
- For several feedings you can mix up 3 pints (1.4 L) low-fat cow's milk (2%); 1 pint (0.5 L) water; ¼ package Sure-Jell pectin or 20 mL of 50% medical glucose.

Limewater (a solution of calcium hydroxide in water) has traditionally been used to provide extra calcium but isn't always available in an emergency. Most drugstores no longer sell it. Instead, you can use tap water unless it is soft (i.e., low in minerals) or high in fluoride. In these cases, use distilled water.

Two percent cow's milk is the best basis for a homemade foal formula because it has the proper calcium-phosphorus ratio. For increasing the carbohydrates, add ½ cup (0.1 L) honey or corn syrup or ¼ package of Sure-Jell pectin to a gallon (3.8 L) of milk. Glucose (simple sugar), which is easy to digest, is an ideal source of extra carbohydrates if you can get it. Pectin (the thickening agent for homemade jelly or jam) is also a good carbohydrate that has a protective effect on the foal's stomach and intestines to prevent ulcers or diarrhea. If you use pectin as the carbohydrate source, add a few tablespoons of honey for palatability.

Bottle or Bucket?

If you are feeding with a bottle, get one with a nipple that doesn't run too fast; otherwise the foal may choke or get digestive upsets. A lamb nipple is a better size for the foal than a calf nipple. Be sure the bottle is clean, and remember to warm the formula or milk replacer to body temperature (100°F [37.8°C]) before you feed him. A foal can also learn to drink from a bucket, which is easy to clean. Start him on the bucket within the first few days or he may not want to switch from the bottle. If you get him drinking from a bucket, you can put in enough milk for 4 to 6 hours at a time.

Amounts to Feed

A young foal will drink 2 gallons (7.6 L) in a 24-hour period. He should be fed every half hour for the first few hours, then once an hour for the first 5 days of life (about a half pint [0.2 L] at each feeding, then up to a pint [0.5 L]) and preferably for the first 10 days. Young foals nurse small amounts often. It helps if someone can share feeding duty so you can get some sleep. When the foal is 10 days old you can go to 2-hour intervals at night but continue hourly feedings during the day. As you increase the length of time between feedings, you'll need to increase the amount given at each feeding. By the time the foal is a month old you can start cutting down to feeding every 4 to 5 hours once he is also eating solid food. If he is doing well you can wean him by 6 to 8 weeks of age.

Signs of Illness

Because orphan foals easily become dehydrated, make sure the youngster is getting enough fluid, especially when you start him on dry feed and cut back on the number of fluid feedings. Also make sure fresh water is always available.

Take his temperature daily. Remember that fever can be an early sign of sickness. Any abnormality (e.g., reluctance to eat, weakness, fever, diarrhea or constipation, distressed breathing) should be a signal to call your veterinarian. Many problems quickly prove fatal to a young foal. Modern drugs, transfusions, and intravenous feedings can often save him, but don't let sickness get a head start. Treatment should begin as soon as you suspect the foal is sick.

Diarrhea

If the foal gets diarrhea, temporarily substitute electrolyte formula for milk. A simple mix is 2 quarts (1.9 L) water, 1 package Sure-Jell pectin, 1 tablespoon (15 mL) Lite salt (sodium chloride and potassium chloride), and 1 tablespoon (15 mL) flavored gelatin for palatability. This gives the gut a chance to heal (whereas milk irritates it) while providing fluid and nutrients, and it replaces the electrolytes the foal is losing through diarrhea. If the problem is due to infection, he'll need an oral antibiotic; consult your veterinarian.

Goat's Milk

Goat's milk can be used as a substitute for mare's milk and is usually digested well by premature or sick foals.

Growth and Weight Gain

Foals should gain 1 to 3 pounds (0.5–1.4 kg) per day. If your orphan is not growing well or has a rough coat, pot belly, or loose bowel movements, he needs a better milk replacer. If he feels and looks good but is always hungry, just feed him more of what you're using. Get him eating alfalfa hay, green grass, or foal pellets as soon as he shows interest. Milk replacers are not designed to be a foal's only source of nutrition during his first months of life.

Nursemare

An orphan can be raised on a substitute mare. The easiest time to persuade a mare to adopt an orphan is very soon after she herself has given birth. If her own foal dies at birth or soon thereafter, she is more inclined to accept a substitute than she would be a few days or weeks later. It's harder to fool her after she's had her own baby a while. Also, the hormones in her body at foaling time heighten her maternal instincts and help in the bonding process; they make her more receptive to mothering. A mare that has lost her own foal generally makes the best nursemare. If you can't locate a nursemare you can sometimes convince a mare to raise an orphan along with her own foal. This takes more work and careful monitoring.

Introduce the orphan as soon as possible after the mare loses her own foal, rubbing the mare's nose with Vicks VapoRub (to disguise other smells so she can't tell the orphan is not hers), and thoroughly rub the mare's fresh afterbirth over the orphan — especially rump and tail area, where the mare will be nuzzling and smelling as the youngster nurses. It also helps to put some Vicks on the foal, especially on his hind end.

Use Hobbles

Hobble the mare's hind legs so she can't kick. The hobbles should have sheepskin lining or some other soft material that won't chafe her pasterns in case you have to leave them on for several days. Keep her in a stall. She can still move around with hobbles but can't hurt the foal. For the first nursing, tie her to the side of the stall, with the rope not too short, allowing some freedom of head and neck. Alternatively, have someone hold her so she can still reach around and smell the foal but can't hurt him or knock him down. Introduce her to the foal and let her smell him.

If the orphan is critically in need of milk, accomplish the first nursing quickly with the mare tied, hobbled, and twitched (or restrained with a Stableizer); then allow bonding to take place afterward. After the foal has nursed a few times the mare may mellow in her attitude. Many mares are more inclined to accept a foal after it has nursed. To make sure the mare doesn't hurt the orphan, tie her in the box stall with her hind feet hobbled, and watch closely until you are sure bonding is complete and she accepts the foal. This may take 1 to 3 days, sometimes longer.

Raising Two

If a nursemare can't be located and you have a gentle mare who gives a lot of milk, you may be able to persuade her to raise the orphan along with her own. Complete success is most likely when you bring the extra foal to her immediately after she gives birth to her own. If you can be there at the birth and present her with both foals at once (using Vicks VapoRub on her nose, and rubbing birth fluids and membranes on the orphan), she may think they are both hers. To make sure the adoption works, hobble and tie the mare, and monitor the nursings. Sometimes a mare adopts both from the beginning, but usually you must supervise the process for a while.

Keep Foals Penned Separate from the Mare

If the mare prefers her own foal and rejects the orphan, keep both foals in a pen by themselves between nursings, but in sight of the mare. Adjacent pens (with a safe fence such as diamond mesh wire, or boards or poles with little space between them so the foals won't be injured while trying to stick their heads through) work best. The bonding process takes longer with two newborns, but after a time most mares will accept both foals.

A. *Sometimes you can raise an orphan on a mare that already has a foal, and she will feed both babies.* **B.** *Keep both foals penned separate from the mare at first, except at nursing time.*

Having the foals separate from the mare keeps her from bonding with only her own. Apply Vicks to her nose several times a day at first. Also put Vicks on the foals — especially their hindquarters — to make them smell the same, confusing the mare as to which is hers. At nursing time, tie the mare before bringing in the foals. Have them nurse together, one on each side. When they are finished, remove them to their pen. Don't leave them with the mare lest she mother her own and reject the other.

After the mare accepts the nursing procedure (it may take several days) she can be left free in the stall during nursing, with just the hobbles on so she won't try to kick the extra foal. You can remove the hobbles later after she accepts her role as mother of "twins." She may never love the orphan as her own, but she will likely come to tolerate him.

For the first few days, the mare should wear hobbles when nursing, so she cannot kick the new foal.

The Foals

The foals should have access to feed (such as Foal-Lac pellets) and water in their own pen. Some foals need encouragement to learn to eat pellets; you may have to put the feed into their mouths at first. They need good bedding, and a heat lamp if weather is cold. Be sure the heat lamp is well away from anything flammable and out of reach of the foals.

Safety Tip

Mares often get upset when you are handling their foals. *Always* restrain the mare *before* putting the foals with her. She may also fuss when you take them back to their pen after nursing. Work quietly and firmly, so as not to upset her and the foals.

Feed the Mare Well

Due to the increased demand on her milk supply, the mare will produce more milk. Thus, you must increase her feed and make sure she always has plenty of water so she can keep up with the demand. Most well-fed mares can easily raise two foals if the foals are given supplemental feed. Keep close watch on the foals to make sure they are doing well. After the mare is well bonded with the foals, they can be allowed to remain with her and no longer need supervision at nursing. However, you should continue to keep the mare and her babies separate from other horses.

The extra time it takes to graft an orphan onto a nursemare will pay off greatly. Not only will the mare feed the orphan (and you won't have to), but the youngster growing up with a "mother" will do better physically and mentally. He'll be better adjusted than the hand-raised orphan that lacks contact with other horses. The mare can teach a foal a lot more about being a normal horse than you can. If you do have to raise an orphan yourself, let him live with a kindly old gelding or older mare, after he is about 1 month old, to teach him manners and "horse sense" while you provide his dinner by bottle or bucket.

Twenty-Four

CARE OF THE YOUNG HORSE

Raising a horse to adulthood is a challenge and a pleasure; there's considerable satisfaction in watching the youngster grow and develop. It also takes good care and conscientious handling. The young horse is vulnerable to problems that are not a worry with adult horses, so you must be prepared to prevent or minimize them.

Because the young foal is most vulnerable to diseases and other problems, he needs the most care and monitoring. If he obtained a good supply of colostrum from his mother he'll be temporarily protected from many diseases that would otherwise be life threatening during his first weeks of life. As temporary immunity wears off, however, he becomes susceptible to disease until his own immune system responds to pathogens in his environment. He will encounter all kinds of dangers — from parasites to respiratory illnesses — but with good care you can reduce risks and stresses, keeping him strong and healthy as he makes the transition from babyhood to yearling.

Handling the Foal

The foal should be handled from the day he's born, not only to instill a trust of people but also to promote ease of managing him when it's time to give vaccinations, medications, or other aspects of health care. When he is very young it's better to restrain him with your arms than to use a halter. He may resist restraint on his head, rear up, and flip over backward. Until he's more accustomed to people and to restraint, there is a risk of injury when using a halter and lead rope.

If you must handle the foal and restrain him, hold him with one arm around his chest and the other around his hindquarters.

It's best to handle the foal with one arm around his front and one around the hindquarters; this is much safer and a lot less traumatic than a halter. If the foal must be treated, given an enema, or moved from stall to pasture, it's easiest with two people — one to hold the foal and one to treat him. If the foal is rambunctious, the most effective way to restrain him is the tail twitch — the person holding him places one arm around the front of the foal and with the other hand holds his tail straight up, over his back. This tends to make him stand still. When working by yourself, crowd the foal into a corner, use your hip to hold his shoulder and ribcage in a body block, and hold his tail straight up with one hand while you use your free hand to administer the enema, injection, or other treatment.

If you must move a mare and foal by yourself, use an adult horse halter as a foal restraint. By putting the big halter on the foal as a body harness, you can hang on to him with one hand and have the other hand free to lead the mare.

Putting On a Foal Harness (Adult Halter)

Using a full-size nylon web horse halter with round buckles, hold the halter upside down and slip the noseband over the foal's head like a collar, positioning it around the base of his neck above the shoulders. The halter's jaw strap will then be resting on the foal's withers, aligned with his backbone. The crownpiece (which would normally go behind the ears of the adult horse) will be at his girth. Buckle the crownpiece around the foal's belly like a cinch or girth, but not too tightly — just snug enough that he can't get a hind foot caught in it. This foal "harness" gives you a safe and sturdy handle for hanging on to or restraining the youngster.

An adult horse halter can become a foal harness that allows you to hang onto the foal with one hand.

Leading Lessons

Once the foal is several days old and used to being handled — having been "led" a few times with your arms around his front and rear — you can progress to a rope around his chest and buttocks with a figure 8 over his back. After he goes willingly with you in this type of restraint, start putting a foal halter on him; he can have his first leading lessons as you take him and the mare from stall to pasture or from pen to pasture. Be sure to keep using a rope around his buttocks during his early leading lessons. Then if he pulls back or tries to get away, you have control over his movements and he cannot pull back hard on the halter or be as likely to pull away or flip over backward. A few leading lessons, going along with the mare, will be invaluable later, even if he doesn't get much formal training until he's a weanling or older. Early handling will make your later work with him much easier.

In the foal's early leading lessons, use a rope around his buttocks so he can't hang back on the halter.

Feeding the Foal

If mare and foal are at pasture while he grows up, he will get all the nutrition he needs if the pasture is good and the mare milks well. If you live in a climate or region where pastures are dry or of poor quality, or if the mare and foal are kept in a paddock, they will need good hay and perhaps some grain. By the time the foal is a week old he'll be nibbling the mare's hay; make sure it's of good quality.

Mare's Milk

A good milking mare will provide her foal with proper and adequate nutrition without the foal needing any supplemental feed other than good pasture or good hay — as long as the mare is well fed on a well-balanced diet. Broodmares usually produce about 3 percent of their body weight daily in milk during early lactation. A 1,200 pound (544.3 kg) mare gives about 36 pounds (16.3 kg) of milk daily (more than 4 gallons [15.1 L]). This is more milk than the average beef cow produces. Clearly the lactating mare needs adquate food to meet her body's needs for milk production.

Nursing Rate

Foals are aggressive eaters, nursing more often than calves do. A young calf nurses about four times in a 24-hour period. A young foal nurses about every 30 minutes; when he's a few weeks old, about once an hour; then later, about eighteen times per 24-hour period.

Milk production in the mare increases during the first 2 months and then begins to decline. The energy and protein content of the milk decline also. The foal's greatest need for energy and protein from milk is during those first 2 months; then as he starts eating more solid foods his nutritional needs are partly met by grass or hay and grain. By the time he is weaned, his need for mother's milk is past as long as he has an adequate and well-balanced diet.

Grain

By 3 weeks of age the foal can be started on grain. If you are graining the mare, the foal will have already sampled hers and learned how to eat it. After all, foals learn best by mimicking their mothers; the easiest way to teach him to eat grain is to let him try some of his dam's. Most horsemen feed mares and foals grain, but not all of them need it. You should never underfeed a lactating mare and growing foal, but don't overfeed them either. Some rapidly growing youngsters develop serious leg deformities and soundness problems if they are overfed on grain and rich feeds; this problem is addressed later in this chapter.

When feeding grain, make sure the mare doesn't eat the foal's. You can make a foal creep — putting his grain tub or manger in a corner of the pasture, with a pole he can go under but the mare cannot (4–5 feet [1.2–1.5 m] high, depending on the size of your horses) — or make an enclosure the foal can go into. Tie the mare and foal while they eat in their separate end of a manger in a stall.

Whether you use grain or foal pellets, feed only the amount the foal can clean up within 20 minutes; do not overfeed. Don't let any uneaten grain accumulate in corners of a manger or in a feed tub. Grain that becomes moist from saliva can spoil. It can also be contaminated by birds. Always clean out leftover crumbs or kernels.

Health Care and Problems of the Foal

Pay special attention to cleanliness of the foal's environment so he is free of possible infection and parasites. You should regularly deworm the mare so she won't pass internal parasites to the foal, and he himself should be dewormed starting at 1 to 2 months of age. At 3 months he should be vaccinated with tetanus toxoid and started on other immunizations, including those for eastern and western encephalomyelitis. He may need other vaccinations depending on your locale and situation; discuss this with your veterinarian. Foot care should begin at 2 or 3 months of age — earlier if the foal needs special trimming. Handle his feet regularly, and teach him to shift his weight and pick up his feet. (See chapters 4 and 5 for information on picking up a horse's foot.)

Diarrhea

Diarrhea is a serious problem; the younger the foal, the more critical it can be. Indeed, the young foal does not have much body reserve and can quickly dehydrate. He also has little resistance to infection. He may lose body fluid more rapidly than it can be replaced by nursing. If he is sick he may not feel like nursing, which compounds the problem.

Diarrhea can best be treated if it is discovered early. It's important to try to determine the cause and treat accordingly. You need to know if it's the result of something the foal has eaten, too much milk, infection, or the mare's first heat cycle.

Symptoms

The normal healthy foal is bright and alert, active and sassy. If he doesn't feel well he becomes listless and dull, perhaps with drooping head or ears. The young foal stays close to his dam. If you find him off by himself or the mare standing over him with a full udder and a worried attitude, something's wrong. In early stages of diarrhea the foal may not have obvious evidence on his hindquarters. He may be squirting liquid, but he lifts his tail high and squirts it out in a stream. Even a close check in his pen, stall, or pasture may not reveal the liquid feces because they sink into the ground, grass, or straw. If you fail to find any fresh normal foal manure, be suspicious.

Effects of Diarrhea

After the foal has had diarrhea for a few hours it takes a toll on him. He feels so poorly that he no longer bothers to lift his tail when passing manure, soiling tail and buttocks. The high acid content of the feces burns and irritates the skin; if he's had diarrhea for any length of time, he'll lose hair from his buttocks. Each flick of the tail takes off more hair; the tail is matted and abrasive.

Gut Pain

If diarrhea is severe the foal will also have gut pain. His intestines are cramping as they work overtime; the gut is irritated and sore. If pain is mild he may just stand listlessly, shifting his weight from one foot to the other. If it worsens he may paw or look around at his flank. He may tread his feet or back into a corner in his attempt to get away from the pain. If it becomes severe he may lie on the ground paddling with his feet or rolling. A foal's gut is so delicate and sensitive that a few cramps from diarrhea may put him in such misery he won't feel like nursing. The pain may be so debilitating that he seems weak, dragging his feet and staggering. You may think he is already severely dehydrated and weak, when in fact the true culprit is pain from the cramping in his gut.

Merely easing the pain may make him feel so much better that he'll go ahead and nurse — if the diarrhea is still in the early stages and the foal is not yet too dehydrated and ill. An injection of flunixin meglumine (Banamine) may help him feel better. If he's dehydrated he'll need additional fluids given orally by stomach tube or intravenously.

Dehydration

Dehydration, if not corrected, will ultimately kill the foal because his body cannot function without fluid. At first the affected foal is merely dull and his skin loses elasticity. If you pinch or pull out the skin on his neck, it doesn't spring right back into place but goes back slowly. As dehydration progresses the skin becomes less and less elastic, and his eyes seem sunk back into their sockets; he has lost fluid in the fatty tissues around the eyes. As he goes into shock his legs become cold and pulse weakens. If the condition is not reversed, his kidneys fail. Because there is no longer enough circulating blood volume (too much fluid has been lost into the gut), he slips into a coma and dies.

Too Much Milk

In the very young foal the most common cause of diarrhea is too much milk. Many mares are heavy milkers, producing 3 to 5 gallons (11.4–19 L) a day. This is too much for a young foal. It will cause scouring. To avoid this, mares should not be overfed during the first weeks of lactation — no grain for the first 2 weeks or until after the first heat cycle — unless they are not giving enough milk. If that is the case, their rations can be increased.

Foal Heat Diarrhea

The most common and least dangerous diarrhea in young foals occurs around 4 to 14 days of age, coinciding with the mare's first heat. Almost all foals scour a little at this time, with diarrhea lasting 1 to 3 days. Bowel movements may become soft, loose, or squirting, but the foal keeps nursing and doesn't seem sick. This diarrhea usually requires no treatment other than soothing the foal's buttocks with petroleum jelly or saturating his tail and rear end with mineral oil to keep the acid feces from irritating his skin and taking all the hair off. Wash (and dry) his buttocks gently before applying the soothing coating. Keeping his buttocks clean and well greased will make him more comfortable and prevent hair loss.

Bismuth Subsalicylate (Pepto Bismol) for Diarrhea

If the foal has watery feces more than a day, Pepto Bismol (30 to 40 mL — about 2 tablespoons [30 mL], the adult human dose) can be given orally by syringe. Place the syringe (without needle) into the back of his mouth and gently squirt the Pepto Bismol in, a little at a time, as he swallows it. This will soothe his gut and help slow the diarrhea, making him more comfortable. Kaolin and pectin mix (Kaopectate) is another gut soother that can also be given by mouth. The usual dose for foals is 2 to 3 ounces (59–89 mL).

You can minimize foal heat diarrhea by deworming the mare with iver-mectin immediately after foaling; there is a definite correlation between *Strongyloides* worms and foal heat scours.

A foal with diarrhea during the mare's first heat usually needs no drastic treatment but should be watched closely to make sure the diarrhea does not worsen. If the problem persists or he becomes dull, he may need fluid and elec-trolytes. Make sure he bounces back after the mare goes out of heat. If he seems at all sick, he will need medical attention. The stress of foal heat diarrhea may set him up for an infection and more serious problems.

Diarrhea Caused by Gut Infection

Diarrhea that develops after the third week of life is often caused by infection (as are diarrheas that occur before foal heat — and these are most deadly, because the pathogens reached the gut before the protective antibodies from colostrum did). Most cases occur in foals that did not absorb enough antibodies from colostrum. The severely infected foal may have fever or reddened gums, because the infection produces toxins that circulate throughout his body. In this case the foal needs immediate antibiotic therapy. Some types of infectious diarrhea (including salmonella, rotavirus, and some caused by protozoa) kill a foal swiftly if not treated; consult your veterinarian at once for diagnosis and treatment.

Pneumonia

Foal pneumonia can be caused by many airborne organisms and is most common in youngsters 2 to 6 months of age, although cases can occur at any time after birth. Pneumonia is usually caused by a virus or bacteria, and some-times by fungi or parasites. Bacterial pneumonias are more likely to be fatal. A virus may invade the lungs, allowing bacteria to gain entrance. Any disease that lowers the foal's resistance and is accompanied by stress can open the way for pneumonia; and it is often the pneumonia, rather than the primary disease, that kills the foal.

Symptoms

Typical symptoms are increased respiration (breathing rapidly and shallowly), fever, cough, increased pulse rate, dullness, and lack of appetite. Some foals continue to nurse and are fairly bright, maintaining body weight and nursing until just before they die. Others stop nursing and become progressively dull and weak. The foal may stand with front feet wide apart to try to breathe more easily, or he may try to breathe with his mouth open. There may be nasal dis-charge — thin and runny or thick and pus-filled.

Health Tip

Pneumonia can be caused by foreign substances — contamination and subsequent infection — in the lungs. For example, giving a foal a large volume of liquid medication is dangerous if some gets into the windpipe and lungs.

Stress is also a factor. Wet, stormy weather, overcrowding, dusty conditions, and prolonged emotional or physical stress from injury or illness all deplete the foal's energy reserves and make him more vulnerable.

If a foal develops pneumonia, stress should be avoided at all costs until he recovers fully. Foals with pneumonia should not be weaned, shipped, or handled much. Total rest is crucial.

Treatment Requires Veterinary Help

With prompt and adequate treatment in the early stages, most bacterial pneumonias respond quickly and completely to antibiotics. However, infections involving more than one organism are not uncommon, and treatment aimed at one may not affect the other. Sometimes it is hard to tell if the foal has pneumonia or an upper respiratory infection such as a cold. If this is the case, have your veterinarian check the foal. Don't wait to see if he gets better or worse. The foal may need supportive treatment — additional fluids and medications to help him breathe more easily.

Permanent Damage

If he is treated early and diligently, the affected foal has a chance for recovery. If the infection is neglected, treated too late, or not treated long enough, it may overwhelm the foal. Lung tissue is destroyed, the foal becomes weaker, and his chances of recovery diminish. Foals that do recover may never regain total lung capacity.

Preventing Pneumonia

Dust often carries bacteria that cause pneumonia. This is why foal pneumonia often hits in summer if foals are confined in dusty paddocks. The dust irritates their airways and lung lining, and the bacteria aboard the dust particles invade the tissues. If your foals spend time in a dusty paddock, reduce the dust by sprinkling regularly. Avoid confinement in stalls. The dusty, sometimes damp ammonia-laden air in the lower part of the stall (and bacteria in the stall bedding) make a very unhealthy environment. Being kept in a moist barn (saturated bedding and high humidity) can make a foal susceptible to pneumonia. A clean, grassy pasture is the best place for him.

Prevention includes good sanitation and close observation. Where shelters have adequate ventilation and mares with foals have plenty of space, the problem is seldom encountered. Overprotection of winter foals (i.e., keeping them in barns from January to March) and confinement in dusty paddocks are the prime causes of pneumonia. Parasite control is also important; foals subjected to heavy parasite infestation are quite susceptible to pneumonia.

Gastric Ulcers

Overfeeding a foal or having the mare on a heavy grain ration can lead to foal ulcers. Feeds containing anabolic substances (such as rice oil, which is commonly used as a body builder for foals that will be shown at halter) can cause digestive problems. So can stress. Older horses have usually learned to live with stress, but foals are prone to develop ulcers — raw spots in the digestive tract — from too much stress. If the problem persists, the ulcer may penetrate the gut lining and rupture the wall, causing peritonitis and death.

Stress

Temporary stress, such as fright during a thunderstorm, is not a problem; but prolonged, unrelieved tension and anxiety caused by an environmental condition or a medical problem can produce ulcers. For example, long-term confinement in a box stall increases the risk of ulcers. When a foal is confined in a stall all his youthful energy is bottled up with no outlet, and he becomes irritable, unhappy, and bored — which leads to tension and stress. Overuse of certain drugs such as phenylbutazone (Butazolidin) or other painkillers, as well as dietary deficiency of selenium, also cause ulcers.

Symptoms of Ulcers

A foal with an ulcer usually has gut pain. He may grind his teeth, show symptoms of colic, or lose weight. The pain tends to increase at night and between meals, as in human ulcers. The discomfort can be temporarily relieved by eating because food in the digestive tract tends to buffer the ulcer and keep digestive fluids from irritating the raw spot in the gut lining. Symptoms may be mild and subtle at first; the foal may just grind his teeth and slobber. He may try to eat wood or bedding, lie on his back, or prop himself in an odd position to try to relieve the discomfort.

Diagnosis and Treatment

A veterinarian can diagnose ulcers by means of physical examination (pressing deeply behind the edges of the ribcage to see if this causes pain), lab tests, blood samples, looking into the stomach with an endoscope (a long flexible tube), and evaluating abdominal fluid. If the foal has an ulcer he can be treated with cimetidine (Tagamet) and other medications to keep the gut lining from being damaged.

Parasite Control

Of utmost concern is minimizing a foal's exposure to internal parasites. The first step is to keep the mare dewormed during pregnancy. Her last treatment should be given 1 month before she's due to foal, using ivermectin (approved for use on pregnant and lactating mares because it will not hurt the foal). Worm her again right after she foals, sometime that same day, to protect the foal against threadworms *(Strongyloides westeri)*. These are more damaging to the young foal than to older horses; foals develop immunity to them by 4 to 6 weeks of age. The parasites enter the foal through the mare's milk or even through the foal's skin if he has contact with infested ground, grass, or bedding. Threadworms can damage the lungs and liver of the young foal. They also cause diarrhea.

Health Tip

Three types of internal parasites cause problems for young horses from birth through the first year of life: intestinal threadworms *(Strongyloides westeri),* large roundworms (ascarids, *Parascaris equorum*), and blood-worms *(Strongylus vulgaris).* The mare is usually the primary source of infestation, so she should be kept wormed and her manure picked up regularly.

Young foals are especially vulnerable to parasite damage in body tissues. Early damage in lungs, blood vessels, liver, and other tissues adversely affects the youngster's potential as a future athlete. A foal with serious parasite infestation may never reach his true potential.

Even if a mare is regularly dewormed during pregnancy, she can still pass threadworms to her foal in large numbers right after she foals. This is because threadworm larvae remain dormant in the tissues around the mammary glands for several months and are out of reach of dewormers (even ivermectin). During labor the larvae break their dormancy, perhaps due to hormonal changes in the mare's body. They migrate into the milk and enter the foal as he nurses. If the mare is not dewormed within 12 to 24 hours after the birth, larvae will appear in her milk 4 to 20 days after foaling, with the highest number appearing at days 10 to 12. The larvae mature quickly, becoming adult worms approximately 2 weeks later in the foal's gut. They cause infection and diarrhea when he's 2 to 3 weeks old. The mare should be dewormed the day she foals, as soon as you have attended to the newborn.

A young foal eats his dam's fresh manure to obtain the micro-organisms needed for breakdown and digestion of roughages. If the mare has not been dewormed, he picks up parasite eggs at the same time. The best way to protect the foal is to make sure the mare is dewormed and her udder clean at foaling time. During the first weeks of his life the mare is the primary source of parasites.

Ways to Minimize Parasite Problems in Young Horses

- ◆ Remove manure from stalls daily.
- ◆ Feed horses in mangers, troughs, or other feeders that hold hay off the ground, so it can't be contaminated with manure and worm eggs.
- ◆ Never spread fresh manure over a pasture. Compost it first; heat from the compost pile will kill eggs and larvae.
- ◆ Rotate pastures. If you always use the same pasture for mares and foals or young horses, there will be lots of ascarid eggs there, waiting to infest foals.

For more information on parasites, see chapter 10.

Weaning the Foal

Foals can be weaned at 2 to 6 months of age. It is detrimental to leave a fast-growing, heavy-bodied foal on a heavy milking dam past 2 or 3 months if he is developing leg problems from overfeeding. Other foals do better physically and mentally if they are allowed to stay with their mothers until about 5 or 6 months of age. Foals can be left on their mothers longer than 6 months, but this is hard on the mare if she is pregnant again.

Weaning Methods

Horses are herd animals, happiest with other horses. The worst way to wean a foal is to separate him from the dam and leave him all by himself. He should be where he can at least see other horses, not feel alone and isolated. Left by himself in a stall or corral, he may become so frantic he will literally climb the walls.

Traditional weaning methods involved putting the foal in a stall and taking the mare away, or putting several foals together in a weaning pen. This is stressful for both mare and foal and can lead to injury and illness. The foal doesn't need his mother's milk anymore but still depends on her emotionally and is insecure without her. Foals that are put into weaning pens often run themselves to exhaustion and whinny themselves hoarse in frantic efforts to get out and find their mothers. All of them are desperate, pacing the fence, running back and forth, and stirring up dust that irritates lungs and opens the way for pneumonia. Any frantic activity by one sets off the others, and then they start pacing again, rarely stopping to eat or rest. After a few days they resign themselves to life without their mothers, but this abrupt and traumatic separation is very hard on them. Well-fed foals may be physically ready to wean at 4 to 5 months of age but are not yet emotionally prepared for total separation from mother or from other horses.

Easy Weaning in Adjacent Pens

A better way to wean is to put the mare and foal into separate but adjacent pens where they can still see, hear, and smell each other. The pens should be built so the foal cannot reach through to try to nurse and high enough (6 feet [1.8 m] or higher) that neither mare nor foal will try to jump over. A safe mesh fence works well if it's tall and has a pole at the top so the mesh cannot be jumped or mashed down; diamond mesh wire works best. With this arrangement the foal still has the emotional security of being near his mother.

Foals weaned in pens next to their dams usually spend most of their time near the fence that separates them, but they are not overly worried. There's little whinnying or pacing. They may sniff each other through the fence, but the foal is unable to nurse so the mare starts drying up. After a few days the mares can be moved to a more distant pen or pasture with little fuss or protest, and the foals will not be frantic if they have each other or other horses for company. If you only have one foal to wean, let him live with a gentle, easy-going gelding or some other tolerant "baby-sitter" horse after weaning or at least in a pen or pasture next to another horse; he'll be a lot happier than if he were all by himself.

Weaning the foal with mare and foal in separate but adjacent pens minimizes stress to the foal.

Health Tip

Minimize stress at weaning to avoid illness. Worry can also lead to injury if the foal spends a lot of time running or trying to get through a fence. If he works up a sweat he may chill during cold nights. Cold, windy, or wet weather at weaning often leads to illness if foals are stressed.

Safe Fencing

Pens or pastures used for weaning should have safe and adequate fencing, and no hazardous obstacles for a foal to run into by accident. There should be no places where a foal could be tempted to put his head through or get a foot caught. If a group of foals are in a weaning pen together, it should be large enough that the timid ones can get out of the way of more aggressive ones. The emotional trauma of weaning may make the aggressive foals into bullies; they take out their frustrations on the timid ones. The pen should not be so large that a frantic foal could get up a lot of speed, trying to jump or crash the fence because he is running too fast to stop and turn in time. This kind of behavior is usually not a problem if foals are weaned next to their mothers.

Drying Up the Mare

The mare's udder fills and becomes swollen and uncomfortable after she is separated from her foal, and this fullness signals her body to stop producing milk. Any milk that remains gradually resorbs. Watch the mare closely to make sure she dries up properly and does not get mastitis. If you try to milk her out it will just prolong the drying-up process.

The best way to help her is to limit her feed and water for a few days. A day or two before weaning, cut down or eliminate her grain ration if she's been on grain; and on the day of weaning, stop feeding grain entirely. Keep her on plain grass hay (or cut down the amount of legume hay if that's all you have) and no grain until she has dried up (about 2 or 3 weeks). You can also limit her to approximately 10 gallons (38 L) of water daily, unless weather is hot, until her udder is less tight and she is drying up.

Feeding Weanlings and Yearlings

Once the foal has been weaned, he should be on a balanced diet that contains adequate protein and energy for his growing body. His needs can be met with a mix of good grass and alfalfa hay, and grain if necessary. He needs more protein and total calories than a mature horse does, but he should not be overfed. This is a common mistake. More young horses are ruined by overfeeding than by underfeeding. The damage often begins even when they are foals.

Overfeeding and Leg Problems

Many breeders push young horses to rapid growth with grain and high protein feeds, but this can lead to contracted tendons, epiphysitis (i.e., inflammation at the ends of growing leg bones), and osteochondritis dessicans (OCD), a developmental disorder in the growing skeleton that involves inflammation of cartilage and bone. The growth plates near the end of the horse's bones are soft, producing cartilage that turns to bone. As it produces more cartilage, the portion nearest the main bone shaft hardens and calcifies, causing the bone to grow in length. Growth plates are fragile; too much stress leads to permanant damage. Inflammation can result in bony enlargements, leg deformities, or contraction of the associated tendons.

Leg problems from overfeeding sometimes occur in foals under 2 months of age but may happen in foals up to 16 months of age. The youngster may appear to have swollen joints, but the problem is not in the joints themselves — it's due to abnormal growth patterns at the ends of the long bones next to the joints. Degree of lameness may vary from stiffness and reluctance to move to a definite limp.

A young horse with OCD has weak spots in the bone where the cartilage at the bone ends and joints does not turn into bone properly. These areas are vulnerable to inflammation, twisting, breaking, or tearing because they are weaker than normal. This can result in twisted bones, compressed vertebrae, shoulder cysts, swollen fetlock joints, and separated pieces of cartilage and bone within the stifle joints. It also results in youngsters developing spavin (i.e., hock problems) at an early age.

Dietary Imbalance

Too little or too much calcium, phosphorus, or protein disrupts normal cartilage-to-bone conversion. Thus you should avoid excessive grain in the ration (high in phosphorus) and rich alfalfa (high in calcium). The younger the animal, the more sensitive he is to calcium excess, because most of the growth of long bones takes place at an early age. The faster he grows, the greater the damage. The daily calcium requirement for a growing horse is approximately 30 grams, but a typical alfalfa and sweet feed diet gives a horse about 154 grams of calcium.

Most breeders feed high energy diets to youngsters, with large amounts of grain and concentrates containing starches and nonfiber sugars in much higher levels than would be available in natural feeds. This overfeeding stimulates hormones that trigger cartilage growth. Immature cartilage at the bone ends is produced more rapidly than normal but matures more slowly, resulting in permanent damage to the equine skeleton. This leads to lameness and unsoundness when the horse grows up and is expected to work.

Preventing Skeletal Problems

The best approach is to allow young horses to grow up lean and active on natural feeds or on a diet carefully balanced in terms of minerals, energy, and protein. Natural feeds include pasture and grass hay (with only enough alfalfa added to give sufficient protein for normal growth, or alfalfa that is not too rich) and a minimum of grain, if any. Confinement, lack of exercise, and overfeeding, along with a genetic tendency toward rapid growth and early maturity (which many breeders select for) all work together to cause crippling bone disease. Adult horses are not as adversely affected by overfeeding or nutritional imbalance because these bone problems only occur in animals whose bones are still growing.

To protect the youngster from leg and joint damage, grow him a little slower. Give him room to run and exercise. Raising young horses in large pastures at total liberty reduces the risk of OCD and epiphysitis. The increased exercise level converts the sugars (obtained from grain feeding) to energy instead of increasing the hormone production that interferes with growing cartilage. Foals, weanlings, and yearlings should never be confined to box stalls or small paddocks. They need room to run to develop strong legs and lungs and burn off energy.

Exercise Is Important

The less exercise a youngster gets, the less energy he can handle in feed. Stuffing a confined animal with grain is a sure way to make him an unsound cripple by the time he's grown up. The horse that grows up at pasture without any grain may take a little longer to fill out than his pampered counterpart, but because of regular exercise the pasture-raised youngster will stay more sound.

Care of Weanlings and Yearlings

Because a growing horse needs more protein and energy than a mature horse, feed weanlings and yearlings separately. If they are kept with older horses they won't get their share, for they don't eat as fast as the adults do. Weanlings and yearlings (and sometimes 2-year-olds) eat smaller portions more often. They get full more quickly than the mature horse and stop eating. Then when they go back to eat some more, the other horses have eaten it all up.

If you have several young horses, keep them with their various age groups and sexes. Don't have fillies and colts in the same pen or pasture, especially if the colts are not yet gelded, because young males tend to roughhouse more than fillies. Unless a colt is to be used later for breeding purposes, he should be castrated as a foal or during the spring of his yearling year.

Castrating Colts

The testicles are usually descended at birth or soon thereafter, so the first 3 weeks of life is a convenient time to geld. After a few weeks the testicles may draw up against the abdominal wall and not descend again until the colt is a year old or older, but most stay down. Thus most colts can be gelded at any time. It's best to do it at some point prior to weaning while he still has his mother for security.

Gelding at this age not only means the foal has his dam for comfort, but it also means she'll help exercise him. This is an advantage for you. There will be no aftercare (as with an older colt, which must be forced to exercise to prevent swelling and encourage drainage) — just a daily check to make sure the incision is draining properly and there is no swelling and infection. When gelding a colt this young there is rarely much swelling or drainage, compared with gelding a larger animal. Also, if colts are gelded young they don't have to be separated from the fillies at weaning time; they can all live together as weanlings and yearlings.

Years ago most draft colts were castrated in the first weeks after birth, but in recent years most colts have generally been left entire until they became yearlings or 2-year-olds — the rationale being to let them develop more fully. Yet most of the traits a colt acquires by age 2 are the ones horsemen seek to eliminate by gelding.

The animal with a heavy, muscular, cresty neck is less balanced (i.e., less able to flex and collect) than the horse with a lighter front end. Also, the ungelded colt develops the stallion's habits and attitudes and may become aggressive. Waiting until age 2 also means the veterinarian has to deal with a larger, stronger, and more aggressive patient.

Castrate the Colt on Clean Grass

Choose a clean, grassy area and hold the mare nearby as the veterinarian gives the colt an anesthetic, steadies him, and eases him to the ground. Grass is a practical surgical surface, being a clean place to work; the incision is not likely to be contaminated with dirt and dust. Moreover the mare can graze near her colt, not likely to worry much about him. Because the surgery takes only a few moments, the pair can simply be turned loose (if it's done in their pasture) or taken through a gate to their own pasture or paddock after the anesthetic wears off.

Castrating a Colt

After the sedative begins to take effect, steady the young colt prior to gelding.

Ease the colt down on clean grass.

The veterinarian disinfects the scrotal area, while one helper holds the colt's hind leg out of the way, and another helper protects the colt's eye from sunlight.

The testicles are removed.

As the sedative wears off, the colt wakes up.

Soon, the colt is back on his feet with his mama and ready to be turned loose again at pasture.

Advantages to Gelding the Colt

Early castration does not adversely affect the animal's athletic ability and is less risky to his health than castration later. Also, colts gelded young grow taller (by as much as 1 or 2 inches [2.5–5.1 cm]). Castration is minor surgery at this young age; the testicles are small, as are the blood vessels and supportive structures. There is much less trauma and stress to the colt (mentally and physically) when he is very young; the tissues removed and size of the incision are less extensive. The castrated colt recovers from the surgery much more quickly than the older colt and with fewer complications.

Disadvantages

The main disadvantage of early castration is a slightly greater chance of scrotal hernia (i.e., the intestines slip out through the incision if the colt has an inherited weakness) after the surgery. Looseness in the abdominal rings is a hereditary condition that can be corrected with a stitch or two during the surgery. The veterinarian doing the castration should check for this condition, regardless of the age of the horse. Advantages of early castration far outweigh potential risks. If a colt is to be gelded, there is no point in waiting.

Cryptorchid

Occasionally a colt has one or both testicles undescended; the missing testicle (usually only one) is partway up the passageway it came through during fetal development or clear up in the abdomen. If so, the colt cannot be gelded by ordinary means; major surgery is required to find and remove the missing testicle. If the testicle is partly descended, the veterinarian may wait until after the colt reaches puberty. The enlargement of the testicle at that time and its extra weight sometimes bring it on down. If it's retained in the abdomen there's not much chance it will descend. The colt will still be a stallion even if one descended testicle is removed in normal fashion, and although he appears to be a gelding (with no testicles in the scrotum) he will have a stallion's attitude, be able to breed mares (although he is infertile because the testicle is too warm for sperm viability), and likely be dangerous to handle. The best solution is to have him gelded by means of abdominal surgery.

Epilogue

For the horse lover, there's no greater pleasure than raising a foal and seeing him grow up to be the horse you've always wanted. Raising a foal to maturity may cost more in the long run than buying an equivalent adult animal, when you consider stallion fee, cost of feed for the mare and the young horse as it grows up, vaccinations and dewormings, veterinary services, visits from the farrier, and so on — but there are other factors that can't be measured in dollars. The experience alone is priceless. Moreover the young horse is special because he is uniquely yours, raised and handled the way you want him to be. Because you've handled him from birth he trusts and respects you; the two of you can be a team from the beginning. If you are conscientious in your handling of him, he will be a joy. If you spoil him, it's your own fault.

Raising a foal is time-consuming and costly. It's also fun, satisfying, and emotionally rewarding — giving hours of pleasure. As with any worthwhile venture, there are risks in breeding a mare and raising a foal. Things don't always go as planned or turn out with a happy ending. However, risk and challenge are part of what makes any venture worthwhile. The emotional rewards are all the greater if you succeed, and the accomplishment is special.

It is my hope that this book can help you along the way. No matter how much you learn through reading, however, experience is the best teacher. Good luck.

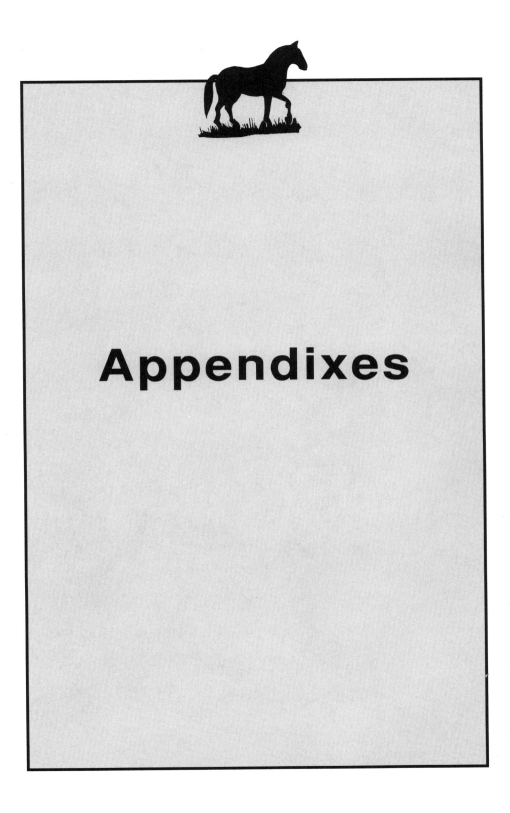

Appendixes

A. Horse Anatomy

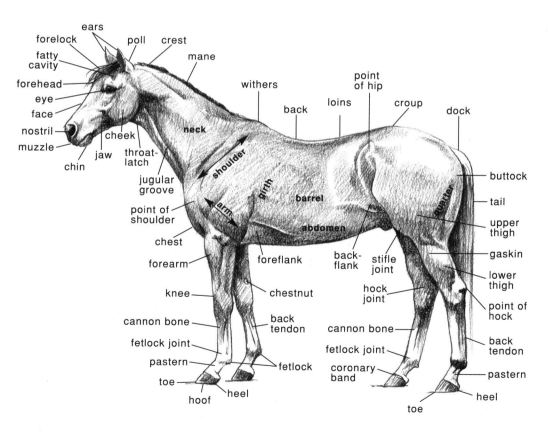

ears
forelock
poll
crest
fatty cavity
mane
forehead
withers
point of hip
eye
back
loins
croup
face
dock
nostril
neck
muzzle
cheek
chin
jaw
throat-latch
jugular groove
shoulder
girth
barrel
quarter
buttock
tail
point of shoulder
arm
abdomen
upper thigh
chest
foreflank
back-flank
stifle joint
gaskin
forearm
chestnut
lower thigh
knee
hock joint
point of hock
cannon bone
back tendon
cannon bone
back tendon
fetlock joint
pastern
toe
fetlock
fetlock joint
coronary band
pastern
hoof
heel
toe
heel

B. Resources

Breed Associations

American Connemara Pony Society
2360 Hunting Ridge Road
Winchester, VA 22603
540-662-5953

American Morgan Horse
Association
P.O. Box 960
Shelburne, VT 05482-0960
802-985-4944

American Paint Horse Association
P.O. Box 961023
Fort Worth, TX 76161-0023
817-834-2742

American Quarter Horse
Association
P.O. Box 200
Amarillo, TX 79168-0001
806-376-4811

American Saddlebred Horse
Association
4093 Iron Works Parkway
Lexington, KY 40511-8434
606-259-2742

American Shetland Pony Club
81-B E. Queenwood
Morton, IL 61550
309-263-4044

American Shire Horse Association
P.O. Box 739
New Castle, CO 81647-0739
970-876-5980

American Suffolk Horse
Association
4240 Goehring Road
Ledbetter, TX 78946-5004
409-249-5795

American Trakehner Association
1520 W. Church Street
Newark, OH 43055
740-344-1111

American Warmblood Society
6801 W. Romley Avenue
Phoenix, AZ 85043-6906
602-936-6621

American White/American Creme Horse Registry
RR 1, Box 20
Naper, NE 68755-2020
402-832-5560

Appaloosa Horse Club
P.O. Box 8403
Moscow, ID 83843-0903
208-882-5578

Arabian Horse Registry of America
P.O. Box 173886
Denver, CO 80217-3886
303-450-4748

Belgian Draft Horse Corporation of America
P.O. Box 335
Wabash, IN 46992-0335
219-563-3205

Clydesdale Breeders of the USA
17346 Kelley Road
Pecatonica, IL 61063
815-247-8780

Foundation Quarter Horse Registry
P.O. Box 230
Sterling, CO 80751
970-522-7822

International Andalusian & Lusitano Horse Association
101 Carnoustie North, #200
Birmingham, AL 35242
205-995-8900

International Arabian Horse Association
10805 East Bethany Drive
Aurora, CO 80014-2605
303-696-4500

The Jockey Club
821 Corporate Drive
Lexington, KY 40503-2794
606-224-2700

Lippizan Association of North America
P.O. Box 1133
Anderson, IN 46015-1133
765-644-3904

Missouri Fox Trotting Horse Breed Association
P.O. Box 1027
Ava, MO 65608-1027
417-683-2468

National Foundation Quarter Horse Association
P.O. Box P
Joseph, OR 97846
541-426-4403

National Pinto Horse Registry
P.O. Box 486
Oxford, NY 13830-0486
607-334-4964

National Show Horse Registry
11700 Commonwealth Drive,
   Suite 200
Louisville, KY 40299-2344
502-266-5100

North American Mustang
Association & Registry
P.O. Box 850906
Mesquite, TX 75185-0906
972-289-9344

North American Trakehner
Association
P.O. Box 12172
Lexington, KY 40581
502-867-0375

Palomino Horse Association
HC 36, Box 24
Dornsife, PA 17823
717-758-3067

Palomino Horse Breeders of
America
15253 East Skelly Drive
Tulsa, OK 74116-2637
918-438-1234

Part Thoroughbred Stud Book
(American Remount Association)
P.O. Box 28
Mercerburg, PA 22117-1901
301-791-1238

Paso Fino Horse Association
101 North Collins Street
Plant City, FL 33566-3311
813-719-7777

Percheron Horse Association of
America
P.O. Box 141
Fredericktown, OH 43019-0141
614-694-3602

Performance Horse Registry
P.O. Box 24710
Lexington, KY 40524-4710
800-787-0027

Peruvian Paso Horse Registry of
North America
3077 Wiljan Court, Suite A
Santa Rosa, CA 95407-5702
707-579-4394

Pinto Horse Association of
America
1900 Samuels Avenue
Fort Worth, TX 76102-1141
817-336-7842

Pony of the Americas Club
5240 Elmwood Avenue
Indianapolis, IN 46203-5990
317-788-0107

Spanish Barb Breeders Association
International
13862 Gulf Haven Road
Gulfport, MS 39503-1110
228-831-5173

Spanish Mustang Registry
HC 3, Box 7670
Wilcox, AZ 85643
520-384-2886

United States Trotting Association
750 Michigan Avenue
Columbus, OH 43215-1191
614-224-2291

Tennessee Walking Horse Breeders
P.O. Box 286
Lewisburg, TN 37091-0286
615-359-1574

Welsh Pony & Cob Society of
America
P.O. Box 2977
Winchester, VA 22604-2977
540-667-6195

More breed organizations can be found in the AHC *Horse Industry Directory*. These include associations for Akhal Teke, Shagya-Arabian, Azteca Horse, Bashkir Curly, Belgian Warmblood, Blazer Horse, Colorado Ranger Horse, Dartmoor Pony, American Cream Draft Horse, Exmoor Pony, Hackney, Haflinger, Hanoverian, Holsteiner, Icelandic Horse, Miniature Horse, New Forest Pony, Quarter Pony, Racking Horse, Rocky Mountain Horse, Suffolk, Trakehner, Walking Pony, Welara, and others.

The *American Horse Council Horse Industry Directory* is available from the
American Horse Council
1700 K Street, NW, Suite 300
Washington, D.C. 20006-3805
202-296-4031

This directory gives listings for breed registries, educational organizations, equine veterinary schools, equine welfare organizations, federal information sources, organizations of interest to horse owners, guide to interstate health requirements, health and research organizations, equine publications and newsletters, show and sport organizations, state departments of agriculture, Extension specialists and horse specialists, state veterinarians, trail organizations, and more.

Note: There are many good books on all aspects of horse raising, horse care, breeding, training, and so on. Check with your local library or bookstore. Other good sources are the advertisements and book reviews in equine periodicals.

Suppliers

Foaling Alarm Systems

Birth Alarm
200 Harbour Street, Apt. 104
Kincardine, ONT N22 3A3
CANADA
1-800-562-4758

Breeder Alert
P.O. Box 1335
Novato, CA 94948
1-415-883-3218

Foalert, Inc.
P.O. Box 2400
Acworth, GA 30102
1-800-237-8861

Foal Milk Replacers

Equilac
Calva Products, Inc.
P.O. Box 126
Acampo, CA 95220
1-209-339-1516

Foal-Lac
Pet-Ag, Inc.
201 Keyes Avenue
Hampshire, IL 60140
1-847-683-2288

Foal-Mate
Manna Pro Corp.
707 Spirit 40 Park Drive, Suite 150
St. Louis, MO 63005
1-800-690-9908

Grow-n-Glow
Merrick's, Inc.
P.O. Box 620307
Middleton, WI 53562-0307
1-800-637-7425

Mare's Milk Plus
Buckeye Feed Mills
Dalton, OH 44618
1-800-417-6460

Silky Coat Milk Replacer
Royal Mills, Inc.
P.O. Box 548
Lakeville, MN 55044
1-612-469-3492

Restraint Devices

Stableizer
Wheeler Enterprises, Inc.
Rt. 2, Box 26A
Ellendale, MN 56026
1-800-287-4791

C. Glossary

Abortion. Loss of pregnancy; resorption or expulsion of embryo or fetus before it is viable

Abscess. Localized collection of pus

Acupuncture. Ancient healing art that involves pricking the skin or tissues with needles

Alfalfa. A legume used for hay, high in protein and calcium

Alternative therapies. Treatments such as acupuncture, massage, chiropractic, and herbal remedies used instead of or in conjunction with traditional Western medical remedies

Amino acids. Group of organic compounds forming the basic structure of proteins

Ammonia. Strong alkaline gas with pungent odor, formed by breakdown of urine and manure

Amnion. Innermost membrane surrounding the fetus in the uterus

Amniotic fluid. Fluid surrounding the fetus

Anaphylactic shock. Hypersensitivity of tissues to an antigen; reaction against a previous dose, putting the animal into shock

Anaphylaxis. Hypersensitivity in which exposure to a specific antigen results in life-threatening respiratory distress, circulatory collapse, and shock (see *anaphylactic shock*)

Anestrus. Period during which a mare does not come into heat

Anhidrosis. Inability to sweat

Antibody. A blood protein produced in response to an antigen for the purpose of counteracting it

Antigen. A foreign substance (toxin) that causes the body to produce antibodies to destroy or neutralize it

Antihistamine. Substance that counteracts the effects of histamine, used in treatment of allergies and other conditions

Antiseptic. Disinfectant used to prevent growth of disease-causing microbes

Antiserum. A blood serum containing antibodies against specific antigens, injected to treat or protect against a disease

Antitoxin. Antibody that counteracts a specific toxin

Arthritis. Inflammation of a joint

Artificial insemination (AI). Manually placing semen into the genital tract of the mare

Ascarid. Large white intestinal parasitic worm

Awns. Stiff bristles that protrude from a grain sheath or grass seed head

Azoturia. Term for cramping of the large muscles; "Monday morning disease"

Banamine (flunixin meglumine). An anti-inflammatory, painkilling drug used for colic

Bar. Support structure of the foot between the heel and the frog

Barren mare. Nonpregnant mare

Base narrow. Feet too close together

Base wide. Feet too wide apart

Bean. Small, hard deposit of debris that collects near the end of the horse's penis

Benzimidazoles. Group of deworming drugs that have similar action

Betadine (povidone iodine). Tamed iodine

Bighead. Condition characterized by enlargement of bones in the skull due to calcium-phosphorus imbalance caused by eating too much bran

Birth canal (vagina). The passage from the uterus to the outside

Bladder stone. Concretion of mineral salts in the bladder

Blister beetle. Small flying insect containing a potent toxin, often baled in hay

Bloodworm. Large strongyle; parasitic intestinal worm of which the larval stage damages the horse's arteries

Body rope. Rope that goes around the horse's girth and through the halter ring, used for safely training a horse to tie

Booster. Repeat immunization to increase immunity level

Bot. Parasitic grub that lives in the horse's stomach; the adult fly lays eggs on the horse's hairs

Botulism. Poisoning caused by a toxin produced by bacterium (*Clostridium*) sometimes found in decaying organic matter

Bran. High-protein roughage supplement consisting of the outer husks of wheat kernels

Breech. Foal positioned rump first (no legs in the birth canal)

"Bute." Butazolidin, trade name for phenylbutazone, an anti-inflammatory, painkilling drug

Buttock. Rounded back portion of the upper hind leg muscles

Caesarean. Surgical removal of fetus from the uterus

Calf knees. Back at the knees; a conformational flaw in which the knee bends backward

Canine tooth. Small sharp tooth behind the incisors, present in most male horses and in some females

Cannon bone. Bone between the fetlock joint and knee (or hock)

Cap. Remains of a baby tooth covering the permanent tooth

Carrier. (Genetic) individual having a recessive gene that is not expressed but that could be passed to offspring; (disease) an animal that appears healthy but harbors pathogens or parasites that can be shed and passed to other animals

Caslick repair. Surgical technique in which the lips of the mare's vulva are cut and sutured so they grow together, making the opening smaller

Cast horse. Horse that is down in the stall (or next to a fence) on his back, unable to get up

Cecum. Large pouch ("fermentation vat") between the small and large intestine where roughage is broken down by microbes

Cellulose. Fibrous part of roughage

Cervix. Constricture between uterus and vagina that opens during estrus and during labor (to allow the foal to be born)

Chiropractic therapy. Treatment involving manipulation of the horse's spine

Chiroptic mange. Leg mange; skin irritation caused by mites

Choke. Obstruction in the horse's esophagus

Chronic obstructive pulmonary disease (COPD; heaves). Respiratory ailment characterized by forced expiration and difficult breathing; caused by allergic sensitivity to dust and other irritants

Club foot. Upright, stumpy foot with short toe and long heel

Coffin bone. Bone inside the hoof (pedal bone, third phalanx)

Coggins test. Laboratory test for equine infectious anemia

Colic. Abdominal pain

Colostrum. First milk, containing antibodies that give the newborn foal temporary disease resistance

Colt. Male horse under 4 years of age

Combined immunodeficiency (CID). Recessive genetic defect in some Arabians (affected foals cannot make antibodies and soon die)

Concentrates. Feeds low in fiber and high in food value, such as grains and oilmeals

Concussion. Jarring impact on feet and legs when traveling

Conformation. Body proportions and angles; how the horse is built

Conjunctivitis. Inflammation of mucous membrane covering the front of the eyeball and inside of the eyelids

Contracted feet. Condition in which there is narrowing of the hoof, especially at the heels

Corn. Bruised area on sole of the foot, often located where the bars meet the heel, due to pressure from poorly fitted shoe

Cornea. Transparent front of the eyeball

Coronary band. Top of the hoof at the hairline, where hoof growth takes place

Corticosteroids. Steroids produced by the body or synthetically (such as dex-amethasone), used in treating allergy, pain, and inflammation

Cow hocked. Having hocks too close together

Cribbing. The act of grabbing a surface (such as fence or manger) with the teeth and sucking in air at the same time

Crossbred. Horse with parents of different breeds

Cross tie. To tie a horse from two sides (from the halter to the walls or posts on either side)

Croup. Highest point of the horse's rump

Cryptorchid. Male horse with one or two retained testicles that did not descend into the scrotum

Curb. Enlargement at the back of the hock

Cushing's syndrome. Metabolic disease that sometimes affects older horses (often due to pituitary tumor); symptoms include failure to shed winter hair

Dam. Mother of a horse

Dermatitis. Inflammation of the skin

Dexamethasone. Steroid used to reduce swelling, inflammation, pain, fever, shock, and allergic conditions

Diestrus. Period between heat cycles when the mare is out of heat.

Digital cushion. Spongy area inside the foot above the frog

Dimethyl sulfoxide (DMSO). A solvent that has many medical uses

Disposition. Temperament and attitude

Distemper. Old term for **strangles**

Dominant gene. One that produces the expressed trait in offspring when more than one gene for a certain trait are present

Eastern equine encephalomyelitis (EEE; "sleeping sickness"). Viral disease of horses affecting the brain, spread by mosquitoes

E. coli. Bacteria that sometimes cause gut infection

Edema. Swelling caused by excessive fluid in the tissues

Electrolytes. Important body salts that can be lost through sweating or diarrhea

Embryo. The developing foal in earliest stage of pregnancy

Encephalitis. Inflammation and infection in the brain

Endorphins. Chemicals produced in the brain, having pain-relieving effects

Endotoxemia. Presence of endotoxins in the blood, which may result in shock

Endotoxic shock. Shock caused by body systems shutting down from reaction to bacterial poisons

Endotoxin. Poison created when bacteria multiply in the body

Enteritis. Inflammation of the intestines

Enteroliths. Concretions of mineral salts ("stones") that build up in the intestinal tract around a foreign object

Epinephrine (adrenaline). A hormone affecting circulation and muscle action, used as a stimulant

Epiphysitis. Inflammation of the growing end of the long bones, usually caused by too much rich feed and dietary imbalance in the young horse

Epsom salts (magnesium sulfate). Used as a purgative or for soaking a foot to draw out infection

Equine infectious anemia (EIA; "swamp fever"). Incurable viral disease spread by bloodsucking insects

Equine protozoal myeloencephalitis (EPM). A parasitic infection spread by opossums, affecting the horse's spinal cord and the nerves that control his movements

Equine viral arteritis (EVA). Viral disease that can cause respiratory disease, abortion, and infertility in horses

Estrogen. Female hormone produced by developing follicles in the ovary; stimulates the mare to come into heat

Estrous cycle. Recurring period of sexual receptivity

Estrus. Heat period, during which the mare ovulates and is receptive to the stallion

Euthanasia. Act of ending a horse's life with merciful death

Farrier. Professional horseshoer

Fecal count. Examination of manure under a microscope to count worm eggs so as to estimate degree of infestation

Fetlock. Tuft of hair at the back of the fetlock joint

Fetlock joint. Joint between pastern and cannon bone

Fetus. Developing foal after 55 to 60 days of pregnancy

Filaria. Tiny threadlike worms spread by biting flies

Filly. Female horse up to 4 years of age

Flat bone. Term used to define depth of cannon, with the flexor tendon set well back of the bone

Flatulent colic. Abdominal pain caused by excessive gas in the gut

Flexor tendon. Large tendon behind the cannon bone

Float. File used to rasp off sharp edges of a horse's teeth

Foal. Newborn horse of either sex; the act of foaling

Foal heat. Mare's first heat period after foaling

Follicle. Fluid-filled sac on an ovary, containing the egg

Founder (laminitis). Inflammation of laminae within the hoof

Frog. Spongy V-shaped cushion on the bottom of the foot

Gelding. Castrated male horse

Gestation. Length of pregnancy

Girth. Area behind the front legs; circumference of the horse's body at this location

Girth itch. Skin disease caused by a fungus in the girth area

Granulation tissue. The tissue that grows to fill in a wound

Granulosa cell tumor. Cancer of the ovary

Grub. Fly larvae (such as the botfly or heel fly)

Guttural pouch. Sac between the ear and mouth that sometimes becomes infected

Habronemiasis ("summer sores"). Sores caused by larvae of a stomach worm trying to develop in the horse's skin instead of in the stomach, after being brought to the skin by flies

Heat cycle. See **estrus**

Heat stroke. Circulatory collapse and shock caused by exposure to high temperatures (especially while exerting)

Heave line. Ridge of muscle along the abdomen caused by the extra effort of forcing air out of the lungs; common in a horse with chronic obstructive pulmonary disease (heaves)

Heaves. Breathing problem caused by respiratory irritation or allergy (see also **chronic obstructive pulmonary disease**)

Heterosis. Hybrid vigor resulting from crossbreeding

Heterozygous. Genes of a specific pair that are different

Histamine. A chemical compound that dilates capillaries, constricts the smooth muscle of the lungs, and increases secretions of the stomach

Homeopathy. Treatment of disease by using tiny doses of substances that in a healthy horse would produce the same disease

Homozygous. Genes of a specific pair that are the same

Hoof tester. Tool used to put pressure on various areas of the bottom of the foot to check for pain or sensitivity

Hook. Elongated portion of a tooth

Hyperthermia. Use of heat to burn off a growth or tumor

Immunotherapy. Prevention or treatment of disease with substances that stimulate an immune response

Impaction. Blockage of digestive tract with food material

Incisors. Front teeth

Influenza. Respiratory disease caused by a virus

Interdental space. Space between the incisors and molars where there are no teeth

Isoerythrolysis. Condition in which antibodies in the mare's colostrum destroy the foal's red blood cells after he nurses; newborns with this condition are called Rh foals

Ivermectin. Deworming drug that paralyzes the parasites; they cannot eat and eventually die

Joint ill. Inflammation in the joints due to bloodborne infection, often from bacteria entering the navel of the newborn foal

Lactating. Producing milk

Laminae. Attachments inside the hoof that bind the hoof wall to the inner tissues

Laminitis (founder). Inflammation of the sensitive laminae

Lampas. Tenderness and inflammation of the soft tissue in the roof of the mouth, usually caused by eating hard feeds

Larvae. Immature stage of a worm or fly

Legumes. Plants (such as alfalfa and clover) that make their own nitrogen while growing; they have more protein than grasses

Leptospirosis. Infectious disease caused by bacteria; may cause abortion in mares or eye problems (moonblindness)

Lethal genes. Genes that if doubled up cause serious defects leading to death of the foal before or soon after birth

Lice. Tiny insects that feed on the skin or suck blood

Lip chain. Chain portion of lead shank placed over the gum above the top teeth (in the front of the mouth); pressure applied here sedates and immobilizes the horse like a twitch does

Maiden mare. Mare that has never been bred

Mammary glands. Milk-producing tissues

Mange. Skin disease caused by mites

Mare. Female horse after her fourth birthday

Mastitis. Inflammation of the udder

Meconium. Dark material passed in the foal's first bowel movements

Melanin. Skin pigment

Melanoma. Skin cancer; tumors in the pigment-producing cells

Microbes. Micro-organisms, germs; refers especially to bacteria that cause disease or fermentation

Mites. Tiny insects that feed on the skin

Molars. Grinders; back teeth

We'd love your thoughts . . .

Your reactions, criticisms, things you did or didn't like about this Storey Book. Please use space below (or write a letter if you'd prefer — even send photos!) telling how you've made use of the information . . . how you've put it to work . . . the more details the better!

Thanks in advance for your help in building our library of good Storey Books.

Pamela B. Art

Publisher, Storey Books

Book Title: _____

Purchased From: _____

Comments: _____

_____

_____

_____

_____

Your Name: _____

Mailing Address: _____

E-mail Address: _____

☐ Please check here if you'd like our latest Storey's Books for Country Living Catalog, (or call 800-441-5700 to order).

☐ You have my permission to quote from my comments and use these quotations in ads, brochures, mail, and other promotions used to market Storey Books.

Signed _____ Date _____

e-mail=thoughts@storey.com   www.storeybooks.com   PRINTED IN THE USA   11/99

From: _____

BUSINESS REPLY MAIL

FIRST-CLASS MAIL   PERMIT NO. 2   POWNAL VT

POSTAGE WILL BE PAID BY ADDRESSEE

STOREY'S BOOKS FOR COUNTRY LIVING
STOREY COMMUNICATIONS INC
RR1 BOX 105
POWNAL VT   05261-9988

"Monday morning disease." See *azoturia*

Moonblindness (iridocyclitis; periodic ophthalmia). Inflammation of internal parts of the eye; recurrent attacks eventually cause blindness

Mucous membrane. Mucus-secreting tissue that lines many body cavities

Mucus. Secretion from a mucous membrane

Navel ill. Disease caused by bacteria entering the foal's navel

Navicular disease. Lameness due to inflammation of the navicular bone that lies behind the coffin bone inside the foot

Night blindness. Recessive genetic defect most commonly found in Appaloosas

Nolvasan (chlorhexidine). All-purpose disinfectant

Oilseed meal. High protein concentrate obtained after extraction of oil from flax, soybeans, or cottonseed

Osteochondritis dessicans (OCD). Developmental disorder in the growing skeleton that involves inflammation of cartilage and bone

Ovulation. Process by which a follicle ruptures on the ovary and releases an egg into the oviduct (the tube connecting the ovary and uterus)

Oxytocin. Hormone that stimulates milk letdown and contraction of the uterus during foaling

Paddling. Walking with feet flung outward

Palpate. To feel, as when the veterinarian checks the uterus or ovaries through the rectal wall

Parasite. Organism that lives on or within another animal

Parrot mouth. Condition characterized by mismatched jawbones; upper teeth protrude over lower teeth

Passive transfer. Temporary immunity transferred from mare to newborn foal via antibodies in colostrum

Pastern. Area between the hoof and fetlock joint

Patella. "Kneecap" on the stifle joint

Pathogen. Invading organism (bacterium, virus, etc.) that causes disease

Periodic ophthalmia. See *moonblindness*

Phenylbutazone. See *"bute"*

Photodynamic agents. Chemicals in a plant that cause a reaction in the horse's skin tissues when activated by sunlight

Photosensitization. Serious skin disease caused by reaction to sunlight after a horse eats certain plants

Pigeon-toed. Having toes pointing toward one another instead of straight

Pinworms. Parasites that live in the rectum and lay eggs at the anus, causing severe annoyance and itching

Placenta (afterbirth). Membranes that attach the developing fetus to the uterus; it is shed after foaling

Pneumonia. Infection in the lungs

Poll. Top of the horse's head

Potomac horse fever (PHF). Disease caused by a rickettsia spread by flukes that parasitize freshwater snails, causing severe diarrhea and laminitis in horses

Pressure bandage. Bandage used to halt excessive bleeding

Progesterone. Hormone that stops heat cycles and keeps the mare pregnant

Prolapse. Protrusion of an inverted organ, such as the uterus

Protozoa. One-celled animals; some cause disease

Proud flesh. Excessive growth of granulation tissue in a wound

Purebred. Animal with sire and dam of the same breed

Purpura hemorrhagica. Serious condition caused by internal bleeding from small blood vessels

Quicked nail. Nail driven into sensitive tissues when shoeing, causing pain and possible infection

Quidding. Dropping pieces of partially chewed hay from the mouth

Rabies. Fatal viral disease spread by bites of infected animals

Rainrot. Fungal skin disease most common in wet conditions

Recessive gene. One that must be received from both parents for its trait to be expressed in the offspring

Rhabdomyelitis. Muscles tying up (see *azoturia*)

Rh foal. See *isoerythrolysis*

Rhinopneumonitis. Viral disease (a herpes virus) that causes respiratory problems and abortion

Rickettsia. A genus of bacteria that often live in the gut of ticks and insects and cause diseases such as Potomac horse fever, typhus, and Rocky Mountain spotted fever

Ringbone. Arthritis of pastern joint or joint between pastern and coffin bone, causing excessive bone growth and lameness

Ringworm. Fungal infection causing scaly patches of skin

Roundworm (ascarid). A damaging parasite common in foals

Salmonella. Bacteria that causes severe diarrhea

Sand colic. Potentially fatal condition caused by eating sand with the feed

Sarcoid. Skin tumor that sometimes develops after a wound

Scours. Diarrhea

Scratches. Skin disease of the lower legs caused by a fungus or bacteria

Septicemia. Generalized infection in the bloodstream

Settle. Become pregnant

Shaker foal. Progressive weakness in a newborn foal due to botulism

Shear mouth. Condition in which teeth are worn at an angle so top and bottom teeth do not match but slide past one another

Sidebone. Calcification of the cartilage just above the side of the hoof; bony enlargement above the coronary band

Sidelining. Tying the hind leg to the front leg on the same side to prevent kicking

Sire. Father of a horse

Sole. Horny bottom of the foot

Sound. Having no faults or injuries that impair usefulness and ability to travel freely or without pain

Sow mouth. Mismatch of jawbones; lower jaw protrudes farther forward than upper jaw

Spasmodic colic. Acute intermittent gut pain due to spasms

Spavin. Disease of the hock joint

Splay footed. Having feet toe out instead of pointing straight ahead

Splint. Bony growth on the splint bone

Squamous cell carcinoma. Skin cancer usually affecting unpigmented areas of skin

Stableizer. Restraint device that goes over the poll and under the top lip; the loop is tightened to activate pressure points that stimulate release of endorphins, relaxing the horse and blocking out pain

Stallion. Ungelded male horse 4 years old or older

Stifle. Large joint high on the hind leg, next to the flank

Stocks. Chutes for restraining a horse

Strangles. Respiratory disease caused by bacteria, causing swollen and abscessed glands

Strongyle. Bloodworm

"Summer sore." Lesion that won't heal caused by worm larvae in the skin, spread by biting flies; see **habronemiasis**

Surcingle. Belly band; strap that goes around the girth

"Swamp fever." Equine infectious anemia (EIA)

Tamed iodine (povidone iodine). Type of iodine not as harsh on raw tissues as tincture of iodine

Tapeworm. Segmented parasite that infests the intestine

Teasing. Bringing a stallion near a mare to see if she is in heat

Thrush. Foul-smelling foot infection in tissues next to the frog

Tooth bumps. Enlargements under the lower jaw from impacted molars

Total digestible nutrients (TDN). Digestible portion of diet

Toxemia. Condition in which bacterial toxins invade the bloodstream and poison the body

Toxin. A poison created by a living organism such as a bacterium

Toxoid. Vaccine that stimulates immunity to bacteria that produce toxins

Twitch. Tool used to restrain a horse by putting pressure on the nose, which stimulates release of endorphins (noun); to use such a tool (verb)

Tying up. Muscle cramps in the large muscles

Umbilical hernia. Protrusion of tissue by the navel

Unsound. Having weakness or injury that makes a horse unable to perform without pain or disability

Unthriftiness. Poor feed efficiency despite adequate nutrition; often due to illness or parasites; characterized by thin condition and poor hair coat

Urinary stones. Concretions of mineral salts that form in kidneys or bladder, sometimes blocking the urinary tract

Urticaria (hives). Swellings in the skin due to allergy

Venezuelan equine encephalomyelitis (VEE; "sleeping sickness"). A form of encephalitis spread by a virus and transmitted by insects; more deadly than Eastern equine encephalomyelitis or Western equine encephalomyelitis, the Venezuelan variety is sometimes brought north from Central America

Vital signs. Temperature, pulse, and respiration rates

Warble. Lump in the skin of the back where the grub of a heel fly is developing

Warmblood. Any European breed of heavy horse that has some Arabian blood in its ancestry; also used to describe the result of crossing cold blood (draft) with hot blood (Arabian or Thoroughbred)

Water bag. Membrane filled with amber-colored fluid that precedes the foal through the birth canal

Wax. Secretion that coagulates at the end of the teat just before foaling

Weanling. Foal that has been weaned but is not yet a yearling

Western equine encephalomyelitis (WEE; "sleeping sickness"). See *Eastern equine encephalomyelitis*

Windpuffs. Puffy enlargements at the fetlock joints

Winging. Having feet swing inward as the horse travels

Withers. Top of the shoulders of a horse

Wolf tooth. Small vestigial first premolar

Worms. Helminth parasites that live in the body of the host animal

Yearling. A horse between 1 and 2 years old

INDEX

Note: Numbers in *italics* indicate illustrations and photographs; numbers in **boldface** indicate charts.

A

AAEP (American Association of Equine Practitioners), 296
Abdominal pain, 131–33, *132–33*. See *also* Digestive problems
Abortion, 364–68
Abrasions first aid, 281
"Acting funny," 137
Active labor, 380–89, *381–88*
Activity level and feeding, 54, 55, 57–58
Acupuncture, 297–98
Adrenaline (epinephrine), 153
Afterbirth, checking, 394
Age
   for breeding, 343
   feeding and, 54, 55, 56, 58
   fertility of stallions, 340–42
   teeth and, 252–53, *252–53*, 255
Aggressive horse handling, 122–24
Airways, clearing, 390
Alfalfa hay, 44–45, 46, 185, 237
Allergic reactions
   skin problems, 240–41, *241*
   to vaccinations, 151–53
Alternative therapies, 295–98
Alum for proud flesh, *283*, 284

American Association of Equine Practitioners (AAEP), 296
Anaphylactic shock, 152–53
Anatomy
   of eye, *260*
   of feet, 81, *81*
   of horse, *454*
Anestrus, 349
Anhidrosis (without sweat), 78
Anthrax, 174–75
Antibodies
   from colostrum, *406*, 407–11, 422
   vaccinations and, 139–53, **143**
Antifreeze toxicity, 238
Antigens, 139
Antitoxins, 156, 188
Arteries versus veins, 281
Artificial insemination, 358
Artificial respiration, 391
Ascarids (roundworms), *201*, 201–2
Aspiration pneumonia, 162, 288
Automatic waterers, 28

B

Baby Buzzer, 377
Back at the knee (calf-kneed), *310*, 311
Backward foals, *387*, 387–88

Bad manners, handling, 115–22
Barbed wire fences, 17–18
Barley, 47–48
"Barn blind," 303
Barn itch (itching dermatitis), 221
Barns, 3–14
  bedding, 11
  black walnut cautions, 11, 233
  building a barn, *2*, 3–9, *4–6*, **8**
  cast horse, 9
  cleaning stalls, 12, *12*
  conversion caveats, 4
  feeders, 12, *12*
  feed storage, 13–14, *15*
  flooring, 5, *5*, 11
  floor plan, *10*
  foundations, 5
  hay storage, 13–14, *15*
  heated barns, 7
  lights in, 4, *4*
  location, choosing, 3–4
  louvers for ventilation, 6, *6*, 7
  manure pile, 12
  materials (building), **8**
  mats, 5, *5*, 11
  needs for, 3
  permits, 4
  shavings (wood chips), 11
  site for building, 4–5, *5*
  stalls, 9–12, *10*,*12*
  straw bedding, 11, *372*, *373*
  tack rooms, 13
  ventilation, 6, *6*–7, 9
  windows, 5, *5*
  for wintry climates, 9
Barren mares, 360
Base narrow/wide, *305*, *311*, 312, 314,
  314
Beans in sheath, 293
Bedding, 11, *372*, *373*

Bench-kneed, *310*
Benzimidazoles, 208, 247
Betadine (tamed iodine), 244
Birth Alert, 377
Birth defects, 414–15
Bismuth subsalicylate (Pepto Bismol),
  436
Biters, handling, 116–17
Black flies, 214
Black locust toxicity, 232
Black patch disease, 234
Black walnut toxicity, 11, 233
Blankets for winter, 66–68, *68*
Bleeding, controlling, 281
Blister beetles (cantharidin) toxicity,
  236–37, *237*
Blood-brain barrier, 415
Blood circulation in foals, 390
Bloodworms (large strongyles), 198,
  202–4, *203*, 273, *274*
Body
  bashers, handling, 123
  condition in winter, 62–63, *63*
  conformation, 307, *308*, *309*
  contact for handling horses, 101–2,
    103, *104*, 107, 110, 117
  language of horses, 99, 117, 118,
    131–37, *132–36*
Bone and horse's size, 312
Boredom cautions for stallions, 338
Borrelia burgdorferi, 177
Bots (*Gasterophilus*), *200*, 200–201, 210
Botulism (forage poisoning), 170–71
Bowel movements (foals), *412*, 412–13
Bowlegged, *310*
Bowling, Ann T., 330
Brackenfern toxicity, 225
Brain fever (equine encephalomyelitis),
  142, 143, 156–57
Bran mash, 49

Breathing after birth, 389–91
Breech presentation, 388, *388*
Breeding fee, 342
Breeding horses. *See* Breeding mares;
    Breeding stock selection;
    Broodmares; Foal care; Foaling;
    Genetics; Stallions
Breeding mares, 343–59. *See also*
    Breeding stock selection;
    Broodmares
  age for breeding, 343
  anestrus, 349
  artificial insemination, 358
  Caslick repair, 347, *347*
  cervix, 343–44, *344*
  clitoris, 343, *344*
  conception, checking for, 358
  diestrus, 349
  estrus, 349
  feeding, 37, 55–56, 360
  first cycles, 350
  foal heat, breeding at, 359
  handling, 355–57, 358
  heat cycles, 348–51
  hobbles for, 355–56, *356*
  imperforate hymen, 346
  infections, 346, 359
  infertility, 359
  ovaries and oviducts, 344, *344*
  ovulation, 348–49
  preparation, 354–57, *355, 356*
  rectal palpation, 345
  reproductive tract, 343–47, *344, 347*
  seasonal breeders, 350
  signs of heat, 349
  silent heat, 349–50
  soundness, 344–46
  spring and summer breeding, 351
  stallion selection, 316
  tail wrapping, 354–55, *355*
  teasing, 351–54, *352*
  uterus, 344, *344*, 345–46
  vagina, 343, *344*, 346
  vulva, 343, *344*
  washing, 355, 358
Breeding stock selection, 302–17. *See
    also* Conformation; Genetics
  "barn blind," 303
  broodmares, 315
  conformation and, 303
  crossbreeding, 317
  fertility record of stallion, 317
  genetic defects and, 303, 326
  goals of breeding, 302
  pedigree obsession, 304
  perfection, visualizing, 303–4
  semen, shipped, 316–17
  stallions, 315–17
Bribery, handling using, 98
Brittle feet, 94
Broodmares, 360–71. *See also* Breeding
    mares; Foaling
  abortion, 364–68
  barren mares, 360
  for breeding stock, 315
  deworming, 399, 437, 440–41
  drugs and abortion, 366
  edema prevention, 362–63, *363*
  exercise, 362–63, *363*
  facilities for, 17, 364
  false pregnancy, 367–68
  feeding, 37, 55–56, 360, 361–62,
    366–67, 429
  fescue grass cautions, 230, 366
  foaling, countdown to, 371
  handling of, *402*, 402–3, 429
  infections and abortion, 368
  lactating mares, 360
  maiden mares, 360
  mares in foal, 360

mastitis, 369–71
pasture for, 362, 367
placental insufficiency, 365
pregnancy care, *361*, 361–62
progesterone and abortion, 366
stress and abortion, 366
twins and abortion, 365
udder problems, 368–71
vaccinations for, 368
Bruises first aid, 280, *280*
Buck-kneed (over at the knee), *310*, 311
Buckthorn toxicity, 226
Buffalo gnats, 214
Building a barn, *2*, 3–9, *4–6*, **8**
Burdock and eye problems, *263*, 263–64
Burns first aid, 289–91, *290*
B vitamins, 39

C

Calcium, 40–41, 50, 445
Calf-kneed (back at the knee), *310*, 311
Cancer, 179–83, *183*
Canine teeth, *253*, 254, *254*
Cannon bone, 311, *311*
Cantharidin (blister beetles) toxicity, 236–37, *237*
Caps (teeth), 256
Captan (Orthocide), 244
"Captive bolt" to the brain, 299–300
Carbohydrates, 36, 50
Cartilage-to-bone conversion, 445
Caslick repair, 347, *347*, 378–79
Cast horse, 9
Castor beans toxicity, 225
Castrating colts, 447–50, *448–49*
Cataracts, 261, *262*
Cecum, 33, *33*, 35
Cellulose and body heat, 34, 36
Cervix, 343–44, *344*
Chain restraint, 111–12, *112*, 122–23

Chemical poisons, 237–39
Chewers and fences, 18, 20
"Chew time," 46, 51
Chiropractic therapies, 296–97
Chiroptic mange, 222
Chokecherry (wild cherry) toxicity, 231, *231*
Choking, first aid for, 287–88
Chromosomes, 318
Chronic obstructive pulmonary disease (COPD), 184–85, *185*
CID (combined immune deficiency), 303, 327–28
Clitoris, 343, *344*
Clostridium botulinum, 170
Clostridium tetani, 154
Club foot, 84–85, *85*
Coats. *See also* Parasites; Skin problems
quality and feeding, 37, *37*
shedding, 71–72, *72*
for winter, 62, 66
Cocklebur toxicity, 225
Coggins test, 166, *166*
Cold weather. *See* Winter care
Colic. *See* Digestive problems
Colon, 35
Color, genetics of, 320–24
Colostrum
collecting, 378
importance of, 406, 407–11, 422
vaccinations and, 141, 143
Combined immune deficiency (CID), 303, 327–28
Commercial grain mixes, 51
Concentrates. *See* Grains
Conception, checking for, 358
Conformation, 304–14
back at the knee (calf-kneed), *310*, 311
base narrow/wide, *305*, *311*, 312, 314, *314*

Conformation (*continued*)
  bench-kneed, *310*
  body, 307, *308*, *309*
  bone and horse's size, 312
  bowlegged, *310*
  breeding stock, 303
  cannon bone, 311, *311*
  cow hocked, 314, *314*
  cresty neck, *307*
  cut in at knee, *310*
  disposition and, 315
  ewe neck, *307*
  feet, 82, 305–6, *306*
  flat bone, *311*, 311–12
  head and neck, 307, *307*
  hind legs, 313–14, *313–14*
  hock, 313, *313*
  horse anatomy, *454*
  knock-kneed, *310*
  legs, front, *310–11*, 310–12
  muscling, 307
  "out in the country," 313, *313*
  over at the knee (buck-kneed), *310*,
    311
  pastern angle, 306, *306*, 309
  pigeon-toed, *311*
  post-legged, 313, *313*
  round bone, *311*, 312
  shoulders and withers, 308–9, *309*
  sickle-hocked, 313, *313*
  splay-footed, *311*
  tied in knees, *310*, 312
  topline, *308*
  underline, *308*
  unsoundness, *314*
Conjunctivitis, 261
Constipation
  colic and, 277–78
  in foals, 412, 413
Contracted heels, 82–84, *83*

Contracts for breeding, 342
COPD (chronic obstructive pulmonary
    disease), 184–85, *185*
Cord, breaking, 391–92, 393
Corn as feed, 48–49, 235
Cornea, *260*
Corneal injury, 262
Corners of fences, 16, *16*, 24, *24*
Corns in feet, 89–91, *90*
Coronary band, 81, 83, 93–94
Cow-hocked, 314, *314*
Crazyweed (locoweed) toxicity, 227, *227*
Cresty neck, *307*
Crossbreeding, 317, 326
Cross ties, 111, *111*
Crotalaria, 226
Cryptorchid, 450
Curb of hock, 313
Cushing's disease, 71, 186
Cut in at knee, *310*

D

Daily deworming (pyrantel tartrate),
    202, 205, 210
Daily feed rations, 34
Death camus, 226, *226*
Deer flies (Tabanids), *212*, 212–13
Dehydration, 65, 420, 424, 436
Delaying foaling, 376
Delphinium (larkspur) toxicity, 225, *225*
Dental care, 250–59
  age and teeth, 252–53, *252–53*, 255
  canine teeth, 253, 254, *254*
  caps, 256
  floating teeth, 259
  Galvayne's groove, 252, *252*, 253
  gingivitis, 250
  gum problems, 250
  impacted teeth, 256, *256*
  incisors, 253, *253*

lampas, 251, *251*
lost teeth, 258
malocclusion, 257–58, *258*
molars, 253, *253*, 257
for older horses, 259
parrot mouth, 257–58, *258*
premolars, 253, *253*
retained caps, 256
seeds, embedded, 252
sharp ridges, 257
shear mouth, 258
signs of trouble, 257
sow mouth, 257–58, *258*
tartar build-up, 251
teeth, growth of, 252, *252*
tongue injuries, 252
tooth bumps, 256, *256*
uneven wear, 257
wolf teeth, 253, *253*, 254, *254*
Dermatitis, 221, 240
Deworming
broodmares, 399, 437, 440–41
fungicides, dewormers as, 247
overdoses, 239
parasites, 199–200, 207, 208–10, *210*
Diarrhea in foals, 425, 434, 436–37
Diatomaceous earth, 215
Dichlorvos, 205, 208, 209
Dictyocaulus arnfieldi (lungworms), 207
Diestrus, 349
Digestion, *33*, 33–35
Digestive problems, 270–78
constipation, 277–78
endotoxic shock, 273–74
enteritis (inflammation of intestines),
275
enteroliths, 277
feeding and, 34, 44, 46, 52, 274–75
flatulant colic, 276
from foaling, 380, 393

impaction, 34, 52, *276*, 277
intestinal obstructions, *276*, 276–78
parasites and, 198, 273, 274
rolling, dangers of, *272*, 273
sand colic, 275–76
signs of, 131–33, *132–33*, 135, *135*,
195, 271, *271*, 272, 276
spasmodic colic, 271–72
twisted intestine, 272, *272–75*
Dimethyl sulfoxide (DMSO), 247, 287
Disease prevention, 138–53. *See also*
Infectious diseases; Non-infectious
diseases; Parasites; Signs of illness;
Vaccinations
antibodies, 139–40
antigens, 139
colostrum and, *406*, *407–11*, 422
disease, defined, 138–39
in foals, 141, 143
guidelines, 153
immune system and, 139–42
immunity, 139, 140–41
pathogens, 139, 142
of respiratory system, 142, *161*
stress and, 141–42
Disposition and conformation, 315
Distemper (strangles), 157–59
DMSO (dimethyl sulfoxide), 247, 287
Dominance and handling, 98, 99, 101
Dominant traits, 319–20, 323, 324
Droncit (praziquantel), 206
Drugs and abortion, 366

E
Early labor, 379–80
Ear problems, 266–69
diagnosing, 267
ear mites, 269
ear ticks, 220–21, 268, *268*
flies, 268

Ear problems (*continued*)
   foreign objects, 269
   infections, 266
   middle ear infection, 266–67
   signs of, 267, 268, 269
   trauma, 266
   warts, 269
Easyboots, 84, 93
Edema prevention, 362–63, 363
Ehrlichia risticii, 167
EHV (equine herpes virus), 160
EIA (equine infectious anemia), 164–66, 166
Electric fencing, 18, 21–23, 21–23, 25, 25
Embedded seeds, 252, 262–63
Emergency situations. *See* First aid
Endorphins, 114
Endotoxemia and founder, 189
Endotoxic shock, 273–74
Enema, giving, 412, 412
Energy-to-protein ratio, 55
Enteritis (inflammation of intestines), 275
Enteroliths, 277
Epinephrine (adrenaline), 153
Epiphysitis, 444–45, 446
EPM (equine protozoal myeloencephalitis), 39, 48, 177–78
Equine Color Genetics (Sponenberg), 330
Equine encephalomyelitis (sleeping sickness), 142, 143, 156–57
Equine herpes virus (EHV), 160
Equine infectious anemia (EIA), 164–66, 166
Equine protozoal myeloencephalitis (EPM), 39, 48, 177–78
Equine viral arteritis (EVA), 138, 163–64

Equisetum toxicity, 226
Ergot toxicity, 235
Estrus, 349
Euthanasia, 298–300
EVA (equine viral arteritis), 138, 163–64
Ewe neck, 307
Exercise
   for broodmares, 362–63, 363
   for foals, 413, 413, 446
   for stallions, 338, 338
   in summer, 78
Eye problems, 260–66
   anatomy of, 260
   burdock, 263, 263–64
   cataracts, 261, 262
   conjunctivitis, 261
   cornea, 260
   corneal injury, 262
   embedded seeds, 262–63
   iris, 260
   lacerated eyelids, 262
   medications, applying, 266
   moonblindness (periodic ophthalmia), 167, 264–65
   night blindness, 265
   pupil, 260
   signs of problems, 261
   squinting, 261–62
   in summer, 73–74
   tearing, 261–62

F
Face flies, 214–15
Facilities, 2–31. *See also* Barns; Fences and paddocks
   automatic waterers, 28
   for broodmares, 17, 364
   for foaling, 372, 372–75
   for foals, 443, 443–44
   hazards, 29, 29–31, 30–31

pasture, 17, 26–27
poisonous plants, 26–27
run-in sheds, *15*, 15–16
for stallions, 17, 331–33, *332*
water, 27–29, *28–29*
weed control, 26–27
windbreaks, natural, 16
in winter, 28–29, *28–29*, 62
False pregnancy, 367–68
Farriers, 84, 85, 95, 103, *103*
Fats, 36–37
Feeders, 12, *12*
Feeding and nutrition, 32–59. *See also*
    Dental care; Grains; Hay
  activity level and, 54, 55, 57–58
  age and, 54, 55, 56, 58
  alfalfa hay, 44–45, 46, 185, 237
  barley, 47–48
  bran mash, 49
  for broodmares, 37, 55–56, 360,
    361–62, 366–67, 429
  calcium, 40–41, 50, 445
  carbohydrates, 36, 50
  cecum, 33, *33*, 35
  cellulose and body heat, 34, 36
  changing feed cautions, 46
  "chew time," 46, 51
  choking, 287–88
  coat quality and, 37, *37*
  colic, 34, 44, 46, 52, 274–75
  colon, 35
  commercial grain mixes, 51
  corn, 48–49, 235
  daily rations, 34
  digestion, *33*, 33–35
  energy-to-protein ratio, 55
  fats, 36–37
  for foals, 37, 56, 428, 433–34, 444–46
  grass hay, 44, 45, 46, 53, 54, 185
  health and, 32

  for idle horses, 55
  impaction, 34, 52, *276*, *277*
  injectable supplements, 53
  iodine, 40, 41
  iron, 40, 41
  laminitis (founder), 34
  large intestine, 35
  lawn clippings, 46, 236
  legume hay, 44–45
  manure and, 47
  minerals, 40–43, *42*
  mixed hay, 44, 45
  molasses, 50
  moldy (fungi) feed, 45, 234–36
  oat hay cautions, 43
  oats, 47
  oilseed meals, 50
  for older horses, 58
  overfeeding, 52, 53, 56, 57
  pancreas, 35
  pasture, 26, 37, 38, 40, 43–44, *44*, 54
  pelleted feeds, *51*, 51–52
  phosphorus, 40–41, 50, 445
  protein, 37, *37*
  ruminants, 33, *33*
  salt (sodium chloride), 41–43, *42*
  selenium, 40, 41
  silage, 43
  small intestine, 35
  soil and minerals, 40
  sorghum grain (milo), 50
  for stallions, 57
  steamed or rolled grains, 47, 48
  stomach, 33, *33*, 34–35
  storage, 13–14, *15*
  in summer, 78
  supplements, 36, 40, 41, *52*, 52–53
  tips, 59
  total digestible nutrients (TDN), 56
  trace minerals, 40, 41, 42

Feeding and nutrition (*continued*)
  tying up, 50, 135, *135*, 194–96
  vitamins, 38–39, 40, 239
  vomiting, 35
  water and, 35
  for weanlings, 444–46
  weighing feed, 45–46, 48
  wheat and wheat bran, 49–50
  in winter, 60, 64–65
  for working horses, *57*, 57–58
  for yearlings, 444–46
Fences and paddocks, 16–25
  barbed wire, 17–18
  chewers and, 18, 20
  corners, 16, *16*, 24, *24*
  electric fencing, 18, 21–23, *21–23*, 25, *25*
  for foals, *443*, 443–44
  gates, 24–25, *25*
  maintenance of, 16, 17–18, 19, 20, 22, 23, *23*
  for mares and foals, 17
  net or mesh wire, 20, *20*
  for pastures, 17
  polyethylene-coated wood, 20
  polyvinyl chloride (PVC) tubing, 19–20
  posts, 24, *24*
  rubber and nylon strips, 20
  smooth wire, 21
  for stallions, 17, 332, *332*
  steel pipe, 23
  wood fencing, 18–19, *18–19*, 22, *22*
Fertility, 317, 340–42, 359
Fescue grass toxicity, 230, 366
Fiddleneck toxicity, 226
Fireweed toxicity, 226
First aid, 279–99
  abrasions, 281
  acupuncture, 297–98

  alternative therapies, 295–98
  alum for proud flesh, *283*, 284
  arteries versus veins, 281
  beans in sheath, 293
  bleeding, controlling, 281
  bruises, 280, *280*
  burns, 289–91, *290*
  "captive bolt" to the brain, 299–300
  chiropractic therapies, 296–97
  choking, 287–88
  cleaning wounds, 282
  emergency situations, 279
  euthanasia, 298–300
  fractures, 286
  friction burns, 290–91
  head injuries, 287
  hemorrhagic shock, 288
  herbal therapy, 296
  homeopathy, 296
  incised wounds, 281
  kick wounds, 285–86
  kit contents, 294
  lacerations, 282–83, *282–83*
  massage therapy, 297
  porcupine quills, 291, *291*
  pressure bandage, 281
  proud flesh, 94, 249, *283*, 283–84
  punctures, 91, 92, 284–85, *285*
  relaxing injured horses, 287
  rope burns, 290–91
  secondary shock, 289
  sheath swellings, 292–93
  shock, 288–89
  snakebite, 291
  stitches, 284
  tetanus, 142, 143, 154–56, *155*, 280
  urinary stones, 292
  veins versus arteries, 281
  weight estimation for dosage, 295
  wounds, *280*, 280–86, *282–83*, 285

First cycles, 350
First foals, 375–76
Flat bone, *311*, 311–12
Flat feet, 84
Flatulant colic, 276
Flies, *211–13*, 211–17, 268
Floating teeth, 259
Flooring, barns, 5, *5*, 11
Floor plan, barns, *10*
Flu (influenza), 142, 159–60, *160*
Fly masks, 74, *74*
Foal Alert, 377
Foal care, 400–451. *See also* Foaling; Foot
   care
  antibodies, *406*, 407–11, 422
  birth defects, 414–15
  blood-brain barrier, 415
  bowel movements, *412*, 412–13
  calcium imbalance, 445
  cartilage-to-bone conversion, 445
  castrating colts, 447–50, *448–49*
  in cold weather, 411, *411*
  colostrum, *406*, 407–11, 422
  constipation, 412, 413
  cryptorchid, 450
  dehydration, 420, 424, 436
  diarrhea, 425, 434, 436–37
  disease prevention, 141, 143
  enema, giving, 412, *412*
  epiphysitis, 444–45, 446
  exercise, 413, *413*, 446
  facilities, *443*, 443–44
  feeding, 37, 56, 428, 433–34, 444–46
  foal heat diarrhea, 436–37
  gastric ulcers, 439–40
  goat's milk substitute, 425
  grains (concentrates), 433–34
  growth and weight gain, 425
  handling, 115, 119, *401–2*, 401–3,
    429, 430–32, *431–32*

harness, 431–32, *432*
hobbles, nursing, 426, 428, *428*
imprint training, 401, *401*
infections, 418–20
injections, 419
IVs, 420
leading lessons, 432, *432*
legs, crooked, 421, *421*
milk replacer, 422–24
navel ill (joint ill), 419–20
neonatal isoerythrolysis (Rh foals),
  416–17
nervous system, 415
nursemares, 425–26
nursing, 404–11, *405–6*, 426, 428,
  428, 433
orphan foals, 422–29, *427–28*
osteochondritis dessicans (OCD),
  444–45, 446
overfeeding cautions, 444–45
parasites and, 198, 202, 205, 437,
  440–41
phosphorus imbalance, 445
pneumonia, 437–39
premature foals, 421–22
rejection, avoiding, 400
ruptured bladder, 415–16
skeletal problems, 445–46
stalls, 414
standing up, 403–4
stomach tubes, 410, 420
tendons, weak, 421, *421*
ulcers, 439–40
umbilical hernia, 418
vaccinations, 141, 143
weaning, 441–44, *443*
weanlings, 444–50, *448–49*
yearlings, 444–50, *448–49*
Foal heat, breeding at, 359
Foal heat diarrhea, 436–37

Foaling, 372–99. *See also* Breeding mares;
  Broodmares; Foal care
  active (second stage) labor, 380–89,
    381–88
  after, 391–93
  afterbirth, checking, 394
  airways, clearing, 390
  artificial respiration, 391
  backward presentation, *387*, 387–88
  bedding, *372*, 373
  blood circulation, stimulating, 390
  breathing after birth, 389–91
  breech presentation, 388, *388*
  care after, 399
  Caslick repair, 378–79
  colic and, 380, 393
  colostrum, collecting, 378
  cord, breaking, 391–92, 393
  countdown to, 371
  delaying foaling, 376
  early labor, 379–80
  facilities for, *372*, 372–75
  feces and, 389
  first foals, 375–76
  gestation, 375
  grooming and, 379
  head-turned-back presentation, *385*,
    385–86
  heartbeat, checking, 390–91
  hemorrhage after, 397–98
  Klebsiella bacteria, 11
  laminitis (founder), 396
  large foals, 383
  malpresentations, 383–89, *384–88*
  navel stump, disinfecting, 392
  normal presentation, 381–82, *384*
  nose-but-no-feet presentation, 385
  one-foot-turned-back presentation,
    384, 384–85
  in pasture, 374–75
  placenta previa, 388–89
  placenta shedding, 393
  prediction, 377
  privacy for, 376, 379
  retained placenta, 394–96
  rotated transverse foal presentation,
    386, *386*
  shoulders, passing, 382–83, *382–83*
  signs of, 377–79, 380
  in stall, 372–74
  straw bedding, 11, *372*, 373
  supplies for, 373
  uterine prolapse, 397
  uterine torsion, 389
  water bag rupture, 381, *381*
  waxing, 378
Foal-Lac, 422
Foal-Mate, 422
Foot care, 80–96
  abnormalities, 82–85, *83*, *85*
  anatomy of, 81, *81*
  brittle feet, 94
  cleaning feet, 85, *85*, 89
  club foot, 84–85, *85*
  conformation, 82, 305–6, *306*
  contracted heels, 82–84, *83*
  corns, 89–91, *90*
  coronary band, 81, *83*, 93–94
  drainage hole, protecting, 93, *93*
  Easyboots, 84, 93
  farriers, 84, 85, 95, 103, *103*
  flat feet, 84
  frog of foot, 81, *81*, 82
  growth of, 81
  handling, 80, 107–10, *108–10*
  hoof abscess, 91–93, *92–93*
  "hoof bound," 83
  hoof cracks, 94–95, *95*
  hoof testers, 90, *90*
  hoof wall, 81, *81*, 82

injuries, 89–95, *90, 92–93, 95*
iodine, applying to sole, 84, 89
lameness, diagnosing, 96
laminitis (founder), 188–91, *189–90*
nails, unclinching, 86, 88
navicular disease, *191–92,* 191–93
proud flesh, 94
puncture wounds, 91, 92, 284–85, *285*
"quicked" hoof, 91
shoes, corrective, 95
shoes, pulling, 87, *87–88*
soaking feet, 92–93
sole bruises, 89–91, *90*
sole of hoof, 81, *81,* 82, 83, *83,* 84
in summer, 73
thrush, 88–89
in winter, 60, 66, 89
Footing safety in winter, 66
Forage poisoning (botulism), 170–71
Forages. *See* Hay
Force, handling using, 98
Foundations, barns, 5
Founder. *See* Laminitis
Fractures first aid, 286
Friction burns first aid, 290–91
Frog of foot, 81, *81,* 82
Fungi (mold) in feed, 45, 234–36
Fungus infections, *243,* 243–47, *245,*
   247
Fusarium moniliforme, 235

G

Galvayne's groove, 252, *252,253*
Garden toxic plants, 225, *225*
Gas colic, 276
Gasterophilus (bots), 200, *200–201,* 210
Gastric ulcers in foals, 439–40
Gates, 24–25, *25*
Gelding (castrating) colts, 447–50,
   448–49

Genetics, 318–30
chromosomes, 318
color, 320–24
combined immune deficiency (CID),
   303, 327–28
crossbreeding, 326
defects, 303, 326
dilution factors, 323
dominant traits, 319–20, 323, 324
genes, 318
gestation and, 375–76
heterozygous, 320, 323, 324
homozygous dominant, 320
homozygous recessive, 320
hyperkalemic periodic paralysis (HYPP),
   303, 328–29
inbreeding, 324–25
inheritance of traits, 318
lethal genes, 326–27
linebreeding, 325
nicking, 325
outcrossing (outbreeding), 325
pedigrees, *319*
recessive traits, 319–20, 323, 324
Gestation, 375–76
Gingivitis, 250
Girth itch, 244–45, *245*
Goat's milk substitute, 425
Goat weed (Saint-John's-wort) toxicity, 229
Grains (concentrates)
balanced diet using, 53
energy-to-protein ratio, 55
for foals, 433–34
nutrition, 46–51, *47*
toxicity, 235, 236, 238
weighing, 48
in winter, 64
Granulosa cell tumors, 182–83
Grass clippings for feed, 46, 236
Grasses toxicity, 230

Grass hay, 44, 45, 46, 53, 54, 185
Grease heel (scratches), 246–47, *247*
Ground rules, handling, 120–21
Gum problems, 250
Gut sounds and illness, 132, *132*

H

Habronemiasis (summer sores), 211–12,
   217–18
Halters
   for breeding stallions, 336
   for handling, 102, 104, 105
   pullers, handling, 118–19
Hand breeding, 339
Handling horses, 97–124
   aggressive horses, 122–24
   bad manners, 115–22
   biters, 116–17
   body bashers, 123
   body contact for, 101–2, 103, 104,
      107, 110, 117
   body language (horse), 99, 117, 118,
      131–37, *132–36*
   breeding mares, 355–57, 358
   by bribery, 98
   broodmares, *402*, 402–3, 429
   chain restraint, 111–12, *112*, 122–23
   communication, levels of, 98–99
   consistency and, 98, 115, 124
   cross ties, 111, *111*
   distance, maintaining, 121
   dominance, 98, 99, 101
   endorphins, 114
   for farriers, 103, *103*
   feet, 80, 107–10, *108–10*
   foals, 115, 119, *401–2*, 401–3, 429,
      430–32, *431–32*
   by force, 98
   ground rules, 120–21

halter-pullers, 118–19
halters for, 102, 104, 105, 336
hand twitching, 114
hard-to-catch horses, 119–20
herd instinct, 97, 98, 121
holding a horse, 102–5, *103–5*
horse sense, *97*, 97–98
kickers, 104, 109, *109*, 117–18
leading horses, 104–5, *104–5*, 122,
   432, *432*
lip chain for restraint, 112
"manger tie," 107, *107*
pecking order, enforcing, 121
punishment, 100–102, 116–17
quick-release knot, 107, *107*
relaxation and, 99–100
restraints, *111–13*, 111–15, *115*
rubbing, discouraging, 121–22
safety tips, 101–2, 107, 110, 113
Stableizer, 114–15, *115*
stallions, 333–38, *338*, 356–57
temperament of horses and, 98
turnout terrorists, 124
twitch restraint, *112–13*, 112–14
tying horses, 105–7, *106–7*, 119
for veterinarians, 102, 103, *103*
voice tone and, 100
Hand twitching, 114
Hard-to-catch horses, 119–20
Harness, foal, 431–32, *432*
Hay
   balanced diet with, 53
   non-infectious diseases and, 185
   nutrition, 44–46
   as safe feed, 54
   storage, 13–14, *15*
   toxicity, 234–35, 236, 237
   weighing, 45–46, 48
   in winter, 64

Head
    conformation, 307, *307*
    injuries first aid, 287
Head-turned-back presentation, *385,*
    385–86
Healing Your Horse: Alternative Therapies
    (Snader), 298
Health care. *See* Dental care; Digestive
    problems; Disease prevention; Ear
    problems; Eye problems; First aid;
    Infectious diseases; Non-infectious dis-
    eases; Parasites; Signs of illness; Skin
    problems; Toxic plants and poisons
Heartbeat (foal), checking, 390–91
Heart (pulse) rate of horses, 127, 129–31,
    130
Heat cycles, 348–51
Heated barns, 7
Heat energy from roughage, 64
Heat stroke, 75, **76,** 76–77, *77*
Heaves (chronic obstructive pulmonary
    disease), 184–85, *185*
Hemorrhage after foaling, 397–98
Hemorrhagic shock first aid, 288
Hepatitis, 186–88
Herbal therapy, 296
Herd instinct, 97, 98, 121
Heterozygous genes, 320, 323, 324
Hind legs
    conformation, 313–14, *313–14*
    as illness sign, 134–35, *135,* 137
Histamine and vaccinations, 152
Hives (wheals), 240–41, *241*
Hobbles, 355–56, *356,* 426, 428, *428*
Hock, 313, *313*
Holding a horse, 102–5, *103–5*
Homeopathy, 296
Homozygous dominant, 320
Homozygous recessive, 320

Hoof. *See also* Foot care
    abscess, 91–93, *92–93*
    cracks, 94–95, *95*
    testers, 90, *90*
    wall, 81, *81,* 82
"Hoof bound," 83
Hormones and shedding, 71
Horn flies, *213,* 213–14
Horse anatomy, *454*
Horse flies (Tabanids), *212,* 212–13
Horse Genetics (Bowling), 330
Horsekeeping. *See* Facilities; Feeding and
    nutrition; Foot care; Handling
    horses; Seasonal care
Horse sense, 97, *97*–98
Horsetail (equisetum) toxicity, 226
Humidity concerns, 75, 78
Hyperkalemic periodic paralysis (HYPP),
    303, 328–29
Hyperthermia, 75
HYPP (hyperkalemic periodic paralysis),
    303, 328–29

I

Idle horses, feeding, 55
Immune system
    disease prevention and, 139–42
    warts and, 247–48, *248,* 269
    worms and, 197–98
Impacted teeth, 256, *256*
Impaction, 34, 52, *276, 277*
Imperforate hymen, 346
Imprint training of foals, 401, *401*
Inbreeding, 324–25
Incised wounds first aid, 281
Incisors, 253, *253*
Infections
    abortion and, 368
    in broodmares, 346, 359

Infections *(continued)*
　ear, 266
　in foals, 418–20
Infectious diseases, 154–78. *See also*
　　Disease prevention; Parasites; Signs
　　of illness; Vaccinations
　anthrax, 174–75
　antitoxins, 156, 188
　aspiration pneumonia, 162, 288
　from *Borrelia burgdorferi*, 177
　botulism (forage poisoning), 170–71
　from *Clostridium botulinum*, 170
　from *Clostridium tetani*, 154
　Coggins test, 166, *166*
　defined, 138–39
　from *Ehrlichia risticii*, 167
　equine encephalomyelitis (sleeping
　　sickness), 142, 143, 156–57
　from equine herpes virus (EHV), 160
　equine infectious anemia (EIA),
　　164–66, *166*
　equine protozoal myeloencephalitis
　　(EPM), 39, 48, 177–78
　equine viral arteritis (EVA), 138,
　　163–64
　influenza (flu), 142, 159–60, *160*
　leptospirosis, 166–67, 265
　lockjaw, 155, *155*
　Lyme disease, 177, *220*, 220–21
　moonblindness (periodic ophthalmia),
　　167, 264–65
　penicillin for strangles, 158
　piroplasmosis, 174
　pneumonia, 162–63, 288, 437–39
　Potomac horse fever (PHF), 138,
　　167–68
　rabies, 171–73
　rhinopneumonitis, 142, 146, 160–62,
　　161
　salmonellosis, 168–70
　from *Sarcocystis neurona*, 177
　from *Septicemia*, 162, 174
　"shaker foal syndrome," 170
　strangles (distemper), 157–59
　from *Streptococcus equi*, 157
　tetanus, 142, 143, 154–56, *155*, 280
　toxoids, 156
　Venezuelan equine encephalomyelitis
　　(VEE), 138, 156
　vesicular stomatitis (VS), 175–76
Infertility, 341–42, 359
Infestation process, 198
Influenza (flu), 142, 159–60, *160*
Inheritance of traits, 318. *See also*
　　Genetics
Injections. *See also* Vaccinations
　for foals, 419
　supplements, 53
Inoculations. *See* Infectious diseases;
　　Vaccinations
Insects, 74, *74*. *See also* Parasites
Intestinal obstructions, *276*, 276–78
Intramuscular injections, *147*, 147–49
Intramuscular shots, 144
Iodine
　for feet, 84, 89
　supplements, 40, 41
　tamed iodine (Betadine), 244
Iridocyclitis (moonblindness; periodic
　　ophthalmia), 167, 264–65
Iris, *260*
Iron, 40, 41
Isolation and stallions, 333
Itching dermatitis, 221
Ivermectin, 199, 202, 204, 205, 207, 208
IVs for foals, 420

J

Jimmy weed (rayless goldenrod) toxicity,
　232

Jointfir (equisetum) toxicity, 226
Joint ill (navel ill), 419–20

K
Kickers
    handling, 104, 109, *109*, 117–18
    wounds first aid, 285–86
Klamath weed (Saint-John's-wort) toxicity, 229
Klebsiella bacteria, 11
Knock-kneed, *310*

L
Lacerations first aid, 282–83, *282–83*
Lactating mares, 360
Lameness, diagnosing, 96, *314*
Laminitis (founder)
    disease, 188–91, *189–90*
    from feeding, 34
    from foaling, 396
Lampas, 251, *251*
Lantana toxicity, 226
Large foals, 383
Large intestine, 35
Large strongyles (bloodworms), 198,
    202–4, *203*, 273, 274
Larkspur (delphinium) toxicity, 225, *225*
Lawn clippings for feed, 46, 236
Leading
    foals, 432, *432*
    horses, 104–5, *104–5*, 122
Lead poisoning, 239
Legs. *See also* Foot care
    conformation, *310–11*, 310–14,
        313–14
    crooked in foals, 421, *421*
    as illness sign, 133–37, *134–36*
Legume hay, 44–45
Leptospirosis, 166–67, 265
Lethal genes, 326–27

Lice, 218–19
Lights in barns, 4, *4*
Linebreeding, 325
Lipomas, 183
Live foal guarantee, 342
Lockjaw, 155, *155*
Locoweed toxicity, 227, *227*
Louvers for ventilation, 6, *6*, 7
Lungworms (*Dictyocaulus arnfieldi*), 207
Lyme disease, 177, *220*, 220–21

M
Maiden mares, 360
Malocclusion, 257–58, *258*
Malpresentations, foaling, 383–89,
    384–88
Mange mites, 221–22
"Manger tie," 107, *107*
Manure pile, 12
Mares. *See* Breeding mares; Broodmares
Mare's Milk Plus, 422
Massage therapy, 297
Mastitis, 369–71
Materials (building), **8**
Mats, stall, 5, *5*, 11
Medication overdoses, 239
Melanomas, 180–81
Mesh wire fences, 20, *20*
Middle ear infection, 266–67
Midges (no-see-ums), 214
Milk replacer, 422–24
Milkvetch (locoweed) toxicity, 227, *227*
Milo (sorghum grain), 50
Minerals, 40–43, *42*, 445
Mites, 221–22
Mixed hay, 44, 45
Molars, 253, *253, 257*
Molasses in feed, 50
Mold (fungi) in feed, 45, 234–36
Monday morning disease, 194–95

Monensin (Rumensin) poisoning, 239
Moonblindness (iridocyclitis; periodic ophthalmia), 167, 264–65
Mosquitoes, 217
Mouth problems. *See* Dental care
Moxidectin (Quest), 199, 208, 209
Mud and health concerns, 70–71
Muscling and conformation, 307
Mycotoxins, 236

N

Nails of shoes, unclinching, 86, 88
Navel ill (joint ill), 419–20
Navel stump, disinfecting, 392
Navicular disease, *191–92*, 191–93
Neck conformation, 307, *307*
Needles for vaccinations, 145–46, *146*, 148, 149, *150*
Neonatal isoerythrolysis (Rh foals), 416–17
Nervous system of foals, 415
Net or mesh wire fences, 20, *20*
Nicking, 325
Night blindness, 265
Nightshade family toxicity, 228, *228*
Non-infectious diseases, 179–96. *See also* Disease prevention; Signs of illness
    antitoxins, 156, 188
    cancer, 179–83, *183*
    chronic obstructive pulmonary disease (COPD), 184–85, *185*
    Cushing's disease, 186
    defined, 139
    endotoxemia and founder, 189
    granulosa cell tumors, 182–83
    hay and, 185
    hepatitis, 186–88
    laminitis (founder), 34, 188–91, *189–90*, 396

lipomas, 183
melanomas, 180–81
navicular disease, *191–92*, 191–93
rhabdomyelitis (muscles tying up), 50, 135, *135*, 194–96
serum hepatitis (Theiler's disease), 187–88
squamous cell carcinomas, 181–82
toxoids, 188
Normal birth presentation, 381–82, *384*
Nose-but-no-feet presentation, 385
No-see-ums (midges), 214
Nursemares, 425–26
Nursing, 404–11, *405–6*, 426, 428, *428*, 433
Nutrition. *See* Feeding and nutrition

O

Oat hay cautions, 43
Oats, 47
OCD (osteochondritis dessicans), 444–45, 446
Oilseed meals, 50
Older horses
    dental care, 259
    feeding, 58
    stallion education by, 337
    winter care, 63, 65
Oleander toxicity, 232
One-foot-turned-back presentation, *384*, 384–85
Orphan foals, 422–29, *427–28*
Orthocide (captan), 244
Osteochondritis dessicans (OCD), 444–45, 446
Outcrossing (outbreeding), 325
"Out in the country," leg conformation, 313, *313*
Ovaries and oviducts, 344, *344*

Over at the knee (buck knees), *310*, 311
Overfeeding, 52, 53, 56, 57, 444–45
Ovulation, 348–49
Oxibendazole, 205
Oxyuris equi (pinworms), 205–6

P

Palisade worms (bloodworms), 198,
    202–4, *203*, 273, 274
Pancreas, 35
Parascaris equorum (roundworms), *201*,
    201–2
Parasites, 197–222
    benzimidazoles for, 208, 247
    black flies, 214
    bloodworms (large strongyles), 198,
        202–4, *203*, 273, 274
    bots *(Gasterophilus)*, 200, 200–201,
        210
    buffalo gnats, 214
    chiroptic mange, 222
    colic from, 198, 273, 274
    daily deworming (pyrantel tartrate),
        202, 205, 210
    deworming, 199, 207, 208–10, *210*,
        239, 399, 437, 440–41
    diatomaceous earth for, 215
    dichlorvos for, 205, 208, 209
    digestive problems from, 198, 273
    drug resistance, 208–9
    ear ticks, 220–21, 268, *268*
    face flies, 214–15
    flies, *211–13*, 211–17
    foals and, 198, 202, 205, 437, 440–41
    grooming for, 219
    horn flies, *213*, 213–14
    horse and deer flies (Tabanids), *212*,
        212–13
    immune system and worms, 197–98
    infestation process, 198
    itching dermatitis, 221
    ivermectin for, 199, 202, 204, 205,
        207, 208
    lice, 218–19
    lungworms *(Dictyocaulus arnfieldi)*, 207
    mange mites, 221–22
    midges (no-see-ums), 214
    mites, 221–22
    mosquitoes, 217
    moxidectin (Quest) for, 199, 208, 209
    oxibendazole for, 205
    paste dewormers for, 209–10, *210*
    pasture management for, 207
    pinworms *(Oxyuris equi)*, 205–6
    praziquantel (Droncit) for, 206
    predator wasps for, 215
    psoroptic mange, 222
    pyrantel tartrate (Strongid C) for, 202,
        205, 210
    roundworms (ascarids), *201*, 201–2
    sand flies, 214
    sarcoptic mange, 221
    small strongyles, 204–5
    stable flies, *211*, 211–12
    strongyles, 198, *203*, 203–5, 273, 274,
        440–41
    summer sores (habronemiasis) from,
        211–12, 217–18
    tapeworms, 206
    ticks, *220*, 220–21
    trombidiform mites, 221
    unthriftiness from, 200
    vinegar as fly repellent, 216
    warbles, 207
Parrot mouth, 257–58, *258*
Paste dewormers, 209–10, *210*
Pastern angle, 306, *306*, 309
Pasture
    breeding, 339–40
    for broodmares, 362, 367

Pasture (*continued*)
  as facility, 17, 26–27
  feeding, 26, 37, 38, 40, 43–44, *44*, 54
  fences for, 17
  foaling in, 374–75
  parasite management, 207
Pathogens, 139, 142
Pecking order, enforcing, 121
Pedigrees, 304, *319*
Pelleted feeds, *51*, 51–52
Penicillin for strangles, 158
Pepto Bismol (bismuth subsalicylate),
  436
Periodic ophthalmia (moonblindness),
  167, 264–65
Permits, 4
Pesticides toxicity, 238
PHF (Potomac horse fever), 138, 167–68
Phosphorus, 40–41, 50, 445
Photosensitization, 73, 229, 230, 236,
  242–43
Pigeon-toed, *311*
Pinworms (*Oxyuris equi*), 205–6
Piroplasmosis, 174
Placental insufficiency, 365
Placenta previa, 388–89
Placenta shedding, 393
Pneumonia, 162–63, 288, 437–39
Poison hemlock, *228*, 228–29
Poisons. *See* Toxic plants and poisons
Poisonvetch (locoweed) toxicity, 227,
  227
Polyethylene-coated wood fences, 20
Polyvinyl chloride (PVC) fences, 19–20
Porcupine quills first aid, 291, *291*
Post-legged, 313, *313*
Posts of fences, 24, *24*
Posture as illness sign, 131–33,
  132–33,160
Potomac horse fever (PHF), 138, 167–68

Praziquantel (Droncit), 206
Predator wasps, 215
Prediction of foaling, 377
Pregnancy care, *361*, 361–62
Premature foals, 421–22
Premolars, 253, *253*
Pressure bandage, 281
Privacy for foaling, 376, 379
Progesterone and abortion, 366
Protein, 37, *37*
Proud flesh, 94, 249, *283*, 283–84
Psoroptic mange, 222
Pulse (heart) rate of horses, 127, 129–31,
  130
Puncture wounds, 91, 92, 284–85, *285*
Punishment, 100–102, 116–17
Pupil, *260*
PVC (polyvinyl chloride) tubing fences,
  19–20
Pyrantel tartrate (Strongid C), 202, 205,
  210

Q
Quest (moxidectin), 199, 208, 209
"Quicked" hoof, 91
Quick-release knot, 107, *107*

R
Rabies, 171–73
Rainrot (rain scald), *245*, 245–46
Rayless goldenrod toxicity, 232
Rearing, discouraging, 337
Recessive traits, 319–20, 323, 324
Rectal palpation, 345
Red mange (itching dermatitis), 221
Red maple toxicity, 232, *232*
Red worms (bloodworms), 198, 202–4,
  203, 273, 274
Rejection (foal), 400
Relaxing injured horses, 287

Reproductive tract, 343–47, *344*, *347*

Respiration rate of horses, 127, 131

Respiratory system of horses, 142, *161*

Restraints, *111–13*, 111–15, *115*

Retained caps, 256

Retained placenta, 394–96

Return-in-season, 342

Rhabdomyelitis (muscles tying up), 50, 135, *135*, 194–96

Rh foals (neonatal isoerythrolysis), 416–17

Rhinopneumonitis, 142, 146, 160–62, *161*

Rhizoctonia leguminicola, 234

Ribs and body condition, 63, *63*

Ringworm, *243*, 243–44

Rolled grains, 47, 48

Rolling, dangers of, *272*, 273

Rope burns first aid, 290–91

Rosea (rayless goldenrod) toxicity, 232

Rotated transverse foal presentation, 386, *386*

Roughage. *See* Hay

Round bone, *311*, 312

Roundworms (ascarids), *201*, 201–2

Rubber-and-nylon-strip fencing, 20

Rubbing, discouraging, 121–22

Rumensin (monensin) poisoning, 239

Ruminants, 33, *33*

Run-in sheds, *15*, 15–16

Ruptured bladder in foals, 415–16

Russian knapweed toxicity, 229

Rusts toxicity, 236

S

Saint-John's-wort toxicity, 229

Salmonellosis, 168–70

Salt (sodium chloride), 41–43, *42*

Sand colic, 275–76

Sand flies, 214

Sarcocystis neurona, 177

Sarcoid tumors, 249

Sarcoptic mange, 221

Scouring rush (equisetum) toxicity, 226

Scratches (grease heel), 246–47, *247*

Seasonal breeders, 350

Seasonal care, 60–79
    spring, 70–72, *72*
    summer, 73–79, *74*, **76**, *77*, *79*
    winter, 60–69, **61,** *63*, *68*

Secondary shock first aid, 289

Second stage labor, 380–89, *381–88*

Seeds, embedded, 252, 262–63

Selenium, 40, 41

Semen, shipped, 316–17

Senecio toxicity, 229

Septicemia, 162, 174

Serum hepatitis (Theiler's disease), 187–88

Shade, need for, 79, *79*

"Shaker foal syndrome," 170

Sharp ridges on teeth, 257

Shear mouth, 258

Sheath swellings first aid, 292–93

Shedding in spring, 71–72, *72*

Shelters. *See* Facilities

Shock first aid, 288–89

Shoes, horse, 87, *87*–88, 95

Shooting horses, 299, *300*

Shoulders, passing in birth, 382–83, *382–83*

Shoulders conformation, 308–9, *309*

Shrubs, toxic, 230–33, *231–33*

Sickle hocked, 313, *313*

Signs of illness, 126–37. *See also* Infectious diseases; Non-infectious diseases; Parasites; Skin problems; Toxic plants and poisons
    abdominal pain, 131–33, *132–33*
    "acting funny," 137
    body language, 99, 117, 118, 131–37, *132–36*

Signs of illness (*continued*)
colic, 131–33, *132–33*, 135, *135*, 195, 271, *271, 272*, 276
dental trouble, 257
ear problems, 267, *268, 269*
eye problems, 261
gut sounds, 132, *132*
hind legs, 134–35, *135*, 137
leg position, 133–37, *134–36*
posture, 131–33, *132–33, 160*
pulse (heart) rate, 127, 129–31, *130*
respiration rate, 127, 131
temperature, 62, 127–29, *128–29*, 131
temperature, pulse, respiration (TPR), 126, 127, 131
three-legged stance, *133*, 133–34
tying up, 50, 135, *135*, 194–96
urine, brown, 196
vital signs, 126–31, *128–30*
warning signs, 137
weight loss, *183*
Silage for feed, 43
Silent heat, 349–50
Site for building barns, 3–5, *5*
Skeletal problems in foals, 445–46
Skin problems, 240–49. *See also* Parasites
allergies, 240–41, *241*
dermatitis, 221, 240
fungus infections, *243*, 243–47, *245, 247*
girth itch, 244–45, *245*
hives (wheals), 240–41, *241*
photosensitization, 73, 229, 230, 236, 242–43
proud flesh, 94, 249, 283, *283–84*
rainrot (rain scald), *245*, 245–46
ringworm, *243*, 243–44
sarcoid tumors, 249
scratches (grease heel), 246–47, *247*
urticaria, 241

warts, 247–48, *248*, 269
Sleeping sickness (equine encephalomyelitis), 142, 143, 156–57
Small intestine, 35
Small strongyles, 204–5
Smooth wire fences, 21
Smut toxicity, 236
Snader, Meredith L., 298
Snakebite first aid, 291
Snow for water, 65
Soaking feet, 92–93
Sodium chloride (salt), 41–43, *42*
Sole bruises, 89–91, *90*
Sole of hoof, 81, *81*, 82, 83, *83*, 84
Sorghum grain (milo), 50
Sorghum grass toxicity, 230
Soundness. *See* Conformation
Sow mouth, 257–58, *258*
Spasmodic colic, 271–72
Spavin of hock, 313
Sperm production and life, 341
Splay-footed, *311*
Sponenberg, Philip, 330
Spring care, 70–72
breeding, 351
Cushing's disease, 71
grooming, 72, *72*
hormones and shedding, 71
mud and health concerns, 70–71
shedding, 71–72, *72*
Squamous cell carcinomas, 181–82
Squinting eyes, 261–62
Stable flies, *211*, 211–12
Stableizer, 114–15, *115*
Stallions, 331–42
boredom cautions, 338
breeding, 336, 339–40, 355, 356–57, 358
breeding fee, 342

for breeding stock, 315–17
contracts for breeding, 342
equine viral arteritis (EVA), 164
exercise, 338, *338*
facilities for, 17, 331–33, *332*
feeding, 57
fertility, 317, 340–42
grooming, 337
halters for breeding, 336
hand breeding, 339
handling, 333–38, *338*, 356–57
infertility, 341–42
isolation and, 333
live foal guarantee, 342
older horses to educate, 337
pasture breeding, 339–40
rearing, discouraging, 337
return-in-season, 342
sperm production and life, 341
for teasing, 351–54, *352*
Stalls
barn, 9–12, *10*, *12*
bedding, 11
cleaning, 12, *12*
foaling in, 372–74
for foals, 414
mats, 5, *5*, 11
Standing up, foals, 403–4
Steamed grains, 47, 48
Steel pipe fences, 23
Steroids, 153
Stitches, 284
Stomach, 33, *33*, 34–35
Stomach tubes for foals, 410, 420
Strangles (distemper), 157–59
Straw bedding, 11, *372*, *373*
Streptococcus equi, 157
Stress
abortion from, 366

disease from, 141–42
vaccinations and, 141–42
Strongid C (pyrantel tartrate), 202, 205,
210
Strongyles, 198, *203*, 203–5, 273, 274,
440–41
Sudan grass toxicity, 230
Summer care, 73–79
anhydrosis (without sweat), 78
breeding, 351
dry climates, 73
exercise, limiting, 78
eye irritation, 73–74
feeding, 78
fly masks, 74, *74*
foot care, 73
grooming, 74
heat stroke, 75, **76,** 76–77,
77
humidity concerns, 75, 78
hyperthermia, 75
insects, 74, *74*
photosensitization, 73, 229, 230, 236,
242–43
shade, 79, *79*
sunburn, 73
trailering cautions, 75
water intake, insufficient, 77
Summer sores (habronemiasis), 211–12,
217–18
Sunburn, 73
Supplements, 36, 40, 41, 52, 52–53
Swamp fever (equine infectious anemia),
164–66, *166*
Swamp maple (red maple) toxicity, 232,
232
Swellings from vaccinations, 148
Syringes for vaccinations, 145–46, *146*,
148, 149, *150*

T

Tabanids (horse and deer flies), *212*, 212–13
Tack rooms, 13
Tail wrapping, 354–55, *355*
Tall fescue toxicity, 230, 366
Tamed iodine (Betadine), 244
Tapeworms, 206
Tartar build-up, 251
Tar weed toxicity, 226
TDN (total digestible nutrients), 56
Tearing eyes, 261–62
Teasing, 51–354, *352*
Teeth. *See* Dental care
Temperament of horses, 98
Temperature, pulse, respiration (TPR), 126, 127, 131
Temperature of horses, 62, 127–29, *128–29*, 131
Tendons (weak) in foals, 421, *421*
Tetanus, 142, 143, 154–56, *155*, 280
Theiler's disease (serum hepatitis), 187–88
Threadworms (*Strongyloides westeri*), 440–41
Three-legged stance as illness sign, *133*, 133–34
Thrush, 88–89, *89*
Ticks and Lyme disease, 177, *220*, 220–21
Tied in knees, *310*, 312
Tongue injuries, 252
Tooth bumps, 256, *256*
Topline of horses, *308*
Torpedo grass toxicity, 230
Total digestible nutrients (TDN), 56
Toxic plants and poisons, 223–39
    antifreeze, 238
    black locust, 232

black patch disease, 234
black walnut, 11, 233
blister beetles (cantharidin), 236–37, *237*
brackenfern, 225
buckthorn, 226
castor beans, 225
chemical poisons, 237–39
cocklebur, 225
corn poisoning, 235
crotalaria, 226
death camus, 226, *226*
equisetum, 226
ergot, 235
facilities and, 26–27
fiddleneck, 226
fireweed, 226
fungi, 234–36
Fusarium moniliforme, 235
garden plants, 225, *225*
grains (concentrates), 235, 236
grasses, 230
hay, 234–35, 236, 237
lantana, 226
larkspur (delphinium), 225, *225*
lead poisoning, 239
locoweed, 227, *227*
medication overdoses, 239
monensin (Rumensin), 239
mycotoxins, 236
nightshade family, 228, *228*
oleander, 232
pesticides, 238
photosensitization from, 73, 229, 230, 236, 242–43
poison hemlock, 228, 228–29
rayless goldenrod, 232
red maple, 232, *232*
Rhizoctonia leguminicola, 234

Russian knapweed, 229
rusts, 236
Saint-John's-wort, 229
senecio, 229
smut, 236
sorghum grass, 230
sudan grass, 230
tall fescue, 230, 366
tar weed, 226
torpedo grass, 230
treated grain, 238
trees and shrubs, 230–33, *231–33*
white snakeroot, 229
wild cherry, 231, *231*
yellow burr weed, 226
yellow star thistle, 230
yew, 233, *233*
Toxoids, 156, 188
TPR (temperature, pulse, respiration), 126, 127, 131
Trace minerals, 40, 41, 42
Trailering cautions, 75
Treated grain toxicity, 238
Trees, toxic, 230–33, *231–33*
Trombidiform mites, 221
Turnout terrorists, handling, 124
Twins and abortion, 365
Twisted intestine, *272*, 272–75
Twitch, restraint, *112–13*, 112–14
Tying horses, 105–7, *106–7*, 119
Tying up (rhabdomyelitis), 50, 135, *135*, 194–96

U

Udder problems, 368–71
Ulcers in foals, 439–40
Umbilical cord, breaking, 391–92, 393
Umbilical hernia, 418
Underline of horses, 308

Uneven wear of teeth, 257
Unsoundness, 96, *314*
Unthriftiness from parasites, 200
Urinary stones, 292
Urine, brown, 196
Urticaria, 241
Uterine prolapse, 397
Uterine torsion, 389
Uterus, 344, *344*, 345–46

V

Vaccinations, 140–53
anaphylactic shock from, 152–53
antibodies and, 140–53, **143**
blood, testing for, 149–50, *151*
in breast, 148
for broodmares, 368
in buttocks, 149
colostrum and, 141, 143
desensitizing area before, 149, *150*
epinephrine (adrenaline) for anaphylactic shock, 153
equine encephalomyelitis (sleeping sickness), 142, 143, 156–57
equine viral arteritis (EVA), 138, 163–64
failure of, 141–42
for foals, 141, 143
histamine and, 152
immune system and, 139–42
influenza (flu), 142, 159–60, *160*
by injections, 144–53, *146–47*, *150–51*
intramuscular injections, *147*, 147–49
by intramuscular shots, 144
in neck, 148
needles for, 145–46, *146*, 148, 149, *150*
Potomac horse fever (PHF), 138, 167–68

Vaccinations (*continued*)
reactions to, 151–53
rhinopneumonitis, 142, 146, 160–62, *161*
schedule for, 142, **143**
sites for injections, *147*, 147–49
steroids for anaphylactic shock, 153
storage of vaccines, 147
stress and, 141–42
swellings from, 148
syringes for, 145–46, *146*, 148, 149, *150*
tetanus, 142, 143, 154–56, *155*, 280
Venezuelan equine encephalomyelitis (VEE), 138, 156
Vagina, 343, *344*, 346
VEE (Venezuelan equine encephalomyelitis), 138, 156
Veins versus arteries, 281
Venezuelan equine encephalomyelitis (VEE), 138, 156
Ventilation in barns, 6, 6–7, 9
Vesicular stomatitis (VS), 175–76
Veterinarians, handling horses for, 102, 103, *103*
Vinegar as fly repellent, 216
Vital signs, 126–31, *128–30*
Vitamins, 38–39, 40, 239
Voice tone and handling, 100
Vomiting, 35
VS (vesicular stomatitis), 175–76
Vulva, 343, *344*

W

Warbles, 207
Warmbloods, 317
Warts, 247–48, *248*, 269

Water
facilities and, 27–29, *28–29*
feeding and, 35
in summer, 77
in winter, 65–66
Water bag rupture, 381, *381*
Waxing as sign of labor, 378
Weaning foals, 441–44, *443*
Weanlings, 444–50, *448–49*. *See also* Foal care
Weed control of facilities, 26–27
Weighing feed, 45–46, 48
Weight
gain of foals, 425
loss as illness sign, *183*
and medication dosage, 295
Wheals (hives), 240–41, *241*
Wheat and wheat bran, 49–50
White snakeroot toxicity, 229
Wild cherry toxicity, 231, *231*
Windbreaks, natural, 16
Wind-chill index, **61**
Windows in barns, 5, *5*
Winter care, 60–69
barns for, 9
blankets, 66–68, *68*
body condition, 62–63, *63*
coats (horse), 62, 66
dehydration, signs of, 65
facilities, 28–29, *28–29*, 62
feeding, 60, 64–65
for foals, 411, *411*
foot care, 60, 66, 89
footing, safe, 66
grains (concentrates), 64
hay, 64
heat energy from roughage, 64

manure and, 65
older horses, 63, 65
ribs and body condition, 63, *63*
snow for water, 65
temperature of horses, 62
thrush, 89
water, 65–66
wind-chill index, **61**
Withers conformation, 308–9, *309*
Wolf teeth, 253, *253*, 254, *254*
Wood fencing, 18–19, *18–19*, 22, *22*
Working horses, feeding, *57*, 57–58

Worms. *See* Parasites
Wounds first aid, *280*, 280–86, *282–83*,
    285

Y

Yearlings, 444–50, *448–49*. *See also*
    Foal care
Yellow burr weed toxicity, 226
Yellow star thistle toxicity, 230
Yew toxicity, 233, *233*
Young horse care. *See* Foal care

OTHER STOREY TITLES YOU WILL ENJOY

A Guide to Raising Beef Cattle by Heather Smith Thomas. A bovine expert, Thomas explains facilities, breeding and genetics, calving, health care, and advice for marketing a cattle business. 352 pages. Paperback. ISBN 1-58017-037-4.

Horse Handling & Grooming: A Step-by-Step Photographic Guide by Cherry Hill. This user-friendly guide to essential skills includes feeding, haltering, tying, grooming, clipping, bathing, braiding, and blanketing. 160 pages. Paperback. ISBN 0-88266-956-7.

Horse Health Care: A Step-by-Step Photographic Guide by Cherry Hill. Explains bandaging, giving shots, examining teeth, deworming, and preventive care. Exercising and cooling down, hoof care, and tending wounds are depicted, along with taking a horse's temperature, and determining pulse and respiration rates. 160 pages. Paperback. ISBN 0-88266-955-9.

Horsekeeping on a Small Acreage: Facilities Design & Management by Cherry Hill. The essentials for designing safe and functional facilities whether on one acre or one hundred. Hill describes the entire process: layout design, barn construction, feed storage, fencing, equipment selection, and much more. 192 pages. Paperback. ISBN 0-88266-596-0.

Horse Sense: A Complete Guide to Horse Selection & Care by John J. Mettler Jr., D.V.M. Covers selecting, housing, fencing, and feeding a horse, plus immunizations, dental care, and breeding. 160 pages. Paperback. ISBN 0-88266-545-6.

Safe Horse, Safe Rider: A Young Rider's Guide to Responsible Horsekeeping by Jesse Haas. Beginning with understanding the horse and ending with competitions, this book includes chapters on body language, pastures, catching, and grooming. 160 pages. Paperback. ISBN 0-88266-700-9.

Stablekeeping: A Visual Guide to Safe and Healthy Horsekeeping by Cherry Hill. Explains essential features of the barn, stall, tack room, and storage, work, and turnout areas; recommends tools; and offers valuable information and helpful tips on sanitation and pest control, healthy feeding practices, safety, and emergency precautions and practices. 160 pages. Paperback. 1-58017-175-3.

Trailering Your Horse: A Visual Guide to Safe Training and Traveling by Cherry Hill. An essential guide to safe training and traveling, with detailed information on selecting an appropriate truck and trailer, training a horse to load and unload, and helpful traveling advice. 160 pages. Paperback. 1-58017-176-1.

These books and other Storey Books are available at your bookstore, farm store, garden center, or directly from Storey Publishing, Schoolhouse Road, Pownal, Vermont 05261, or by calling 1-800-441-5700. www.storeybooks.com